# WILDLIFE ON THE EDGE

## ADVENTURES OF A SPECIAL AGENT IN THE U.S. FISH & WILDLIFE SERVICE

TERRY GROSZ

WOLFPACK PUBLISHING

**WOLFPACK**
**PUBLISHING**
— EST 2013 —

Print Edition

© Copyright 2018 Terry Grosz (as revised)

Wolfpack Publishing
6032 Wheat Penny Avenue
Las Vegas, NV 89122

ISBN: 978-1-62918-387-9

Library of Congress Control Number: 2018951484

# CONTENTS

# DEDICATION

ON NOVEMBER 11, 2010, the World of Wildlife lost one of its finest warriors, who was killed in the line of duty protecting the Nation's National Heritage. Pennsylvania Wildlife Conservation Officer David L. Grove was killed by a poacher while protecting the wildlife resources near the Town of Fairfield in the Commonwealth of Pennsylvania.

Many men and women of the Thin Green Line, quietly and without fanfare, serve their fellow citizens at their own peril protecting this Nation's natural resources for those folks "yet to come." Such were the selfless, dedicated acts of Wildlife Conservation Officer Grove who, as a result of his actions, made the ultimate sacrifice for his fellow Americans and those in the World of Wildlife who have little or no voice.

However, the critters of this great land of ours are created by our Maker. And it was into those hands Officer Grove stepped upon the very moment of his ultimate sacrifice. In so doing, he stepped into the good Lord's hands and the eons of love, compassion and history they represented of His own sacrifices for those of us souls left behind.

An old game warden once told me, "We are all travelers in this world. From the genesis of our being, to the exodus of our souls. Between those two great eternities, we all travel looking for the helping hands of God." David Grove found "those hands of God" and now is in His Kingdom with the rest of the angels.

Rest in peace, my brother, for the critters and the rest of us left behind have truly lost a fellow warrior, family member and dear friend. Thank you for your times with us and we look forward to our "reunion."

It is to fellow officer David L. Grove of the Thin Green Line that this book is respectfully dedicated. Dedicated in the hope that David is long remembered for being one of God's warriors for the world of wildlife.

# WILDLIFE ON THE EDGE

## TERRY GROSZ

# CHAPTER ONE

## THE BEAR TRAP AND
## A MORGAN SILVER DOLLAR

I AM A COUNTRY BOY born and bred. I was raised in the Sierra Nevada Mountains in northeastern California near the small lumber and logging community of Quincy. Quincy is a Sierra Mountain hamlet, located 80 miles due west of Reno, Nevada. I caught my first rainbow trout at age three and my first 21-inch German brown trout at age eight. I killed my first California mule deer at age eleven, gutted him out with a straight razor (I didn't own a pocket knife), and brought him home on my bicycle. I was into bird hunting as well, killing my first mourning dove at age eleven, and my first mountain quail and band-tailed pigeon at age twelve. I don't remember when I killed my first Canada goose, but I do remember he was so large, when carrying him by his neck, his tail feathers dragged on the ground.

I was raised by a single, hard-working, divorced mother, and did my share to help out at an early age. I took every job I could find after age seven, pulling weeds, mowing lawns, digging graves and pits for septic tanks, shoveling deep, heavy Sierra snows, along with the ever-present splitting and stacking of firewood in my mountain community.

At age 15, I lied about my age one summer and went to work at the local box factory (one had to be at least 18 years of age to legally do such work). At age 16, I lied about my age the following summer once again and went to work in the woods as a logger. Even at that young age, I knew I wanted to go to college so I could leave the hard work, employment uncertainty, and near poverty levels associated with the lumber industry, for a better lifelong profession. As it turned out, being driven in such a manner was a very prophetic move. Today, most of the small lumber mills and logging shows I once knew, are all gone, as is most of the remaining big timber in Plumas County in the Sierra Nevada Mountains.

From age seven, my mother banked half of the money I earned at my various odd jobs. The other half of the money earned went towards keeping the "wolf" from our door. When I graduated from high school in 1959 and left for college, I had $7,000 saved in the bank! That doesn't sound like a lot in today's money but one could buy a nice three-bedroom, two-bath home for that amount in those days!

During those hard times when I was not working, I was either trout fishing, shooting ground squirrels with my .22 rifle, or hunting deer, ducks, geese, mountain quail, or band-tailed pigeons with a borrowed rifle or shotgun (I couldn't afford one of my own). Yeah, I guess one could say I was a dyed-in-the-wool country boy.

Having saved my money all those earlier years, I put myself almost completely through seven years of college. First at a junior college in Santa Barbara and then at Humboldt State College in Arcata in northwestern California. Donna, my new Bride, helped me through the last year of my master's program, where I earned a second degree in wildlife management.

After graduation in 1966, I competed with 1,400 other applicants for a California State Fish and Game Warden position, and I was one of 25 applicants selected and hired. My first game warden duty station was in the small lumbering and logging town of Eureka. My last position with the California Department of Fish and Game, ending in 1970, was in Colusa, a small agricultural town in the northern Sacramento Valley. I guess one could say both duty stations were in keeping with my love of the country and dislike for the largeness of a city. My quicker readers can see where this happy-as-a-clam, country boy thing is leading...

In 1970, I resigned my position as a state fish and game warden to accept my lifelong dream job with the U.S. Fish and Wildlife Service as a U.S. Game Management Agent. Unfortunately, my new duty station was in the San Francisco Bay Area's Town of Martinez. Not my ideal place to live or in keeping with my country boy principles, being that I was now located in a large and noisy city. A place of constant police sirens, sounds of gun shots, highway noises, where one had to stand in line to use a public restroom, and where one could not see the stars at night because of all the city lights. But I had plans of eventually moving back into my country environment as I moved my way up through the ranks of my new agency. So with those plans, I swallowed the 'country' in me and moved to a city, in order to further my career, increase the challenges in my chosen profession, and eventually provide a better life for my family.

Then in 1973, the Fish and Wildlife Service made an administrative change in my Division of Law Enforcement. Instead of being titled U.S. Game Management Agents, we were reclassified as Special Agents. That meant now being classified as criminal investigators (wildlife) and as such, we were no longer responsible for the wildlife management duties that had been a major part of my previous profession. Henceforth, Special Agents of the Fish and Wildlife Service would be performing criminal investigator duties in the world of wildlife. So, in addition to enforcing the wildlife laws of the land overseeing sportsmen, we would also be investigating commercial market hunters, smugglers of wildlife, those involved with the illegal sale of wildlife parts and products, and other such serious investigations.

My new federally assigned district of operations ran from the City of Monterey inland to the center line of California, then north to the Oregon line. Everything west of that imaginary line was my field of operations and what a boar's nest of illegal activity that turned out to be. Illegal importation of plants and animals; smuggling of wildlife; illegal sale of wildlife parts and products; illegal take of marine mammals off the coast of California by commercial fishermen; illegal interstate commerce of game trophies by so-called sportsmen; and everything in between!

However, my roots were still country and since there was plenty of 'country' in that huge area, that is where I concentrated my federal law enforcement efforts. To that end right out of the box, I ran across one of the strangest and potentially deadly wildlife investigations in my budding federal career.

In the fall of 1973, based on a stream of intelligence reports from Humboldt and Trinity Counties, I made plans to head for the north coast of California. Intelligence reports from numerous citizen sources, mainly the hunting public, advised that band-tailed pigeons, a federally protected migratory game bird frequenting those two counties, were being taken illegally in large numbers. Band-tails were a species of wild pigeon whose population numbers were on the decline. Since that species' life history behavior was to sit or roost high in the treetops, many times out of shotgun range, frustrated shooters were reportedly taking the 'high-sitting' birds illegally with .22-caliber rimfire rifles with scopes. And in so doing, they were violating state and federal hunting regulations. Additionally, said intelligence was that those shooters (note that I didn't call them hunters), were taking over-limits of the birds every time the opportunity arose during the hunting season.

Loading up my Labrador retriever named Shadow and the needed field gear for the detail, I headed north. Arriving in the north coast city of Arcata, I turned and headed for the Kneeland area in the mountains to the east. There I stayed with my old friends, the Ragons. Glenn Ragon was a Humboldt County Deputy Sheriff and a dear friend from my earlier days when I worked the area as a game warden in Humboldt County. By staying with Glenn and his wife, I figured if anyone had a handle on the wildlife pulse of the area, it would be him. After a wonderful dinner of fresh Dungeness crab, sourdough French bread, and green salad, followed by one of Dariel's (Glenn's wife) fresh homemade blackberry pies, we got down to the business at hand. Glenn had good numbers of band-tailed pigeons in his Humboldt County patrol area but because of the vast amounts of private lands they inhabited (no public hunting allowed), had little information on their illegal take. It seemed that whenever someone 'popped a cap' on those private lands without permission to do so, the landowners or Glenn were on the illegal shooters like a bunch of Rhode Island Red chickens on a June bug!

But the information Glenn had on the band-tail problems in the adjoining Willow Creek area in Trinity County was a horse of a different ride. The resident deputy and a hard-working game warden named Hank Marak stationed in Willow Creek, were running across wildlife violators of every make and model. But the area was so rugged and remote, catching anyone breaking the wildlife laws of the land was extremely difficult. In short, work of that kind required a lot of hours of hard work in the band-tailed infested backcountry either afoot or horseback. With only the two officers in that entire rugged, mountainous area, that was almost an impossibility. Between the problems created by the loggers and Hoopa Indians in the area, those men had their hands full of violators of every "creep" and kind known to humankind.

With that relevant information from Glenn, I decided I would add my considerable bulk, and of course extreme beauty, to the fray and see if I could also spread some additional hate and discontent among that breed of lawbreakers as well. Plus, I had worked some of that area as a young game warden in the late '60's. So with that "landed" knowledge, I felt my services would be better served in that arena. With that, my law enforcement plans for the next several days of adventure were set.

Daylight Saturday morning, found dog and me slowly winding up curved, narrow highway 299, behind numerous, heavily loaded logging trucks. Stopping at the top of one of the Pacific Coast Range's many summits, I let dog out to take care of her usual business, as I headed for the ever-present ice chest in the back of my patrol truck. Taking out and opening up a cold Coke, I stood there in the cool of the temperate rain forest environment, breathing in the clean moisture-laden air. Sure not like breathing in the often re-used and polluted air of the City of

Martinez, I mused. It was good to be back once again in one of my old stomping grounds — a place I had long ago quickly fallen in love with. Allowing myself to drift back into the past, I soon found myself thinking about old times, places, and events from the area, especially when I was a rookie game warden chasing loggers, Indians and other rank and file of wildlife outlaws.

Craack went the unexpected sound of a light rifle some hundred yards or so down the old logging road leading off the summit on which I was now standing! That sound was immediately followed by the wing-slapping sounds of numerous band-tailed pigeons leaving a dense stand of nearby Douglas fir trees. Then I heard the soft 'swishing' sounds those feathered rockets made as a flock of about 50 birds sailed noisily overhead.

"In the truck," I told my dog Shadow as I quickly headed for the cab! A resounding thump told me Shadow had bounded into the bed of the truck as I spun its engine to life. Keeping my eyes focused towards the densely forested area from whence came the obvious rifle shot (game wardens quickly learn the difference in sounds between rifle, pistol and shotgun shots), I quietly slipped "her" into second gear (I was driving a four-speed to you younger set not familiar with stick shift vehicles). Then I let the truck more or less idle silently down the old dirt logging road towards where I had last heard the shooting and observed the fleeing band-tailed pigeons.

Rounding a turn in the dirt road moments later, I spied a battered old blue pickup parked in the middle of the road. Its driver's side door had been flung wide open and the driver was nowhere to be seen. Killing my engine and pushing in the clutch, I quietly rolled to a stop and grabbed my binoculars off the front seat. First, I examined the truck and upon confirming its emptiness, quickly glassed the area on the upside of the road. That was the timbered area from whence I had observed the band-tailed pigeons leaving in fright earlier. That is also where I figured my shooter would be, especially if he had killed a bird.

Then down off the brushy side hill came a man running for the pickup's location like he was "hell-bent for election!" In one hand he carried what appeared to be a .22-caliber bolt action scoped rifle, and in the other a wildly flopping dead pigeon as my shooter fought for his balance. Roaring off the steep bank at such a high rate of speed, my shooter misjudged the bank's steepness, lost his footing, and smashed headfirst into the roadbed below in a huge puff of flying dirt! Rolling halfway across the road from the force of his impact, my shooter quickly jumped to his feet, then ran and leaped onto the front seat through the open door of his pickup like nothing out of the ordinary had occurred. Then all of a sudden, he grabbed his head with both hands realizing the road had done a violent, not to be misunderstood number, on his nose and noggin.

Taking that as my cue, I fired up the truck and pulled up alongside my shooter's pickup in such a manner so as to cut off any escape attempt. Stepping out from my truck, I soon saw there was no need for hurry on my part. My shooter was still holding his head in his hands and softly moaning, as he rocked back and forth in pain in the cab of his truck. Below his hands streamed a river of blood, splattering all over his shirt and pants. That stream of blood soon pooled on the floorboard of his pickup in a manner like that found on the floor of a slaughterhouse!

"Damn, Man! Are you all right?" I asked. Rats, that had to rank right up there as another one of my most brilliant statements made when under a stressful moment in time, I thought.

"Hell, no! Can't you see I just broke my nose and now, I am bleeding like a stuck hog," came an angry reply from my shooter who had just tested the gravel road's surface for its Brinell hardness. And in so doing had found it to be "diamond" harder than his noggin or nose.

"You got a rag or something I can get to help stem that flow of blood?" I asked.

"Nah. I can just use the tail-end of my shirt until I can get to a doctor," he replied as he blew a mess of blood off his lips and chin, which instantly splattered all over the inside of his windshield like someone had just been head-shot from behind.

"Well, this probably isn't a good time, but my name is Terry Grosz and I am a federal agent. As such, I would like to check your hunting license and pigeon if I may," I said.

"You got that damn right! Now is not the time. Can't you see I am bleeding to death and you want to see my hunting license and stinkin' damn pigeon! That is the trouble with you federal bureaucratic parasite son-of-a-bitches! Idiots one and all! You can just wait until I get this here flow of blood stopped," he growled.

Detecting a slight bit of unhappiness and lack of cheer at my untimely presence and, suspecting I had an out-of-work, soon-to-be really angry logger in hand, I made my move. Reaching through the open door of his truck, I quickly removed his ignition keys to preclude any future funny business on his part. I had been here before and had shooters run in such situations. And a high-speed chase on a dirt road is never good. Especially, if one is chasing someone hell-bent for leather over a dead pigeon...

"Hey, what the damn hell in 'tarnation' are

you doing? Them is my truck keys and I want them back if you know what is good for you," he growled menacingly.

"Let me try this one more time. I am a federal agent and would like to check your pigeon and your hunting license. Since I saw you come from the side hill carrying a dead pigeon, I just figured it was within my purview to follow through with my field inspection. And to be quite frank with you, removing your ignition keys precludes you from doing something really stupid, especially if you are in the wrong."

"The hell you say. That dead bird was a 'state' quail not a 'federal' pigeon, so there Mr. Federal Agent," he continued in a tone of voice reflecting he was not too happy at seeing me during this moment of time in his life.

"Well, Sir, that doesn't really make one whit of difference. I know for a fact from its overall looks and coloration that the bird was a band-tailed pigeon. However, if you want to claim it was a quail, that is OK with me as well. I am also a credentialed state fish and game warden and would like to see your bird and check your rifle as well," I said in a tone and tenor of voice now matching his.

"Oh, all right," he grumbled as he handed me the pigeon off the front seat of his truck.

"Next, I would like to check your rifle as well,"

I said. "And if you would, Sir, please hand it to me butt first so we don't have an accident," I continued.

Watching the man closely, I saw him take his rifle, turn it over and try to remove a live cartridge from the chamber with one hand in such a manner so I would not see his actions. To say the least, that dog didn't hunt. In a second, I reached in and removed the rifle from his hands before any more dangerous "tomfoolery" continued. Flipping the rifle upright, I then removed a live round from the chamber. "Having a loaded rifle in a motor vehicle on a way open to the public here in California is a violation," I quietly advised.

My chap said nothing, as he tried once again to stem the flow of blood from his nose with the dirty sleeve from his shirt. Finding that insufficient, he took an even dirtier rag from the passenger side floorboards and used that to stem the rampant bleeding from his nose. When he did, I observed two more dead pigeons lying on the floorboards that the rag had been covering.

"I would like those two pigeons from the floorboards as well," I said, "since it appears they were taken with your .22 as well" (he did not have a shotgun in the truck).

With that, the man erupted from the pickup and squared off with me. It was obvious he had more than enough from this extremely beautiful but obviously rude, 'federal person.' First, the bad luck of testing the roadbed with his nose and noggin at a high rate of speed had made its mark on his level of sweetness. Then the untimely arrival of a law enforcement officer and finally being caught red-handed with three illegal pigeons was the icing on the cake to his way of figuring. Having tangled with loggers and lumbermen in the past as a game warden, I was aware of their surprising strength gained from having such hard menial jobs. Most were as stout and wiry as a brick outhouse and just as "smelly" when aroused, temperament-wise.

"That is not what you want to do," I cautioned. "First, I have identified myself to you as a federal officer. Laying a hand on me will get you five years automatically in the federal penitentiary. Additionally, lay one hand on me and that big dog in the back of my truck taking in all your moves will be out of there in a flash and in the middle of your hide, with a mess of the biggest set of teeth you ever saw. And let me tell you from previous situations, that is something you do not want to experience. And lastly, this isn't my first 'walk in the park' with ill-tempered folks, if you get my drift," I said in a very meaningful tone of voice. Of course, my sweet and kind voice was backed with 320 pounds of the most beautiful mass one ever laid eyes upon.

It was at that moment my aroused logger decided he had enough. Between his still brightly gushing nose and his hind end soon to be bleeding if the dog took him on, he lathered back down to common- sense size. Once I could see the fight had left my shooter, I got down to the business at hand. Inspecting the three pigeons, it was obvious from a single entry and exit hole in their bodies that they had been taken with the .22 rifle. That was followed with the non-production of a hunting license and the fact that he had a loaded rifle in his vehicle. One state citation for the loaded firearm and two federal ones for lack of a hunting license and taking a migratory game bird illegally, to wit with a rifle, and the issue on the mountainside was settled. By the time I had finished with all the paperwork, the man's nose had plugged up with crusted blood and he was no longer bleeding like a "stuck hog." However, he was still pissed over having his day going so sour. Especially when he discovered that court for the pigeon violations was clear down in Sacramento, several hundred miles distant!

My chap with the road rash on his nose and noggin eventually paid $250 in federal court for the pigeon violations, and another $50 in state court for the loaded rifle in the Willow Creek Justice Court. I don't know what he had to pay to fix that handsome broken nose but I would bet it was a pretty penny as well...

And that wasn't all. While attending to my 'broken-nosed one,' I had heard more shooting further down the road on which we now sat. Only this time, it sounded like a brace of shotguns making it a hard day for the migrating band-tailed pigeons. Hang in there, pigeons. Help is on the way, I thought as I hurried through my paperwork with the "bloody-nosed one."

Finishing up with my logger chap, dog and I headed down the old logging road for our next bunch of shooters and the adventures that event would generate. Rounding a turn about a mile from my "broken-nosed shooter," I observed another pickup sitting alongside the road. Pulling in behind that rig, I got out and observed several empty shotgun shell boxes lying in the front seat. Checking the grill of the pickup, I found it to still be warm, indicating its recent use. Then further up on the mountainside, I heard several more shotgun blasts. That was followed by a small flock of band-tailed pigeons flying away from the shooting area and sailing over my head into the forested canyons below.

Letting Shadow out from the back of my truck, the two of us quickly headed up the brushy timbered mountainside towards the sounds of the latest shooting. After climbing upwards for a hundred yards or so, dog and I stopped for a breather and to listen for more shooting.

Moments later, I heard another shot ring out still further up on the mountain, so off we went once again. Coming to a small, watered draw, dog and I leapt the stream.

That was when I observed a plastic garden hose running from a small dammed-up area in the stream, down the hill from where I now stood. *What the heck?* I thought. No one lives up here out in the middle of nowhere. So why the watering hose running from the small stream down towards a massive stand of Douglas fir trees? Forgetting my shooters further up on the side hill for a moment and letting my curiosity get the better of me, I followed the garden hose. Soon I discovered a small well-used trail leading from the creek and garden hose towards a dense grove of fir trees. Moving down the narrow trail, I discovered a small, hidden campsite off to one side in the brush. The area had a well-built log lean-to, covered with old, long-dead, tree limbs and brush. Off to one side was a small dumpsite containing cans, bottles, and other garbage from the camper who had occupied the site. *What the hell?* I thought. *Why this old campsite way the hell and gone out here in the brush?* I asked myself. Then I discovered another well-worn trail leading towards a small clearing some thirty or so yards from the campsite area. Still trying to figure out what the heck was going on, I began following the trail leading to the small clearing.

That was when my guardian angels finally woke up and began fluttering around inside me like a large mink was on their trail. I had learned early on that when my guardian angels were on the "prod" inside me, something bad on the outside was close at hand or in the air. In the air either in the form of a bad guy capture or some other dose of trouble. Either way, those feelings meant I needed to be on the alert for any sudden signs of danger. Otherwise, if the facts from such previous happenings and feelings in my life meant anything, a wreck was close at hand or at least in the wind...

Sliding my coat back from the side on which I carried my side arm so I could get to it in a hurry if necessary, I slowed my pace on the trail leading to the small clearing. I now noticed that the warning signs from my guardian angels were now signaling "high alert!" *What the hell?* I thought. I could see nothing in the form of danger, even though I was looking every which way but loose. All I saw was a small trail leading to a clearing that appeared to be devoid of anything signaling danger. Then I thought maybe my angels were foretelling the presence of a black bear close at hand. That could be since I was in great bear country but damned if I could see, hear or smell any sign of that kind of danger. Slowing my pace to almost a stop, I now could see where a space had been cleared out in my clearing and it was obvious something had been grown there earlier. You know, like a small garden or something.

Then the almost physical presence of my angels indicated they now had their flight feathers really in an uproar. So much so, that the feeling I was now getting from them was almost physically felt throughout my miserable carcass. Having had such feelings before throughout my short law enforcement career and finding them to be real signs of danger or action, I stopped and stood dead still in my tracks. Then once again, I carefully examined my surroundings. I had a small clearing to my front in which something had been recently grown. There were smaller hoses laid out across the small clearing attached to the main garden hose running from the creek. There appeared to be signs of four old rows of woody plant stems on the previously cultivated ground but other than that, I observed nothing out of the ordinary or anything appearing dangerous...

Figuring my guardian angels were just awakening from a late-night hangover, I started to walk down the trail once again. It was then I swear I could feel my angels being so upset they were trying to fly around inside me! Stopping once again, I took another look around for the danger they were now signaling was dangerously close at hand. Damned if I could see anything or anybody out of the ordinary. Then looking down, I spotted a shiny silver dollar lying in the middle of the trail not three feet in front of me. *Damn, what a great find out here in the middle of nowhere,* I thought. Starting to take another step to claim my silver prize, my angels went nuts! Stopping before I started, I looked around once more for any sign of danger, which they clearly foretold.

Then, I spotted it! Running through the leaf litter to the right of the trail on which I stood was a heavy steel wire. It led from the edge of the trail over to the top of what appeared to be the end of a partially buried metal "T-post." *What the heck?* I thought. Then I followed the wire with my eyes to where it led onto the trail just in front of the area where I was standing. Once it got to the trail in front of me, it disappeared into the dirt of the well-trodden path. Then I noticed something out of the ordinary. The center of the trail on which I was walking had been slightly disturbed. Reaching down, I grabbed hold of the heavy wire running into the disturbed dirt of the trail and gently lifted it up. It did not move like it was attached to something solid and very heavy.

Turning, I had Shadow who had been following me mash her hind end on the center of the trail. Then I walked back off the trail and picked up a stout Douglas fir limb. Returning to my mystery wire, I began lifting it with the limb as if it were attached to a grizzly bear or something that could bite. Feeling pretty stupid at how I was behaving, I continued poking at my wire in the dirt of the trail. Nothing. Then I gave it a really hard poke. *Wham!* Up from the center of the trail jumped one huge bear trap with my Douglas fir limb firmly clenched in its steel teeth!

Man, you folks have never seen a 320-pound man jump so high in his life! Coming down about a thousand yards away (well, maybe not quite that far), I gathered my beautiful self up along with my shattered composure. Man, you talk about jumping up into the air like a bug on a hot rock. I had completely exploded like a car-hit chicken when that trap had exploded near my feet! Or to you North Dakota farm boy readers, like a red fox does when it is mousing and the mouse runs under the fox's feet. When one sees how high and quickly a red fox can move under such circumstances, you can picture my beautiful airborne carcass in flight... Then realizing just how damn foolish I must have looked, I "eyeballed" around to see if anyone else had witnessed a man my size leap sixty feet into the air as quickly as a snake can strike. Thank heavens, the only witness to my 'embarrassing explosion' was an unconcerned white-crowned sparrow rustling around in the leaves looking for a bug dinner under a small redwood. That plus my dog Shadow. And if she knew what was good for her, she had better never say anything to anyone about my high-flying act when the trap snapped shut... It was then that I noticed my guardian angels had now settled down and were a lot more relaxed than they had been just moments previously.

Walking back to the spot in the trail where the giant bear trap had exploded just inches from the toes of my boots, I got my first real look at what could have been a really bad wreck had I continued walking towards that shiny, luring silver dollar... Especially if I had not listened to my guardian angels, stepped forward, and greedily stooped over and picked up that obvious silver dollar decoy. If I had, my foot would have been over that trap when I reached for that coin! There I would have probably died. Died because the trap had been wired with a heavy wire to that buried T-post off to one side on the trail. Once with a smashed leg bone in that bear trap, I doubt I could have managed in such pain to somehow untangle that heavy wire and crawl back to my truck located a half-mile distant...

Closer examination of the device revealed that it was a heavy, New House #5 bear trap complete with steel dagger-like teeth under the jaws in which to hold its prey! The mouth of the trap was a good twelve-to-fourteen inches across and once a foot was placed therein, there would be hell to pay! Digging the trap the rest of the way from out of the trail, I managed to twist off the heavy wire tied to the trap's heavy steel 'O' ring on its chain. Man, one more step and I would have been in life-threatening trouble. Because of the Grace of God go I, I slowly and thankfully thought. Well, for that and a couple of assigned, now getting more feather-worn on a daily basis, guardian angels that is...

Nothing but all eyes now, I continued down the trail after picking up that shiny decoy Morgan silver dollar as I walked into the garden area. There I discovered why the presence of water hoses and a bear trap for those not wanted in the area. Old marijuana plants, some 60 in number now represented by cut stems, were the center-piece of my hidden garden. No wonder it was set back out in the middle of nowhere! Then I continued surveying the scene for any other clues I could turn over to the Willow Creek Deputy Sheriff. This being my very first find of such a growing site, no wonder I was dumber than a box of rocks and hadn't recognized what I was getting myself into. I had heard of such gardens early on in my career but that was the very first I had the fortune to stumble into. Or misfortune if you will, had I walked forward one more step to pick up that obvious decoy silver dollar. Today both Humboldt and Trinity Counties are known for their numerous illegal growing sites for marijuana and the "druggies" who tend them. Also, as all of us have come to realize, because of their value and the news media, most are guarded with live, "shoot on sight" growers. Or decorated with many forms of booby traps like my previously discovered by dumb luck and a couple guardian angels, buried bear trap.

Slinging the antique bear trap over my shoulder and richer by not only that learning experience but with a silver dollar now resting safely in my pocket, dog and I sought out my earlier pigeon hunters. But alas, they had gone by the time dog and I continued our 'pigeon shooters cold tracking' adventure. Not finding my lads, dog and I had returned to our parked truck only to find my suspect pigeon shooters' pickup long gone.

For the next ten days, dog and I walked many miles over the cutover and virgin-timbered forests of Humboldt and Trinity Counties checking pigeon hunters. Partially due to low pigeon numbers and because there were a lot of good sportsmen out there, dog and I only made seven more pigeon cases during that escapade. However, we did catch several chaps with illegal deer and closed-season quail, so all was not lost. Plus, I had grown up some more and had learned several lessons in the offing. One of the main lessons learned from this adventure was to continue listening to the nervous rattling of my two guardian angels (I have two because I am twice as big as the usual Yogi Bear). Over the many following years in law enforcement, they have saved me from suffering much grief...

Because of other pressing enforcement issues, dog and I finally returned to our duty station in Martinez. Man, after only one hour being back in the closed, stuffy confines of my office, did I realize just how much I missed the outdoors... But that was part of my learning curve in the federal service and I was glad for that opportunity. An opportunity that led me up the ladder until I became a Special Agent in Charge in Denver (the youngest such supervisor in the Nation at that time) several years later. All in all, a career and vision quest that was to prove more than life fulfilling.

By the way, that New House #5 bear trap now resides proudly in my den (no pun intended). And the real treat is this: Until these lines were written, no one in my family ever figured out where I had procured such an antique. I always told them I had picked it up in an antique shop for $45. If they only knew... You know the really great thing about that adventure was the validation once again of the value of my guardian angels and their warnings. Plus, there was another silver lining in that adventure (no pun intended). That decoy Morgan silver dollar turned out to be one from Carson City, Nevada—a scarce mintage that today is worth about $1,750. Not a bad morning's work after almost falling into a trap, eh?

# CHAPTER TWO

## TWO WEEKS IN "HELL"

WRITING UP CASE REPORTS one morning in my Colusa office early in the spring of 1974, my phone irritatingly rang. Not happy with the interruption because I was trying to get my paperwork done and get back out into the field, I still managed a polite answer for my caller.

"Good morning, U.S. Fish and Wildlife Service. May I help you?" I responded.

"Well, I am glad to finally find your miserable carcass working in the office for a change," boomed out the voice from my new boss in Sacramento, Agent in Charge Jack Downs.

"Dang it, Chief. I am trying to get this crappy paperwork done so I can get out into the field and do some real work," I shot right back with a "smile" in my voice.

Jack just laughed and then said, "'Tiny,' I

need you to give Ben Crabb a hand. He needs to take some time off and get out from under the Burlingame Office workload. But he can't do that unless he has someone to backfill for him on his import/export port of entry duties. You are the closest agent at hand so that leaves you, my Man," he continued.

*Ugh!* I thought. All I needed was a stint in the Bay Area doing boring inspection duties on wildlife imports coming into the airport and docks from all over the world.

"I need you to hook up with Ben so he can show you the ropes before he leaves for some much-needed vacation time. Can you do that?" he asked.

"No," I replied with a knowing grin over what was coming next.

"Good. Then it is a done deal," he continued, knowing full well I hated city life, especially anything to do with the San Francisco Bay Area and all the fruit loops who lived therein. "I will let Ben know you are willing to give him a hand and then the two of you can work out the arrangements as to what needs doing and how it's to be done before he leaves," he continued as he ignored my earlier smart-assed "no" to his question of my availability.

Knowing my continued grumbling would do me no good, I teasingly advised my boss that as of now, that would be the last time I would take him duck hunting in Colusa County in retribution for the assignment he just gave me. Jack in turn told me I was fired and then after a good-natured laugh by the both of us, the two of us got down to other business issues at hand.

"Hello, Ben. Jack tells me you need some help. What can I do you in for?" I asked over the phone the following day.

"Hi, Terry. Yeah, the wife says I need some time away from this hellhole. So I asked Jack for a couple days off so my bride and I can do something together. Sorry to do this to you, Old Man, but you were the closest agent, so you got stuck with this detail," Ben replied.

"That is OK, Ben, but you owe me one," I said with a grin. To my way of thinking, Ben had a crappy job. He was a Special Agent who was nothing more than a glorified wildlife inspector—a position whose job was to clear all wildlife shipments entering the United States through the official port of entry located in the San Francisco Bay Area. That meant checking airline passengers, shipping docks, truck terminals, U.S. Customs Houses, and the mail facility at Oakland as needed, on all wildlife imports and exports. To you unwashed readers, that could be up to a jillion requests for import/export assistance daily through that very busy port of entry.

Me, I would have rather taken a swim in a San Francisco sewer pond in mid-July than do that kind of work! Mainly because it was a damn city assignment and secondly, I knew nothing about wildlife imports and exports plus all the related paperwork headaches associated with such regulations and duties. But Ben was a friend and a fellow officer in need, so I pitched in and gave him a hand.

Besides, at some point in time in my career, I was going to need to understand this portion of my new chosen profession. Now was as good a time as any...

"First, I need to show you around to all the locations to which I have to make physical inspections on the imports. Then I have to get you cross-credentialed as a Deputy U.S. Customs Inspector so you can gain entry to their closed secure areas. That means a trip to the District Director's Office for the U.S. Customs Service so you can get cleared and sworn in. Then I will have to spend several days teaching you how to identify all the exotic live critters, their parts and products. That way you can clear or refuse them entry under our existing import/export laws. That will also include learning to identify deleterious species and become familiar with those regulations as well."

Man, the more he droned on about the upcoming assignment, the more I realized this was going to be more than a two-week stint in that slophole known as the San Francisco Bay Area! By now, I suspect all you readers can tell I am a country boy just like John Denver used to be before he augured his experimental airplane into the ocean. Plus, in those days if you thought highly of Bay Area city life, all one had to do was take a walk in any San Francisco city park. First, one had to look out for all the used needles lying around on the park's grounds ready to jab any unsuspecting person in the foot. Then one had to learn to ignore all the illegal sweet alfalfa smells of marijuana wafting through the otherwise-polluted city air. That was topped off with various parties of partially clad, bearded chaps cavorting through the underbrush. I think all you readers get my drift this was not to my way of thinking a 'class' detail... Bottom line, I would have rather taken on a grizzly bear with a pocket knife then spend any time in that hellhole! After Ben had finished filling me in on what we needed to do, we set up a meeting time and place in San Francisco for the next day. What a day that would turn out to be!

Rolling out at daylight the next morning, I left my quiet countryside and headed into the noisy and crowded Bay Area. Soon I went from peaceful country roads to highways, then to freeways and finally onto the main heavily traveled traffic arteries into the maze of Bay Area cities. It wasn't long before I was bumper-to-bumper with a mess of crazy people going seventy miles per hour as we flew down the freeway. All someone had to do was look cross-eyed at someone else and we would have been in a 100-car pileup! It seemed it was all I could do to keep the high-speed crazies from driving right up into the back of my new, three-quarter-ton, stick shift, Dodge patrol truck. And then, because I was driving a truck, I had to pay an arm and a leg in tolls as I passed from bridge to bridge.

Then I really got an introduction into the perils of city driving. Roaring down the "90-lane freeway" at speeds approaching that of a Boeing 707 jetliner at takeoff, I saw a bumper fly off and sail high into the air from about twenty cars in front of me! Then more parts of cars began sailing crazily into the air in many of the lanes in front of me. I tried to slow down but the idiots behind me, not having seen the ongoing wrecks occurring to the front, were pushing me for all they were worth. Then about ten cars ahead of me, vehicles began braking wildly, swerving and driving up over an obvious small obstruction in the freeway. I could plainly see each vehicle ahead of me jump up slightly into the air as they sped along, obviously running over something.

Then it came my turn in the bumper-to-bumper crowded lane with nowhere else to go, to drive over the mystery obstruction lying directly in my lane. Arriving at the obstruction, I was horrified to see the mangled and smashed "sail rabbit" shape of a squashed, flatter-than-a-flounder human form, lying right in front of my speeding truck. Still flying along at 800 miles per hour and with nowhere to swerve to miss the human remains, I, like all the rest of the idiots, drove right over what was left! Let me tell you, that was not a good experience to remember... I still can feel the soft-smashing thump when my three-quarter-ton ton Dodge truck with its ten ply tires crunched over the remains of that chap. It was the unique squishing sound like hitting a small deer that one does not ever forget!

Finally arriving at the District Director's Office for the U.S. Customs Service, I parked my truck and wobbled out from the cab, somewhat lamer in the head and soul from my recent city driving experience. Walking around to the front of my truck, I observed splatters of blood and what appeared to be some sort of intestinal fluid splashed all over the front underside of my truck! *Welcome to Ben's world,* I thought. Sadly, there was no way I could stop and try to render any kind of aid to the smashed lad lying on the highway. To do so would have ended up with me being killed by all

the other madmen rushing into the city and causing a 50-car pileup in the process. Not to mention
ending up as a "sail rabbit" as well. Welcome
to apart of the American dream, I disgustedly
thought. Thus began my "two weeks in hell"...

For the next week, I worked with Ben after being sworn in as a Deputy U.S. Customs Officer.
During that time, I learned where I had to go to
do the inspections, where to take live wildlife
specimens seized from illegal shipments, where
to get treatment from venomous snake bites if
bitten, and how to identify every kind of critter
or its parts being imported. Finally, Ben left me
standing there like a great big bullfrog in a lily
pond without any lilies, as he left for a few days
to get away from that ugly rat race.

For the next week, I came as close to living in
hell as any human being can. Every morning, I
would fight my way into the Bay Area cities' traffic jungle, trying to avoid ending up like that lad
I had run over after ten other folks in front of me
had already done so. Then moving from one part
of the Bay Area to another over its many bridges,
each charging an arm and a leg toll fee, I soon
discovered how to put a daily $30 dent in my
wallet. And that was when $30 had some buying
power unlike our money of today in 2011, after
the Congress and our "lap dog," union-trained
President have spent our U. S. Treasury dry.

Then, from the docks, to the airport, to the Oakland Mail Facility, to the docks, and back to the airport I would go on a daily basis. In between, I would be hauling live wildlife to facilities that cared for such seized shipments, or to Service secure warehouses to store illegal seized parts and products. Then, I would do the same all over again the following day or night as requested by the zillions of importers of record (the import/export facilities, especially those handling live wildlife, never slept). That was especially so with live bird shipments. During my first live bird inspection, I was appalled. There were at least 300 small tropical birds all wedged into a small container! There was only one water container (empty and full of thirsty dead birds) and one seed container, long since empty. Lying in the bottom of the cage were about 100 dead birds — to many heartless importers of record, an acceptable loss... I discovered that was the norm for most small live bird shipments, not the exception! Hence, when I had live bird shipments arriving at all hours of the day or night, I rolled to preclude as much misery on the part of the critters as I could.

One day while in route to the Steinhart Aquarium with a sackful of seized poisonous snakes, I had a slight mishap. Approaching a stoplight, I had a stream of underfed hippies vomit out into the street in a fistfight over who was the ugliest and smelliest of the lot. Catching me by surprise when they careened out from between two parked cars, I locked up my brakes to avoid getting "yuck" all over my truck. In so doing, I threw all the boxes and bags of seized items off my front seat, into a spilled-out jumble onto the floorboards. Skidding sideways, I managed to miss my fighting mess of hippies but came to rest violently against a concrete curb next to a fire hydrant. In so doing, the hard collision with the curb blew off my two outside hubcaps. After all, I didn't want to get "sticky hippie" all over the front of my new truck, did I? Pissed-off beyond all get-out, I bailed out from my truck ready to break a mess of hippie heads like a batch of overripe watermelons. Fortunately, one of San Francisco's finest was there writing another chap a parking citation. His active presence precluded me from helping out humanity by eliminating a mess of San Francisco's finest class of hippie citizens. After the marijuana smoke had settled down so I could see and take stock, I returned to my truck full of boxes and sacks of whale ivory, whale meat, live poisonous snakes, and hundreds of live, endangered yellow-spotted side-necked turtles. The turtles, all of silver dollar size, had been imported illegally to be used as a food item in the nearby Asian markets.

Once back at my truck, I discovered that my front seat was now "occupied!" There in all his six-foot magnificence sat a coiled cobra somewhat pissed-off at being jostled all to hell in a near miss wreck with a bunch of hippies. Then that was followed with the indignity of having a mess of wooden boxes of seized sperm whale carvings landing on top of his sack, tearing it open in the process! As I was soon to discover, I now had seven venomous snakes loose in the front seat of my truck! And to top that off, all of them were as pissed-off as a mess of Irishmen without a cool brew in hand on a hot August day. Quickly slamming my front door, I hotfooted it over to my man in blue and requested some assistance. We soon had a snake handler from the aquarium in route, while the cop and yours truly watched my truck for any escapees emerging from the cracks and crevices in the cab. In so doing, I had already reloaded my .44 magnum handgun with six rounds of bird shot that I always carried in a box on my dash as rattlesnake medicine when in the outback. I figured if any of my "pets" escaped, they would just as soon die from a load of lead shot as from some extreme form of death after biting a hippie! Talk about a day from hell. I was now smack dab unhappily in the middle of one.

Especially after my snake handler had arrived

and found all the snakes except a mean-assed, four-foot-long, "will bite you whenever they get the chance just for the hell of it" black mamba! We never did find that snake. Well, I can't say never, not that day anyway. A month later when I had my truck in the shop for some mechanical work, the mechanic found my very poisonous snake. Well, what was left of him anyway. In all the confusion, that damn snake had crawled up under and through the dash of my truck. Liking the warm not-now-running engine, he had crawled out and made himself at home on the still-warm engine block. Unfortunately or fortunately, depending on which end of the snake you are standing by, he ended up getting the "chop." When I had started my truck later after all the other snakes had been captured, my snake had his head and neck near the fan blade. When I started up my truck, it had apparently struck at the whirling fan blade, losing his head and ultimately that battle! Then my "friend" the snake became fried snake, as his headless carcass laid intertwined around the spark plug wires for the next month on the hot engine block as I moved to and fro. Then with all the drying engine heat, he became a mummified Egyptian Pharaoh snake...

Anyway, for the next two days, I spent most of my time at the Oakland Mail Facility. Into that huge area came many of the imports mailed from

Asia. The Vietnam war was in full swing and into that facility came much of the personal military shipments exported home from the battlefield as well. And along with many of the shipments, both illegal and illegal, came many Vietnamese species of small snakes, spiders and such. So once again I found myself running around in foreign territory, as I checked all those wildlife shipments infested with everything that bites or stings under the sun!

Picture this if you will: A massive, stainless steel table along one entire wall of the Oakland Mail Facility with all the foreign packages being dumped through chutes from trucks outside the building. Since it was tilted and high up, all the mailed items just slid down to the bottom of the table. There they were separated by Customs officers as to class of imports. All fruit and plant items went to the Department of Agriculture and were stacked into one huge, fly-blown and maggot-infested pile. That was because many packages were maggot-infested and fly-blown from all the smashed fruit items in the packages. All wildlife parts and products went to the U.S. Fish and Wildlife Service in another huge pile, and so on. Then in my particular case, it was my job to open and inspect every package being imported from the Orient. Shoes, belts, purses, watch bands, coats, fur-trimmed clothing, Asian medicines, raw skins, you name it and I was confronted by it for identification, clearance or denial purposes.

There I would identify the wildlife part or product and check the items imported against what the numerous associated permits or laws allowed. When I wasn't smashing bugs, picking off spiders, removing maggots, squashing ticks, throwing dead snakes into the trash, and the like, I had my hands greasy, smelly and full (we never used rubber gloves in those days) checking the imports. And in between times, I found myself constantly dusting off my clothing and hands from all the DDT powder that accompanied many of the shipments. As you readers can tell, this was a real jewel of an assignment. Oh, but for a pair of hip boots, a cool running mountain stream and a 16-inch rainbow trout on the other end of my fishing line...

It was during that particular time of my wonderful "Two Weeks in Hell" that I ran across a situation that only because of the grace of God did I manage to untangle. Opening up a heavily packed military duffel bag owned by an Army major, I began carefully inspecting the contents. Dirty clothing predominated the duffel bag but in the bottom I made a unique discovery. Rolled up in heavily oiled brown paper was a fur rug of some kind. Unrolling the rug, I was surprised to find myself looking at a complete tanned tiger skin, head, tail and everything in between. Now in those days, the 1973 Endangered Species Act forbid the importation of such items. Being an endangered species of cat, entry into the United States was prohibited. Seizing the tanned skin, I photographed the same and began the seizure and subsequent initial forfeiture paperwork process. Finishing, I placed the tiger skin off to one side in my seizure pile and refilled the major's duffel bag with his cleared items. Clearing that duffel bag's contents for entry, I proceeded checking the next crate of personal military shipments. To my surprise, I discovered another illegally tanned tiger skin. Placing it off to one side on my seizure table next to the earlier skin I had seized, I reloaded the crate of personal military goods for continuance of its journey to its rightful owner's home in the United States.

Finishing my work at that location for the day several hours later, I began gathering up my seizures for transport to a secure locker Agent Crabb had designated for such items. Picking up the like-in-size tanned rugs, I was surprised at how much one weighed over the other. Putting both tanned tiger rugs back down on the seizure table, I once again hefted each rug. Sure as God made crawdads good catfish bait, one rug was considerably heavier than the other! Taking the heavier tanned skin which had been imported by an Army captain, I rolled it over on the sorting table with its hair side down. It was then I noticed the rug's backing was thicker than that from the other seized tiger skin. Patting the padding, I noticed it was filled with a more solid-like material than the lighter tiger skin. Taking out my Buck knife, I cut a small slit in the backing material and lifted it up for a look-see. I was amazed to see tightly wrapped stacks of U.S. $100 bills! Shocked at what I was seeing, I slit open that tiger skin's backing to reveal several dozen tightly wrapped stacks of $100 notes. Thirty-six thousand dollars' worth to be exact! Then I noticed the head of the tiger was also very heavy. Taking my Buck knife once again, I slit open the backing material on the skull, only to find a plastic bag of a dark substance tightly packed into the tiger's skull area! Pulling out the package inside the hollowed-out skull, the contents inside appeared to be some kind of a drug! Calling over a nearby Customs agent, we examined the items just discovered. As I soon discovered, I was in the middle of a major black tar heroin smuggling case, not to mention the smuggling and importation of thousands of dollars of undeclared U.S. currency!

As I was soon to discover, a ring of crooked Army officers in Vietnam were using tanned tiger skin rugs to smuggle back into the U.S. black tar heroin and undeclared U.S. currency. I had inadvertently stumbled onto a clever way of smuggling into the U.S. not only huge amounts of cash but drugs as well. However, the Army lads did not understand that importing tiger skins into the United States was illegal under U.S. laws. Then they had not counted on those goods being intercepted by a sharp-eyed, extremely intelligent, quick-sensed, brilliant Service officer, who was also extremely handsome as well (read clown) to foil their little plot. To all my slower readers, that description of course aptly described me to a "T"! Well, more or less...maybe less, eh? Those six officers discovered that they were soon being given a trip back to the U.S. that could not be described as R&R. Especially when I discovered eight more such tiger skin shipments being imported under their legal names and service ranks throughout my following days of inspection at the Oakland Mail Facility!

After that bit of excitement, I managed to catch up on all my import/export duties for a short period of time. Then to avoid going nuts with all that kind of import/export paperwork I was now faced with, I decided I was going to look into the Bay Area pet trade for illegally imported wildlife that had slipped by both Ben and me. Plus that would give me the opportunity to travel around and sample some of the Bay Area's better Asian eating places. And being a 320-pound Asian food epicure and chow hound, that was the only part of my city assignment I really looked forward to.

Now in those days, good Asian grits aside, it seemed everyone wanted baby sea turtles of every make and model for their personal turtle collections. So, because of the demand and subsequent value, those gentle critters would be smuggled into the country in every manner and number possible. I figured I would seize as many as I could, being illegally imported. Then those I missed, I would run to ground in the Bay Area's pet stores in between my legendary Asian food sampling expeditions throughout the streets of San Francisco.

Now those kind of identify-and-grab details were fairly simple. Just walk into any pet store and look over the turtle collections offered for sale. Those selling for a few cents on the dollar were for the most part legal. However, as I got closer to Chinatown, I had to be on the lookout for the little endangered yellow-spotted side-necked species of turtle as well. It seemed they were one hell of a food item to many of the Asians, hence they were being smuggled into the U.S. by the tens of thousands. However, since that was contrary to the existing Endangered Species Act at the time, that was a "no-no." So I found myself for days on end, bringing to the local authorized aquariums, bags containing numerous seized yellow-spotted side-necked turtles. But that was alright. Because many times I would find $39.95-priced, baby, endangered sea turtles in among the yellow-spotted side-necked turtles as well. Those, too, would be grabbed off and once again trips to authorized aquariums would be the word of the day. That and another $15 in bridge tolls...

I remember one day in which I walked into a rather large pet store in San Francisco. My usual routine was to walk into the store like "Joe Dumbhead" and just be looking around for "that special kind of pet." The shop owner would always try to be accommodating and sell me whatever was at hand. But I would continue just walking all around looking for that "special" kind of pet to take home to my aquarium. After about 30 minutes of not being able to sell me anything, the many-times greedy and frustrated shop owner, wanting to make a sale, would voluntarily drag out his "specials" (thereby allowing me to dodge the Fourth Amendment search and seizure restrictions). Therein that class of show items would almost always be those species of birds or other critters that were illegal. Anyway on that particular day, I hit a gold mine.

Looking for that special turtle critter for my supposed collection, my shopkeeper stepped over the line of legality and good common sense.

"Sir, I may have just what you are looking for. Come with me into the back of my shop where I am keeping those unique species for just the type of discriminating buyer you appear to be," he greedily advised.

Following him into the back of his musty, floor-water dampened shop (to keep the humidity up for certain species), he took me to a huge aquarium behind a drawn curtain. "I think behind this curtain you will find what you are looking for, Sir," said my shopkeeper as he slid back the curtain with a flourish. When he did so, my eyes damn near fell out of my head! In his large aquarium were about 150 baby sea turtles swimming about (actually 167, as I later learned). My shopkeeper had just fed the animals wads of lettuce and what a spectacle the feeding turtles made. Realizing I had just hit the mother lode, I stepped closer for better identification purposes. In that tank were at least four species of protected endangered sea turtles!

"Damn!" I said in genuine enthusiasm. But not the "buying" kind of course...

"I just knew you might find what you were finally looking for," said an excited shopkeeper with dollar signs whirling across his narrowed "snake in the grass" set of eyes.

"Holy cow!" I said. "Man, I will take every one of these sea turtles, cash on the barrel head!

I suggest you get to figuring what I would owe you for this lot," I said in an obviously highly excited tone of voice. "How did you get so many sea turtles?" I once again mentioned their name to make sure he knew what he had to satisfy any subsequent court proceedings when it came to knowledge of the illegal species and the intent "for sale" on the part of the shopkeeper.

"I have a buddy who drugs these sea turtles and smuggles them in the false-floor bottom of crates of heavy machine parts from Hong Kong," my shopkeeper proudly uttered.

"Can he get me more?" I asked, acting like I was still in shock over my good fortune (no longer an entrapment issue—he had voluntarily offered me sale of the endangered species knowing full well what he had).

"Sure can. He is smuggling another shipment into the port tomorrow," he quickly replied, greedily realizing he had a paying sucker on the hook and a big one at that!

"Great," I said. "I will tell you what. I have a need for all he is smuggling in tomorrow as well as for my pet shop. How many more will he be bringing you tomorrow?" I asked.

"I ordered three hundred more green sea turtles. So minus whatever dies in transport, I will have up to that number for sale," he replied as his eyes once again filled up with dollar signs.

"What time tomorrow do you think he will be here?" I asked.

"He usually clears Customs by ten in the morning. He can be here anytime around noon," my greedy shopkeeper replied.

"Great! I will tell you what. You tally up what I owe you here for all these sea turtles. I will have you keep them here tonight for me if you will. And in so doing, I will pay for their storage and keep as well (I did that in order to keep my greedy outlaw on the hook). Tomorrow, I will bring my live transport truck and we can offload these from you, as well as the newly smuggled ones. In the meantime, I need to go to the Bank of America here in town and draw out a mess of cash in which to pay for these turtles. That way, I am not leaving any trail for the Fish and Wildlife Service to pick up on and that protects you from selling endangered species of turtles as well," I said. To my readers, notice how I got rid of any lack of "intent and knowledge" issues if this case ever came to trial. Especially with the use of my words relative to the "Service trail" and the "endangered species of turtles identification as well."

"That is a great idea," he said now getting as excited as I was pretending to be. "I don't want no trouble with those damn feds either," said my new friend the shopkeeper.

Walking over to the shopkeeper, I reached out and sealed the deal with a handshake. "Keep feeding them," I said as I turned to leave the backroom of his store. "I want to make sure all of them are fit for a long trip of travel in my truck," I said over my shoulder.

Around ten the next morning, I entered that same shop. Only this time, I was carrying a briefcase. I was hoping the show of the briefcase, would cement in my shopkeeper's mind that I was a real businessman and the case would be full of money. As expected, my shopkeeper was happy to see me and could hardly keep his eyes off my briefcase.

"You got me a bill of sale made up for all the critters you have currently in the back tank?" I quietly asked.

"Sure do. However, I have listed all the critters as snakes, puppies, kittens, and every other kind of run-of-the-mill legal pet," he smugly advised.

"Great idea!" I said, trying to fluff up his ego. "That way everyone from the Fish and Wildlife Service to the Internal Revenue Service will be thrown off the trail," I continued, with what had to be a cautious gleam in my eyes.

Looking over some more of that pet store owner's stock, I got the high sign from my dealer. Walking over to my pet store dealer, he beckoned me over to the back of his store as he left another employee in charge of the front of the store.

Walking into the gloomy interior of the back of his shop, I was met with a burly, swarthy-looking man with a mustache that would have made any man proud to 'sport.'

"Terry, this is Jacque. He just brought me in another shipment of sea turtles. There are only 233 live ones though. I know I told you I had ordered 300 but the rest died in transit. But you can have all of the new ones as well if you want," said my pet dealer with just a tad bit of excitement creeping into his voice.

"May I see the new turtles?" I asked.

My pet dealer walked back to the curtained aquarium and drew it back. Therein swam 233 green sea turtles. "Where are the others from yesterday?" I asked, fearful they had already been disposed of to other buyers.

"They are over there in those transport tanks ready for loading into your live truck," said my pet dealer.

Relieved, I walked closer to the tank for a better look at my latest "acquisitions." They all appeared to be "fine as frog hair" and I began to relax. "What is your asking price for these green sea turtles?" I asked.

"I will let these go for $30 each or $6,990, since I promised you another 300 and only 233 survived," said my pet dealer.

"Draw up some more phony paperwork and

you have a deal," I said with an ever-widening grin taking place on my, of course, handsome puss and in my "black heart and soul." *Before this is said and done, this will end up as another one of those "lions one, Christians nothing" scenarios,* I smugly thought.

Then the three of us went into the pet dealer's office and he drew up the rest of the paperwork on the recent addition of his 233 green sea turtles smuggled in from Hong Kong. When he handed that bill to me for my examination, along with the earlier paperwork for the mixed bag of 167 sea turtles ($6,671.65), I pushed a small button located alongside the handle of the borrowed Customs undercover briefcase.

Six burly Customs agents outside in a panel truck listening in on our briefcase-recorded conversations, upon hearing the all-clear signal from my discussion, bailed from their undercover truck and thundered into the pet store. Jacque, realizing something evil was now afoot, tried to run. His "running road trip" was short-lived as two burly Customs agents accustomed to working the rough and tumble San Francisco docks, soon had him in by his "stacking swivel." As for my pet dealer, he made a grab for the bill of sale documents I was holding. Well, my mother didn't raise just another turnip in a 320-acre field. My pet dealer only got his hands on a very determined, not to be touched (but still beautiful of course) Special Agent of the Fish and Wildlife Service. My pet dealer's "run for the roses" ended him up in a large "weed patch" for selling endangered species of sea turtles.

Well suffice to say, the Steinhart Aquarium got another load of endangered sea turtles for safe-keeping and eventual release back to the wild. Additionally, the Customs Service got their two lads as well. As I later discovered, Jacque went under the heel of the federal courts for smuggling illegal wildlife. His heavy machinery import business went under shortly thereafter. As for my pet store owner, he was forced, through shall we say some rather steep federal fines, into selling his place of business. As for me, I had one hell of a meal at the China Dragon Restaurant in San Francisco afterwards for my efforts. If any one of my readers ever goes to San Francisco to partake of some of the finest Asian cuisine food in the world (except that found in Hong Kong and Malaysia), try the Peking duck. It is out of this world in most San Francisco Asian eateries. But don't eat any of the endangered yellow-spotted side-necked turtle soup. Yes, that critter is still being smuggled illegally into the United States. However, on good authority many say it has an annoying habit of catching any one eating such a dish in the last part over the fence, if you get my drift...

Sometime later, my "two weeks in hell" came to an end. Man, I was more than happy to get away from the Bay area once Ben returned and I let him quickly have the reins back once again. However, my trip into the endangered species world didn't just end there. Several years later, I found myself in the Washington, D.C., office as a Senior Special Agent in charge of the Division of Law Enforcement's Endangered Species Desk! Can you believe it? Taking a waterfowl field officer and placing him in charge of the Division of Law Enforcement's Endangered Species Desk! Talk about an out-of-control government in their goofy damn assignments.

However, that move ultimately proved out for the better in the long term. During my stint in Washington, I became so frustrated at getting endangered species parts and products from the field officers to be identified (no one would help or was qualified to do so), that I came up with the idea of a wildlife forensics laboratory for the U.S. Fish and Wildlife Service.

When that lab finally came on-line after many battles, legal, political, Service, and ten years of fighting for such a facility, it has proven itself a crown jewel for Service officers. And at the same time, anathema for wildlife violators, both domestic and international. No longer do importers of record or other world-of-wildlife killers, get away scot-free when it comes to exact species identification. When the Special Agents and the National Fish and Wildlife Forensics Laboratory get on the violator's trails, it is a slam dunk. Anyone breaking the national or international Endangered Species Acts or domestic wildlife laws requiring species identification are going to pay for their efforts at circumventing the federal laws of the land once they are scented. Endangered means just that! And that is one word the Fish and Wildlife Service's Division of Law Enforcement takes seriously. Extinction is forever and that is not lost on the field or forensic officers of the Service.

Lastly to all my conservation officer readers, learn from what I learned. No matter the assignment you draw, face it squarely head on and do your best. You will never know where a crappy assignment will take you. In my case, it led to a better knowledge and understanding of what all the wildlife of the land is going through. In the current instance, I was taken out from the "Dark Ages" as a single-minded, waterfowl field officer and placed into an assignment that eventually led to the concept and inception of one of the greatest wildlife forensics laboratories in the world. And in so doing, gave a leg up to those species in the world of wildlife struggling for their very existence.

Until my retirement in 1998, I remembered the lesson so abruptly handed to me in the San Francisco Bay Area those many years ago. That lesson in learning has served me well over all those subsequent years I served the American people. From grass and mud one can build bricks and eventually a strong house will arise. "A true conservationist is an individual who knows that the world of wildlife is not given by his fathers but borrowed from his children." So let it be written, so let it be said...

# CHAPTER THREE

## WHEN HOLDING ONTO A BISON'S TAIL, THERE IS ONLY ONE HOLD SHORTER...

THE AMERICAN BISON is the lifelong symbol of the U.S. Fish and Wildlife Service. For many years I never figured out why such a representative logo. However, after serving only ten years as a law enforcement officer in the Division of Law Enforcement for the Service, I soon came to experience and realize in my mind's eye, why such a symbol.

I have always been a bit of a romantic dreamer and possessed a lifelong love for everything related to American history. As such, lying in the annals of my mind taken from somewhere is a figurative picture of what I speak, relative to that bison logo. In that mind's eye, I can picture a young Native American holding onto the tail of a madly whirling plains bull bison. From behind the animal's shoulder sticks an arrow, soon to be a mortal wound.

In that visualization, I can picture what has just happened. The young Indian lad was on a bison hunt and shot his arrow deeply into the side of the fleeing beast of the plains. The bison in his life-ebbing agony must have dropped to the ground sometime during that melee. The Indian lad in his innocent excitement, dismounted and approached his dying quarry. However, life is sweet to this noble animal of the plains and he objected mightily to his approaching dying moment. Rising with great difficulty from the ground, the bison attempted to defend itself from his lethal antagonist.

The young Indian lad, first in surprise and then in abject desperation, latched onto the tail of the now-enraged bison in order to save his own life. In so doing, he is holding on for dear life as he is being violently whirled to and fro. If he lets go, he will be trampled or gored to death. If he hangs on, there is a chance to escape with his life if the bison soon expires from this mortal wound.

Standing off to one side, too far away to run to for his escape, is the Indian's quietly feeding pony. A pony more interested in eating a clump of prairie grass than the life-and-death struggle being carried out by his nearby owner and the bison. Scattered around on the ground by the bison and the young Indian lad rests his bow and many scattered arrows. They, too, like his horse are too far out of reach to be of assistance.

With that picture, comes to my mind's eye that there is only one hold shorter on the maddened bull bison for the young Indian lad on which to grab...

Having worked for the oft-times poorly led Division of Law Enforcement with its sometimes lacking leadership in the field and the Washington Office, along with being overwhelmed by the magnitude of the profession's demands, brings to mind the mental picture of the Indian and the madly whirling bison. Then throw into that mix the fact that the officers in the Division of Law Enforcement were historically financially strapped for operating funds, saddled with poor equipment provided to the Service from the lowest bidder, and faced with the extreme everyday odds of violator versus under-staffed law enforcement officers. All that quickly brings to mind the "one hold shorter on the bison scenario." In that mind's eye looking at my Division of Law Enforcement and extrapolating all its challenges, I can truthfully say Custer had better odds...

It was into that figurative scenario of man versus bison that I was thrust upon receiving a phone call one early spring day in 1974. Picking up the phone, I identified the voice of my boss Jack Downs.

"'Tiny,' Jack here. Are you sitting down?" he asked.

"Of course, Jack. I am working at the desk in my office. Why do you ask?" I responded.

"As you are well aware, the Division of Law Enforcement is creating a new first-line supervisor position titled Senior Resident Agent. There will be 45 such positions coming open in the very near future and everyone's hat will be in the ring."

Now in those days, the Division of Law Enforcement had various-graded field agents and a single first-line supervisor titled Agent in Charge. Each Agent in Charge had a geographic district and a number of lower-graded agents working for him (we had no females in the Division of Law Enforcement in those days). The division in Region 1 (California, Oregon, Washington and Nevada) also had an unwritten rule: One did not apply for any type of promotion in the ranks unless he had successfully held a field agent's position for at least seven-and-a-half years. At the time and place of Jack's phone call in 1974, I had only been employed as a Service field officer for somewhat less than four years. I previously had four-and-a-half years' experience as a California State Fish and Game Warden. But those years of state service, in the federal promotional scheme of things, did not apply to the arbitrary "seven-and-a-half years rule" in Region 1 when it came to promotions.

"Jack, what does that have to do with me and the high price of crowbars in Korea? I have been with the outfit less than four years. So, that promotional scheme doesn't involve me and won't for at least another three years," I replied.

"Well, that is why I am calling you. I have nominated you for one of those new positions."

In my personal mind's eye, the deadly arrow had just been shot into the fleeing-for-his-life bull bison...

"Well, how can that be? I don't have enough time in grade to compete. Besides, I am happy at my duty station with my new home here in Colusa and the unlimited numbers of bad guys to chase around in the Sacramento Valley," I replied.

"Well, I nominated you anyway. You have an excellent work record under my command, and had a damn good work record with Fish and Game. So, I nominated you," he continued.

Surprised and pleased that my boss thought so well of me and my work, I said, "Well, I am flattered. But I don't have a chance in hell being so junior in grade for any kind of a promotion over the 150 or more senior-graded officers currently in the Service, so suit yourself," I said. Jack just laughed at my promotional chance's negativity. Then we got down to other more pressing operational business. That business being the numerous commercial market hunters in the northern Sacramento Valley and the many illegal ducks still moving in commerce in the San Francisco Bay Area.

For the next few weeks, no one who was qualified for the new positions knew where they would be going if selected. But just for the hell of it, I looked over the possible list of new supervisory duty stations. I completely overlooked the choice western duty stations like those in Idaho, Montana, Oregon and Washington. I did so because I knew those appointments would be going to very senior, experienced, time in grade officers. In so doing, I only saw four remaining locations out of the 45 proposed that I would really like to be assigned if selected. But once again, I realized any quality duty stations would be gobbled up by more senior officers and rightfully so. So being busy as I was, I hit the road and did not let the promotion works in progress cross my mind. Well, to be truthful, not too much... Come later that August, I was notified that I had been selected as a new Senior Resident Agent!

I now found myself personally holding onto that damn bison's tail in my mind's eye as tightly as I could, in order to avoid being gored or trampled to death...

That official promotion notification came one afternoon when I was on the road. As it turned out, my wife got a phone call from the Regional Director in Region 6 (the big boss) who wanted to talk to me. My Bride advised "the voice" on the phone that I was on the road working covertly and could not be reached. The Regional Director was more than perplexed because the timing of my move to the new duty station was of the essence because of the oncoming fall hunting season.

"Can't you tell me where we are being sent?" asked my Bride.

"Well, I am not supposed to because you are not the person selected. But because we can't reach your husband, I will tell you anyway in case he calls in from his present assignment. Mrs. Grosz, I am sorry to tell you, you are going to Bismarck, North Dakota!"

"Bismarck! That is one of the duty stations we wanted to go to if we were selected," squealed my now very happy Bride (we both had family ties in that area).

"Well, I am sorry," continued the Regional Director, ignoring what he had to think was an agent's wife who had gone daft over that Bismarck duty station announcement. Continuing, he said, "but you folks only have 30 days to move because that is where your next paycheck will be sent! So, I recommend that you get moving in the house-selling department."

Now, I found me and the mad wounded bison were really dangerously whirling around...

I had just built a brand-new home, our first in the Sacramento Valley Town of Colusa. We hadn't lived in it for even one year as of that moment in time, and it was my Bride's dream home! The interest rates, because of the OPEC Cartel reducing oil exports going to the rest of the world, were at an all-time high! That had caused the home interest rates to skyrocket from a normal six to thirteen percent! There was no way anyone could sell a home under those conditions, much less purchase another one, unless they were damn lucky! And I have to be at my new duty station in Bismarck in just 30 days...?

Now that damn bison was stepping all over my arrows, which were scattered across the ground all around us, breaking them into uselessness...

Suffice to say, into the teeth of that financial storm, I had to sell my 3,200 square foot, brand-new, all-brick dream home on an acre of ground, with a shake roof and slate floors within the 30-day, must-move period. Following the Regional Director's orders (remember, he is the big boss), I quickly listed the home for sale. Then I sadly learned, even though he was the big boss, he carried no water with the Service's bureaucratic and cumbersome home sale process. I could not sell my home until the Office of Personnel

Management had notified me in writing that I could sell the home! In the meantime, a new local doctor snapped up our new home in a heartbeat, and in the process, gave us our asking price! Then to satisfy the "move-when-I-tell-you-you-can" bureaucracy, I had to break the contract and "unsell" my home because I had not been notified in writing of my selection before placing the home up for sale! Twelve days later (remember the 30-day deadline), I received the hallowed and blessed home sale paperwork. In the meantime, the interest rates had now soared to an unheard of sixteen percent! I later sold our home for exactly what it had cost me to build, namely $34,600. Imagine purchasing a brand-new, 3,200 square foot, brick home in California for only $34,600! A year later, that very home resold for $250,000! Today, that same home carries a price tag in excess of $650,000!

Now that damn bison had stepped on my bow and broken it in half...

Arriving in Bismarck, we were unable to find another home because of the ongoing oil boom in North Dakota. An oil boom that had gobbled up all the living space for their incoming drilling and roughneck crews. Then through one of my Bride's relatives living in Bismarck, we finally managed to find an unfinished new home still being built. In that purchase, we paid a king's ransom of $24,000 more for half the house that we had just left in California! Plus, my wife had to give up her teaching job in California in order to make that move. The loss of money that represented, as one can imagine, was keenly felt. Not to mention my wife's career was placed on hold so that I might move up through the ranks...

Then the family, all four of us and two large dogs, had to live in one room in a Motel 6 for six weeks. That was because that was all we could afford since the government's housing allowance was so low and our new home was not yet ready for occupancy. Leaving my Bride to cook all our meals in the dingy Motel 6 for a family of four and two Labrador retrievers in her electric fry pan and wash all our clothing in the bath tub, I went to explore my new office and all its warts. However, that was not until a trip had to be made to the local Target store to procure a new pair of shoes for my oldest son so he could go to school. It seemed one of my bored at being cooped up in a motel room all day Labrador dogs had eaten his shoes!

Now that damn horse of mine had moved off to such a distance, that I couldn't let go of the bison's tail and make a quick break for it to escape my wounded, one ton, still very much alive, and full-of-fight antagonist...

At my new office, I discovered one of my officers in South Dakota had also applied for one of the new Senior Resident Agent positions. He was a twenty-year Service veteran, a Second World War military hero, and his new boss (me) had less than four years' federal agent experience. Needless to say, that created some unnecessary friction. That took me the better part of a year to smooth out his bruised feelings over his non-selection. Bruised feelings made even worse because he had been passed over by a junior-graded officer in time, grade and experience, for the very same position he had applied for in Bismarck. Hell, to his way of thinking, I was nothing more than a snot-nosed kid who didn't know up from down!

Another officer in my new squad, as I was soon to discover, was a brand-new lad without any wildlife law enforcement experience. However, he was a warrior. Quickly self-teaching himself, John Cooper became one of the Service's finest officers. And since he had been in North Dakota for some time before my arrival, he drew the short straw in teaching his new boss the wetland easement ropes. Ropes that turned out to be a political nightmare within the Service and on the ground with all the short-tempered, hell-for-drainage-bound, pissed-off at the meddling government, dirt-farming community.

The last man in my new squad, located in the

eastern part of North Dakota, had a work record that indicated I did not have a real fireball in that position. That problem was more than exacerbated when I learned the northeastern part of North Dakota was a hotbed of wetland easement drainage problems! In fact, that area turned out to be one of my busiest and most critical of positions. That officer had also applied for one of the new supervisory positions and had not been selected. That left a bitter taste in his mouth as well, especially since he had 16 more years of time in grade than yours truly. And yet, I had been selected over him.

Now that bison was really getting pissed, as he whirled me to and fro trying to dislodge the one hanging onto his tail for dear life, so he could stomp me into the dust like the small dung beetle I was...

Then I got a closer look at my new duty station's equipment. It was basically worn out trash except for my brand-new, one-wheel drive patrol car. Taking out my new Ford sedan patrol vehicle for a test run, I discovered it was a lemon! In fact, it was even baby crap lemon yellow in color. To my way of thinking, the car's bright yellow color made me look like a well-used baby diaper going down the highway. Also, that damn car sucked in dust like a sieve (remember all the Service's equipment was purchased from the lowest bidder). When I came home after a twenty-mile test run on the typical North Dakota prairie dirt roads, all you could see were my now-bloodshot eyes. All the rest of me was covered in an inch of brown road and topsoil dirt from that dust-sucking piece of crap! Whoever had built that one-wheel drive piece of crap without an AM/FM car radio or air conditioner (both items were not purchased in order to save money), had not bothered to seal the vehicle. Taking it back to the local Ford dealer, I put them to work sealing up the car. However, their efforts were a joke. They never did get that car sealed and for the next year when I drove it (I wore it out in one year!), I came out looking like "Pig Pen" in the Peanuts comic strip. Since that time, I have never purchased any Ford products!

Then there was another problem with my "baby diaper-colored" patrol vehicle. The damn thing only burned unleaded gas. That was when the country was just changing over from using leaded to unleaded fuels. Well, that was fine and good except in rural America, where there weren't many unleaded gas stations! Then take very rural North Dakota. The only places one could buy unleaded gasoline were at a very few service stations in the major cities. As for the countryside, they were in the process of slowly changing over to unleaded fuel because everyone living in those areas was still driving vehicles using leaded gas! So, I was more or less relegated to driving forth into the hinterlands of North Dakota until I had used up one-half a tank of my gas. Then since there was no place to refuel with unleaded gasoline, I had to beat a hasty retreat back to Bismarck to refuel. That pattern of brilliant government mandates to use vehicles using unleaded gasoline before their time went on for the better part of one year at my duty station!

Now that pissed-off bison had pooped all over my hands and the front of my legs, as I still held onto his tail for dear life...

While in my office that first day, I removed 25 messages from my "jug-full" answering machine and responded to another 23 phone calls from my North Dakota state officers and federal National Wildlife Refuge counterparts. All were frantically demanding my presence to do battle with the lawless, local dirt farmers of plentiful supply. Then the Customs officers from my numerous nondesignated ports of entry located along the U.S.-Canada border discovered I was on-line. Instantly, I was deluged with phone calls for assistance on wildlife import/export issues. I was a rather large guy but to answer all those frantic requests for assistance, I would have to be spread thinner than butter on a piece of toast during the Great Depression!

Realizing my Special Agent predecessor had left a hat full of hornets and a bucket load of unfinished business for me, I lowered my head and hit the deck running. In fact for the next calendar year, I drove 58,000 miles and flew another 35,000 miles in my two states of responsibility (I was in charge of North and South Dakota at that time). [Ironically, my older son is a Special Agent supervisor for the Service and is today stationed in Bismarck. As such, he is also in charge of the States of North and South Dakota as I was once was, and just as busy.] In so doing, I completely wore out my brand-new baby-diaper yellow patrol car during that year (mostly dirt road and field driving) and was only home for full days on Thanksgiving and Christmas! The rest of the time I was in travel status or going to and from numerous battlegrounds! Travel status not just because I wanted to be out and about running around, but because I was constantly in demand for putting out major fires or taking on the largest-in-size, irate, illegal drainage-oriented, dirt farmers.

Fires dealing with such violations as smuggling of wildlife across the U.S.-Canada border; dozens of Airborne Hunting Act violations involving the illegal gunning of furbearers with the use and aid of aircraft during the winter months when their furs were in their prime; the illegal killing and sale of eagles by the dozens by Native Americans in North and South Dakota; illegal drainage of wetlands either owned or administered by the Service; cattle trespass issues on many of our 180 National Wildlife Refuges within the two states; theft of valuable trout broodstock from our National Fish Hatcheries; directing covert operations in my two states initiated by the Washington Office operatives; and assisting my two states' fish and game departments on resident species law enforcement operations. That did not include my time meeting with U.S. Attorneys, time in state and federal courts, running roadblocks at my ports of entry along the border, assisting the covert operatives in the field, repairing my heavily used pieces of equipment, travel from Point A to Point B, and sleeping in between. In the process of all that running around and fighting with my dirt farmers and hunting the lawless, I lost 80 pounds in six months from my most beautiful and ample carcass! In short, a "diet" of work that I would not recommend. Then throw into that swirl the mixture of wonderful winter weather associated with working in the high plains. As such, my officers worked until it was 55 degrees below zero, then we quit until it warmed up. In so doing, that created a real "brew" of life on the northern plains to "sip!"

Now that damn hull bison was not only poop-ing on my hands but kicking the hell out of me as well...

To you unwashed readers, North Dakota and South Dakota are major duck-producing areas in the spring and summer months. So much so, that they are called "duck factories" by those in the world of wildlife. The reason those two states are called duck factories is because a zillion years ago when the glaciers moved down throughout those parts of the United States, they gouged out a gazillion small depressions in the soil. Today those gouged-out areas are called potholes or wetlands. Then with the winter snows and high water tables, those wetlands filled with water and eventually plant life. In so doing, that created a veritable living paradise for many species of nesting shore birds and waterfowl.

Then came the rub. With the increased drawbar horsepower on tractors, the local dirt farmers with their newer equipment technologies began draining those wetlands with a desperation-like fury. A fury to put as much land into crop production as possible to offset the low cereal grain prices. And with the drainage of those wetlands went the prime habitat the waterfowl species had come to require and enjoy.

Alarmed at the crazy rate of dirt-farmer land drainage, the Fish and Wildlife Service instituted a small wetlands purchase program to save as many wetland acres as possible. That worked fine for a while, but higher grain prices and better drainage technologies soon eclipsed the Service's small wetlands purchase program. To save even more of the rapidly disappearing wetlands, the Service instituted another wetlands program. In that program, the Service paid the landowners to not drain, burn, fill or level certain existing wetlands on their properties. The lands remained in the farmers' hands, but they gave up the right in perpetuity to drain, fill, burn or level those wetland acres identified and paid for on their properties. Soon, over 10,000 North Dakota landowners were enrolled in that wetland easement program and for the moment, the valuable wetlands survived drainage. A like-in-kind program was instituted in the States of Montana, South Dakota and Minnesota as well, in the '60's and '70's.

Then as the original landowners aged out of the farming business, their kin or other land purchasers took over those lands formally protected by previously executed wetland easement contracts. Soon drainage of those protected wetlands once again became the norm! The Service, through an anemic wetlands law enforcement protection program, was soon steamrolled almost into oblivion by the increasing drainage lawlessness. All justified in the eyes and under the battle cry of the dirt farmers that they "had to feed the world."

Then enter the newly appointed Senior Resident Agent in Bismarck with less than four years' experience as a federal law enforcement officer. One look at the easement contracts showed dangerous flaws in the program. First, the Division of Law Enforcement was only a 'weak sister' when it came to helping the hard-pressed refuge officers trying to hold the line against the illegal drainage of the wetlands protected by government easement contracts. Second, if that wasn't enough of a problem, the Department of Agriculture was paying the dirt farmers to drain like-in-kind wetlands across the prairie pothole states of North Dakota, South Dakota, Montana and Minnesota! Third, the Service's regional attorney, a Regional Solicitor named Jack Little, who was as gutless and useless as teats on a side of bacon, was on his own personal, do-nothing tear. In that misguided, self-instituted policy, he mandated that all wetland easement drainage violations would only be processed civilly. That meant a process that took up to five years to adjudicate, followed by a small fine for such violations and the land restored to its original condition. In essence, a slap on the wrist and the wetland was now, after five years of doing nothing legally, farmed out and wiped off the face of the earth! In so doing

making return of the wetland to its original condition impossible. Fourth, the Division of Law
Enforcement was so financially strapped that, for
a time, field officers had to call the Special Agent
in Charge in my region for permission to buy
each and every tankful of gasoline! In short, the
Division of Law Enforcement, under the above
restrictions, was also as useless as teats on a side
of bacon as well!

That had to change! As the new Senior Resident
Agent in North and South Dakota, I began
looking into the law enforcement program and
what it could do to stop all the illegal drainage
on Service-protected lands. First off, I noticed I
had 363 reported, serious drainage violations in
North Dakota, and 181 in South Dakota! That
alone was a mountain of work for me and my
three officers to tackle. Don't forget, that work
had to be undertaken in addition to all the other
criminal investigations I had on going in those
two states! Looking at the federal law covering the
wetland easement program (16 U.S.C. 668dd(c)),
I thought I could see a loophole. A loophole
that allowed me to prosecute my draining dirt
farmers criminally instead of always going the
useless, solve nothing, civil route. Checking with
Assistant United States Attorney Lynn Crooks in
Fargo and Assistant United States Attorney Dave
Peterson in Bismarck, they supported my thesis

that I could take my illegally draining farmers into federal court criminally.

Then it was off to see my Regional Solicitor in Denver. Regional Solicitor Jack Little tried to back me down from going criminally in order to continue trying my dirt farmers civilly. Suffice to say that led to one hell of a clash between me and my gutless and useless Regional Solicitor. But I had an ace in the hole. I had the backing of the United States Attorney's Office in North Dakota, and hell hath no fury like a Special Agent with the backing of the United States Attorney! But then, Jack Little went over my head into his office in Washington, D.C., to try and stop me in my proposed criminal prosecution process and weaken the North Dakota U.S. Attorney's Office involvement and support! But once again, my guardian angels lent me a hand. I, too, had friends in Washington. In fact, friendly attorneys in the Solicitor General's Office were not only my close friends but disliked Little's lack of guts. Because of those issues, they forewarned me of his interference! Well, my mother, bless her soul, never raised a turnip... Soon, I was out in front of my meddling Regional Solicitor once again. In so doing, I headed for the Chief Federal Judge's Office in Fargo for his blessing going criminally on my hell-bent-for-election, illegally draining dirt farmers on federally administered and protected wetlands!

There I received a blessing from Chief Federal Judge Benson and the race was on for the roses. Well, not quite... Because of the negative sensitivity of the Wetland Easement program in the Governor's Office in North Dakota (Governor Art Link supported the dirt farmers — surprise, surprise), Judge Benson required one thing of me and the soon-to-be criminal drainage prosecutions. The chief federal judge required that the service personally contact every wetland easement holder in North Dakota, and let them know we were coming after them if they illegally drained any of our protected wetlands — all 10,000 of them, with just my three officers and little old me!

Now that damn bison in my mind's eye had just left a beautiful bruise mark on my rump with its horn, as it almost caught me during one of its wild lunges. Damn, I should have shot that dang bison with a .50-70 Sharps instead of that useless arrow...

For the next three months, all four of us agents visited every wetland easement holder in North Dakota as per the Chief Federal Judge's orders! Never in my life had I seen or experienced such animosity or treatment on the part of fellow Americans. They had gladly accepted the government handout when it came to setting aside lands they couldn't farm because of the water and the associated high water tables. And in the process criticized the government for purchasing the rights over such worthless wetlands. But when the price of wheat went up, it soon became apparent all bets were off! In short, most landowners associated with and bound by the Wetland Easement contracts just ignored them and drained the wetlands at will. Then as we personally met with all of them to explain the new wetland contract criminal enforcement programs, they exploded! The agents were threatened with firearms, had dogs sicced on them, threatened politically, chased with pickups, threatened with moving farm equipment, and met with bawling, weeping wives, threatening to stab or throw boiling water on us!

But we got the job done as per the judge's orders. Then several days later after being given the green light, I was on patrol in the Jamestown area of North Dakota. There I spotted a dirt farmer in the act of draining a Type II wetland under easement contract. Thirty minutes later, I had issued the first criminal citation for the illegal drainage of a wetland protected under a Service wetland easement agreement! Brother, you talk about a surprised dirt farmer. He couldn't believe his eyes. In his hands lay a criminal citation for his illegal drainage actions issued to him by a monster-sized federal agent! Heretofore, the Service personnel had just issued warnings and more warnings. In fact, so many warnings that the farmers finally came to believe they were issued by no one but "paper tigers." Now the paper tigers had teeth and man, did that single ticket-writing episode upset the North Dakota apple cart! By the time I arrived at the Jamestown Wetland Office some hour or so later, Dave Goeke, the wetland manager, had received over 30 phone calls from the farm community asking what the hell was going on! The dirt farmers now had their entire world turned upside down. Heretofore, they had done as they pleased when it came to their illegal drainage of naturally occurring wetlands. Now, there was a new sheriff in town and all hell had broken loose!

For the next two weeks, my agents cut a wide swath across the wetland areas of eastern North Dakota addressing illegal drainage issues with the hard-headed dirt farmers. Soon, many of those illegal drainage-and-ditching-oriented chaps on Service-owned or -administered lands were trooping into the magistrate's court in Devils Lake. There they were sternly lectured over the error of their ways, fined $500, and given 30 days in jail! Man, you talk about sets of eyes growing larger than garbage can lids when those verdicts were rendered... Then the judge suspended the 30 days in jail providing the farmer never ditched his protected wetlands ever again. The magistrate judge made sure the dirt farmers understood if they did and came before him once again, they would sit out the 30 days in the local jail! Man, nothing more but big eyes once again over those ominous words.

Now I had removed my hand from the tail and had grabbed that remaining one hold shorter on that danged bison. When I did, damned if he didn't keel over...

For the remainder of my stay in North Dakota, and with a similar program being instituted in South Dakota, I still had my violators. But never again did the Service have such widespread illegal ditching across the northern prairies. Don't get me wrong. We continue to still have some of these illegal drainage problems to this day on Service-owned or -administered lands. But when we do, the refuge lads are on the illegal drainers like stink on a dead cow lying out in the hot July prairie sun. We always will have those hardheads attempting to drain our administered wetlands. But at least our waterfowl and other water-loving birds will have a home to migrate to in the spring of the year, now that we have the illegal drainage somewhat under control.

Would you believe it? As I walked by that damn bison lying on the ground, he gave one more death kick as a reminder of that battle and caught me right in the last part over the fence!

That would not be the last time either during my career that damn bison would have a piece of me... I guess the winds coursing over the waving prairie grasses never quit blowing either, do they?

After all, a bison lying on its side can still manage one last kick... But I eventually had the last laugh during my 28-year assignment with the Service. Before it was all said and done in the Dakotas and many other like states with the Service, I managed to eat more than my fair share of "bison steaks," broken arrows or not...

Like I have come to learn, the world of wildlife has two faces: One which we can easily see and the other we have to look for...

# CHAPTER FOUR

## A NORTH DAKOTA WINTER SURPRISE

EXITING MY BISMARCK HOME at the crack of dawn in 1974, I looked eastward. The sun was just beginning to awake and I noticed the forerunner of a brilliant red sunrise. Red skies in the morning, sailors take warning, I thought through sun-squinted eyes. Standing there on my front steps, I sensed the air density and temperature as my weather-wise father had taught me to do so many years earlier in my life as a young man. My stepdad, Otis Barnes, had been raised in northern California with local neighboring Maidu Indians for his friends while growing up. It was during those times "they" had taught dad the "Wisdom and Ways" of their ancestors. He, in turn, had passed those extraordinary senses down to me. And on that North Dakota November morning in the lands of the

Lakota Indians, I was practicing what dad had preached. There was a soft sense of moisture in the air, a slight breeze from the northwest, and the "Way" was a little unsettled to my soul's way of thinking. To my way of thinking, The Great Spirit was in a weather-messing mood.

Walking down my steps, I opened up the back of my patrol car and took stock as I often did every winter morning of my North Dakota survival gear. In the backseat resided an ice chest full of high-energy food such as peanuts, peanut butter, candy bars, crackers, and the like. Off to one side on the backseat sat a box full of candles, pans to melt snow for water, matches, extra flashlight batteries, snowmobile mittens, and several warm caps. Piled in a corner of that seat were also several winter jackets of various thickness, all the way up to a surplus-Air Force hooded bomber parka and snow pants good to 100 degrees below zero. Last but not least, a sleeping bag also good to 100 degrees below zero sat all rolled up as if waiting for use.

Satisfied, I shoehorned my more than ample and beautiful carcass into my baby diaper colored yellow patrol sedan. Turning the key in the ignition, I spun her engine to life as I checked my now coming-to-life gauges. Rolling down my window since the morning was so unusually warm and pleasant, I waited for the oil to reach the top of the engine and my gauges to read normal. Once that was done and I checked to see that my gas gauge read normally, I slipped the car into gear and backed out onto the road in front of my house.

Leaving Bismarck, I headed north on Highway 83. I reached the Town of Minot some time later. Turning east on Highway 2, I headed for the northern plains historical Town of Rugby. The great hordes of lesser snow geese were in country in the Rugby area, and it was to that general destination and because of the open hunting season that I was now heading. Once there, I had intentions of checking my numerous North Dakota resident and Minnesota non-resident duck and goose hunters. Since reports of the great goose hunting, with the usual collateral over-limits of geese had been reported earlier, I figured that would be a good place to spend my day as a federal wildlife enforcement officer doing what I did best...

Reaching over to my breakfast-lunch sack sitting on the front seat, I removed my typical "game warden" breakfast item, namely a stick of hard, dry Italian salami. Still enjoying the unusually warm outside air temperatures (warning number 1), I continued heading east with my window rolled down. And in so doing, found myself comfortably wearing only a lightweight shirt and a wool Navy watch cap. Damn, it was great to be alive, I thought as I fired off a quick 'thank you for my blessings' shot to the Old Boy skyward.

Then a big bite of my garlic-spiced sausage added even more meaning to my wonderful morning. Well, sort of... I had no more than taken a bite of that hard salami when I bit down hard on a small piece of bone. In so doing, I shattered the hell out of one of my molars!

Wincing in pain, I spit out the piece of hard salami and several broken pieces of tooth into the fresh North Dakota air outside my open window! Damn, I thought, that sure turned my day into something smelly, brown and usually disgustedly discovered on the bottom of one's shoes in a public park.

Speeding down the almost-abandoned highway (typical in the wilds of North Dakota), I came up with a Plan B. I figured I would speed into the nearest dentist in Rugby, have him put a temporary cap on the tooth and then I would be back into the wildlife wars without any "hitch in my giddy up." Kicking my baby diaper yellow steed in her flanks, I zipped down the winter highway like a man on a mission.

However, as I passed the country's numerous prairie potholes, I noticed that every one of them were unusually loaded with happily bobbing species of waterfowl (warning number 2). *What the hell?* I thought. Duck season had been ongoing for some time now and most of the local birds had been shot off or driven out of the country. Why so many ducks in country this particular "blue bird" morning? Looking skyward out the front windshield, the day couldn't have been more beautiful. The sky was as blue in color as the good Lord could have made it. Not a single cloud graced the sky, the sun was now burning brightly and the air temperature was warm and glorious. So much so, I decided when I stopped again, I would change into a lighter shirt so I wouldn't sweat so much as I walked across the fields checking goose hunters (warning number 3).

Pulling into Rugby, I noticed the temperature gauge at the local Farmer's Bank read an unbelievable 79 degrees (warning number 4)! Shaking my head at my good luck weather-wise this late in the season, I finally located a dentist. Unlike any big city dentist, when I walked in the "drill doctor" had an opening on his schedule and took me in immediately. Directing me into his dental chair, he headed out to get a needle so he could deaden the pain in my tooth. Backing my beautiful but more than ample carcass into his dentist chair, I discovered I was a "Colt .45 ACP" too wide. Reaching down and without giving it a thought, I removed my .45 from my belt holster and laid it up onto the dentist's tool tray. Then I settled my beautiful carcass comfortably back into the chair and waited for the "sawbones" to return.

When the "doc" returned, he lifted up the needle and squirted out the air in the needle so he could give me a shot to kill the pain. Then he froze like a small bird does when it sees a mongoose several inches distant! "Honest, Mister. This won't hurt. I will be as careful as I can be," he squeaked out.

*What the hell?* I thought, as I looked at my steadily going whiter-in-the-face dentist. Then I saw the reason for his strange behavior. His eyes were glued to his tool table lying alongside my chair like there was a hungry grizzly bear standing there drooling all over the place. Then I realized what was causing the dentist his huge case of "gas." Looking back at the now-terrified dentist I said, "Don't worry about the gun, Doc. I had to put it there because I couldn't squeeze my dainty hind end into your dental chair while I was wearing it. So I took it off and laid it up onto your tray until you were finished. I am a federal law enforcement officer, Doc, and can legally carry such an item," I continued with a grin.

Those words of explanation did little to calm

the now-ragged nerves of my "drill doctor," so needless to say, fixing my tooth took some time. Especially after every time he would drill a little, stop and then ask if I was all right. That ordeal was soon finished and I paid the good dentist his "piece of silver." However, I sure wouldn't have wanted to be the patient following me with the "doc's" spent nerves and all...

Returning outside, I once again was amazed over the late November warm weather (warning number 5). I knew there was a weather change coming because of the red sunrise and now a sprinkling of mare's-tails skyward. But for now, that corner of my world was beautiful. Continuing east on Highway 2 soon found me parked in the harvested wheat and barley fields around Hurricane Lake. There the air was full of thousands of lesser snow geese beautifully back-lighted against the gorgeous North Dakota blue sky. That was a sight I will carry with me until I meet my Maker... Then my daydreaming was cut short with the nearby sounds of several shotguns going off to the north of where I stood alongside my car. Getting back into my patrol car and moving to the sound of the guns, I soon had located a pit blind holding three individual goose shooters. After watching them for a short time, I determined they were having one hell of a good shoot due to all the flocks of birds in the air nearby.

So much so that they probably needed the attention of a wildlife badge carrier. Switching into a lighter shirt and grabbing a shotgun so I would look like any other goose hunter, I ventured forth into the shooters' field. As I did, I noticed that the air was now literally alive with what appeared to be every snow goose known to mankind. Damn, I thought, I had never seen so many aroused white geese in one area. Not even in the great Sacramento Valley of California had I seen such a phenomenon (warning number 6).

Working my way carefully up to my hunters so I would not be seen until the last minute, I finally walked boldly the last thirty yards out from my cover directly at them. They, too, were looking intently skyward at the thousands of white geese milling all around in wonder as well and never saw me approaching. However, once they realized I was there, they were not happy campers. As it turned out, they had 22 snow geese over the limit! Once I had finished with the paperwork and as my guys gutted all the seized geese so they wouldn't spoil, I walked back to my car. Arriving, I realized that it had now cooled down somewhat, so I put on a heavier shirt (warning number 7). Motoring back across the plowed wheat field, I stopped at the goose hunters' pit blind and loaded all their seized geese into the trunk of my patrol car. Since they had no further questions over what they had done and the court consequences to come, they somewhat grumpily left me standing there in the middle of Mother Nature.

Me, I dug out a Diet Seven-Up from my ice chest, sat back and relaxed. It was then I noticed it! The air for as far as I could see was now chock-full of every kind of bird known to man! Skein after skein of migratory birds filled the skies as they now fled southward with a purpose. And when I say skeins, I mean thousands of them! The air was literally alive with hundreds of thousands of ducks, geese, swans, coots, birds of prey, cranes, loons, grebes, and every other feathered kind of winged critter in between! I just stood there in amazement! Never had I seen such a mass migration! It was like Mother Nature was dumping out every bird she had in her apron from the Arctic! Then as if by magic, it seemed every hunter in the world had now arrived on the scene to take advantage of Mother Nature's bounty. Soon I had my hands full of good ethical sportsmen and numerous others walking on the "dark side" who had trouble counting what they had just shot.

The first group of Minnesotans I caught breaking the law had just jump-shot a small deeply watered pothole holding roughly 5,000 mallards. Between those three shooters firing just nine shots, they had killed 77 ducks! Killed in the air because they had been so closely packed trying to escape when initially jumped, that it seemed every shotgun pellet had killed a critter! Writing up my lads in plain view of the pothole they had just shot, I noticed it filled right back up with another 5,000 or so migrating mallard ducks in no time! It was just like a mad wildlife rush at Grand Central Station...

It was also during that time I noticed it was getting physically colder. With that realization, I returned to my car and put on my down vest (warning number 8). Walking back to my errant shooters as they continued gutting all the ducks they had just killed, I once again noticed the air was now even more crowded with hundreds of thousands of migratory birds dumping out from beyond the north wind as they quickly headed south. Plus the clear blue sky to the northwest was now clouded instead of being clear water blue (warning number 9). Finishing up the paperwork on my lads from Minnesota, the four of us just stared skyward in awe, hardly talking or believing the phenomenon we were witnessing.

Then a string of shots to my south brought me back from my daydreaming awe at what I was seeing and I quickly headed in that direction to investigate. Driving down a prairie trail, I rounded a knoll only to see two chaps stuffing ducks behind their pickup seat for all they were worth! Driving up behind them and parking my baby diaper yellow patrol car with its new catalytic converter off to one side of the trail so it wouldn't ignite the tall grasses, I stepped out. Man, you talk about two lads who looked like they had just eaten the canary, eh, mallard. Before that little get-together was over, I had fished 58 mallards from behind the seat of their pickup! While writing up their citations, I noticed the air was cooling even more from its earlier temperature, enough that my fingers were getting colder as was the windward side of my face (warning number 10). Loading their over-limits into the fast-filling trunk of my patrol car, I finished up with my two lads and headed for even more shooting coming now from the Wolford area.

Arriving in the area where I felt my last heavy shooting had come from, I noticed four men loading up their goose decoys into the back of a pickup for all they were worth. Pulling up alongside my fast-loading chaps, I quickly put a crimp in their style. Especially when I noticed a mess of geese lying under a pile of their decoys in one of their pickups. Before that little "come-to-Jesus" meeting was over, I had four Minnesota shooters with 31 geese and 16 ducks over the limit!

Walking back to my car for another cite book, I noticed the clouds from the northwest were now getting more massive, dark and damn ugly. That plus it was getting cold enough that I changed from my lighter coat to a heavier one. However, the air was still wonderfully filled with every kind of bird known to man. In fact the mass migration was so grand that the four bad guys and I just stood there marveling at what we were observing like a bunch of dingbats. No two ways about it, Mother Nature was dumping every bird with a heartbeat out of the north country (warning number 11). And since the government hadn't provided me a patrol car with an AM/FM radio so I could listen to the news or weather, I remained in the dark as to what was happening. And being that this was my first winter in the northern plains, I didn't have a clue as to what was in store for me...

The reason given by our 'leaders' for the lack of a car radio was that if they gave the agents such devices, they would just spend all their time listening to the ballgames and not working... While my Minnesota lads cleaned the illegal birds, I wrote up their citations. As I did so, I noticed the winds from the northwest were picking up. And as they did, they filled with blown topsoil and North Dakota winter, cold as all get-out (warning number 12). Loading the seized birds into the now-filled trunk of my patrol car and putting the rest onto the front passenger seat and floorboards, I made ready to leave.

"But not too soon," said Mother Nature, as my attention was once again drawn to an adjoining harvested barley field where I swear a war was going on. Cranking up my baby diaper yellow patrol car, I headed towards that latest flurry of frantic shooting. By now, it was plain by the way the vehicle handled that I had a load in the car when one took into account all the seized ducks and geese I was transporting.

It was also now plain to me that a winter storm was on its way and a big one at that. Then another revelation was experienced by my now getting-colder carcass. I could sense that the air was getting dryer than a popcorn fart. And as a result of the increasing wind velocity and flying dirt, the air was becoming full of static electricity. That became even more apparent whenever I touched anything metallic.

Moving over to the area where the massive amount of shooting was taking place, I was now aware that something of untold proportions in nature was taking place. The outside temperature was rapidly plunging. Even the birds in the air seemed more frantic now to getting the hell out of North Dakota. The air was full of flying topsoil as well as exhausted ducks looking for places to land and rest. Every drop of water that I passed now had thousands of bobbing ducks on the water and more were dropping into the potholes like there was no tomorrow! And still, even more skeins of birds filled the air until it seemed no more could fit into the airways. I also noticed the great numbers of deer running across the harvested fields as they raced for cover in the nearest coulees and weed-filled ditches. As for me, I was now wearing a down vest, heavy coat, gloves and a warmer hat.

Moving to my group of lads shooting the hell out of everything that flew, I soon got that halted. In their pile of birds I found 4 loons, 32 ducks, 19 geese, 8 cranes, 3 grebes, 2 red-tailed hawks, 4 swans, and an even-dozen coots. My lads had never seen so many low-flying birds and targets of opportunity. And in so doing had tested the Winchester Firearms Company and its ability to produce enough ammunition for those six chaps to discharge into the masses of low-flying birds that frantic day! By the time I had that mess cleaned up, the air was full of flying topsoil, clouds hanging menacingly close to the land at hand, snowflakes whizzing by portending more to come, and more static electricity — it was so cold, I now went to my heaviest winter parka (warning number 13)!

Finishing up with my six lads and now having filled my patrol car clear full of seized birds of every kind, I headed for the town of Rugby. As I did, I noticed it seemed everyone else was doing just like me. Basically, all of us were trying to get the hell out of Dodge before we were blown or frozen off the face of the North Dakota high plains map!

I finally got my government two-way radio working and hooked up with the weather channel. There I learned the first blizzard of the year was enveloping the entire northern plains! Weather warnings flew every which way and never having been in a blizzard before, I called her quits and headed for Bismarck. Besides, I didn't have any more room in my patrol car to put another seized duck or goose!

The wind was now howling up a gale, it seemed all the high plains topsoil was airborne heading for Kansas, and it was now colder than all get-out. Since the wind was coming out of the northwest, I figured returning home the way I came and into the teeth of that storm was out of the question. If I did, I would be driving directly into the storm's fury, while fighting a side wind along with driving snow, and none of that sounded like a real good option to me. So I headed due south on highway 3 from Rugby so I could go with the flow so to speak.

By the time I got to Interstate 94, I couldn't see twenty feet in front of me for all the flying snow and topsoil! Then turning west, I was hit with the full force of the howling blizzard coming down from the northwest. The fastest I could safely drive was 21 miles per hour without skidding all over the place on the now-icy highways. Then I noticed my trusty engine temperature gauge for the first time. It read ice cold! It was so damn cold outside, my engine temperature never got off the freezing cold mark! Pulling off the Interstate to clean off my frost-covered windshield since my defroster did not work, I was hit with the full force of Mother Nature for the very first time in my life. A real northern plains full-force blizzard if you will! It was so cold, I could not take in the outside air to breathe in a normal breath! And for the first time in my life, I felt the edges of panic setting in at not being able to breathe normally! I soon discovered that I had to hold my cupped hands over my mouth to pre-warm the frigid outside air before my body would accept the slightly warmed air into my lungs! Getting back into my car, I now realized why, when I went out onto my front porch that morning, the "Way" inside me had been so unsettled... The old saying, "Red sky in the morning, sailors take warning," now took on a much-fuller meaning than it had ever had when back in my home state of California.

I finally arrived home around nine o'clock that evening to a very worried wife and family. By then having driven all that way without a heater since I had no engine heat to speak of, I was frozen solid. Struggling out from my unheated patrol car (I later learned to drape a canvas covering over the front of my radiator during such weather and use a hotter thermostat), I staggered up the steps to my home trying hard to breathe. Unable to work the door handle because of stone cold hands, my wife had to let me in and then help undress my near-frozen carcass! A hot shower later and something warm to eat finally brought some life back to my now-rapidly-learning carcass about winter weather on the northern plains. That evening I stayed glued to the television watching the weather events unfolding in North Dakota. Many people were missing, several had already died and more trouble was to yet come. Before that blizzard blew itself out, if I am not mistaken, 19 people died in that 1974 storm! Six kids' car had broken down and they tried to walk to a farmhouse one-quarter of a mile distant. They found all six frozen to death in a field 100 yards from that farm home.

The next day once again I stepped out onto my front porch. It was cold but not bad. I only wore a heavy wool shirt and a light vest for that day's weather. The wind was not blowing, snow covered the ground and the sky was clear as a bell. My senses advised all was well, so I unplugged my patrol car (engine blocks had to be plugged into an electrical source in order to keep the oil warm so the engine would turn over), and it started right up. From there I went to my storage shed to put all my seized birds into a freezer. Boy, was that a joke. I just laid the frozen bird blocks on the floor of my unheated shed and they survived in good shape until the court cases had cleared weeks later. In fact, I later learned that during winter, many people in Bismarck who had died naturally, had their bodies placed in an unheated shed during winter by the morticians. It was so cold that they remained frozen throughout until graves could once again be dug into the frozen North Dakota ground. Later, I took my blocks of frozen migratory game birds down to the Bismarck Zoo, and there Mark Christianson, the Director, thawed them out and fed them to his animals to keep food costs down.

Later that day after the storm in the weak winter sunlight on the North Dakota high plains, I took a spin around my new world of white. That trip was unique to say the least... It was cold, not a breath of wind blew and the snow-covered ground was now frozen hard as a block of cement. All the watered potholes were now frozen over hard as a piece of steel and there wasn't a single duck or goose to be seen, anywhere! As for the rest of the country, not a thing was moving.

Driving by a Waterfowl Production Area (like a small wildlife refuge owned in fee title by the

Fish and Wildlife Service) during the late morning, I noticed a lone deer standing at the edge of a frozen pothole. Stopping, I took a hard look at what appeared to be the storm's lone land survivor. Then I noticed it had not moved the entire time I sat there watching it. Thinking the deer may have been frozen standing up, I got out from my patrol car and began walking towards it. It did not move as I approached. It remained standing as I walked right up to the animal not moving one whit. Then as I got closer, I could see the damage caused by the previous day's blizzard. That deer was almost frozen solid! It still had some life but you could tell it was in bad shape and I sensed it would never recover from being partially frozen. I put the animal out of its misery and left it for the coyotes or wolves in the area.

That evening back at my home, I had a chance to think back over the phenomenon I had personally experienced and been fortunate enough to have lived through, being a novice and all. I saw for the very first time how the Arctic wildlife emptied out in front of an oncoming blizzard. What I saw that day in the way of bird migrations has to be one of the wonders in the world of wildlife. I literally saw millions of migratory birds leaving in front of a killer storm. And after that, almost all the wildlife in the northern high plains in the form of many species of migratory birds were absent until spring returned. There were still many species of little birds present but one only found waterfowl where there were sources of open water in abundance, like in parts of the Missouri River. Talk about force and fury, I had a chance to witness and live through one of Mother Nature's very unique cycles. I only saw such an event one other time in my life and consider myself lucky to have done so. That migratory event, occurring just before Old Man Winter arrived in the northern plains in his full force, is one to be experienced and seen if one wants to go full cycle in the world of wildlife wonders.

Having come from North Dakota farm stock and lived through some tough times myself in northern California, I can see why my tiny North Dakota-born mother was such a hardy person. She basically raised two kids on her own after divorcing my biological father during some hard times. It is those life's experiences that I still carry close to my being this very day. Then being able to experience some of the things she lived through in North Dakota as a young woman made me realize just how tough she was and how lucky my sister and I were to have her!

As an aside, the high temperature that fine day so long ago in North Dakota was 79 degrees above zero when I started working my duck and goose hunters. When I arrived home later that evening, with wind chill, it was 79 degrees below zero! A swing of almost 160 degrees in one day. Just another day in the life of Terry Grosz. And what a day of wonder, the full force of Mother Nature, the world of wildlife and magnificence it was. One in which I will happily take to my grave as one of my life's many unique experiences.

# CHAPTER FIVE

## A SPECIAL AGENT'S "STEW"
## ON THE DAKOTA NORTHERN HIGH PLAINS

FINISHING MY MEETING with Jim Gritman, the Fish and Wildlife Service's Area Manager for North Dakota, regarding an Airborne Hunting Act violation by one of his employees, I headed out the door. It had been a good meeting culminating three days of hard field work, and I was anxious to get home for some clean laundry, a shower, some good home cooking by my Bride, and to spend some time with my family. Dream on...

On the way home, I ran the events of my recent meeting with Jim Gritman through my mind. It was 1975 and I was the Fish and Wildlife Service's Senior Resident Agent for law enforcement for the States of North and South Dakota. Jim on the other hand, was the senior manager for all the other Service offices throughout the State of

North Dakota except for the Jamestown laboratory facility. Jim was a very bright, aggressive, far-thinking, tough, no-nonsense fighter for the Service, her programs, its people and the critters. I really enjoyed working with the man because he stood for no nonsense, and had a 'Damn the torpedoes, full speed ahead' aggressiveness that fit the life and times of the wildlife resource problems of the state.

Case in point was our recent meeting. I had just caught one of his employees, a Service Predator Control Agent or 'mouse choker' as I called them, in a clear violation of the federal Airborne Hunting Act. Basically, while on stakeout in the western portion of North Dakota for another Airborne Hunting Act violator I was investigating, I had crossed swords with "one of our own" breaking federal law!

Simply put, the federal Airborne Hunting Act prohibits taking, harming or harassing wildlife with the use or aid of an aircraft. However, it does contain provisions whereby a permit may be issued to a landowner to take depredating predators from the air with the use and aid of an aircraft to protect his or her livestock. Additionally, Service employees during that time period could also be covered by such a permit to assist any landowner in taking such offending predatory animals from the air. However, the terms and conditions of said permit did not allow one to kill every heartbeat in the country! A common mindset which many Service "mouse chokers" of the time seemed to want to do; i.e., the only good coyote is a dead coyote... Also when such permits were issued, they were geographically specific. That is, one could not just kill wherever the pilot or his gunner wanted or found the opportunity. The killing was only authorized on the landowner's specific geographic area where the predation problem was occurring. And that killing field was specifically spelled out in the permit to preclude over-killing and knocking out of balance the predator-prey relationships existing in the area!

Case in point. While on stakeout in Dunn County just south of the Fort Berthold Indian Reservation early one morning, I heard the sound of a low-flying approaching aircraft engine. I had been on stakeout for the last three days in the area cold camping. In other words, I had hidden my patrol vehicle a long distance away to avoid discovery. Then I had walked into an area which had been reported to me by a bunch of sharp-tailed grouse hunters as one visited and gunned over regularly by a red-and-yellow Super Cub with taped-over aircraft numbers (to foil identification). Since I had received that information three different times from several different groups of grouse and deer hunters, I had decided to cold camp in the area and see if I could catch my aerial gunner in the act of breaking federal law.

On my third day of 24-hours-a-day stakeout, eating food out of bags and cans, and sleeping on the ground in heavy brush in a coulee, I now had company. Making sure I was still well-concealed so I didn't get a face full of shot from my illegal flying "Red Baron," I got out my binoculars and waited. Sure as God made a mess of gophers for the State of North Dakota, here came a red-and-silver aircraft heading right at my monster-sized coulee. It was not a red-and-yellow aircraft as reported, but after living out on the prairie eating out of cans and bags for three days without the comfort of a fire, one is not particular whom he catches with their fingers in Mother Nature's cookie jar...

As if on a string, my suspect aircraft flew right at my coulee and then began low-circling the area as if looking for something on the ground. Sure as God makes a gob of worms on a number six Eagle Claw hook pure dynamite for catching German brown trout, my suspect airplane all at once gunned his engine and swung back into a tight circle. Below him, scared from the brush, ran a pair of coyotes for all they were worth. My plane then lined-out on the fleeing 'song dogs,' fired two shots and killed both animals deader than a car-hit chicken!

Damn, that guy is a good shooter, I thought. Then his airplane overflew the two dead coyotes to examine his results and that sunk his goose. When he did, I was able to not only get the numbers off his aircraft but clearly identify the low-flying pilot as well. "God dang it," I uttered under my breath. "I know that guy and he is one of our own 'mouse chokers' working for the Service!" That wasn't the bad part. There was no airborne hunting act permit issued for that area for anyone! The closest that particular chap could shoot coyotes was on another permit holder's property some twelve miles distant! In short, my Service aerial gunner was in clear violation of federal law!

Waiting until my chap had finally left the area, I gathered up my camping gear and headed for where I had seen him shoot the coyotes. I found one right away but it took a little longer to locate the other. Dragging my two coyotes and carrying the rest of my gear, I walked back to my hidden at an abandoned barn patrol car and then headed back to Bismarck.

Arriving at the Bismarck Area Office, I met

with Jim and told him what I had observed that morning. Aside from the good-natured ribbing I took on how badly I looked and smelled after a three-day stakeout, and dragging two dead stinking coyotes back as evidence, the meeting went well. Jim later called in his employee and after hearing his side of the story, he was given two options: Be criminally prosecuted in federal court for his violation of the Airborne Hunting Act or resign. He chose to resign. Me, I chose to go home, shower, get cleaned-up and spend a few precious days with my family. As I said earlier, fat chance of that happening...

Arriving home and after kissing my Bride and being told once again how badly I smelled, I got the bad news. I had a refuge manager who needed to see me just as fast as I could get to his duty station in north central North Dakota over some wetland easement violations. It seemed some of their people working wetland easement violations on the ground had been threatened by two mean-assed farmers. There was also another more ominous call from my Washington Office. It seemed one Bob Halsted, Special Agent in our Special Operations Branch (covert operations), was in town and needed to see me immediately. As my wife related the rest of the story, his call was regarding illegal eagle feather sales in

South Dakota and some mean-assed, threatening Native Americans. Damn, there sure were a lot of 'mean-assed' folks in the Dakotas in those days.

Seeing this was shaping up into another long detail away from home, I so advised my wife and then headed off to a welcome shower. Finishing, I gave Agent Bob Halsted a call where he was staying at a local motel. As I later discovered, he had been working in North and South Dakota on an undercover eagle feather buying trip. During that trip, he had run into a group of South Dakota Indians in the Rapid City area who were selling huge amounts of eagle parts and products. As he advised, he was to meet this group ten days hence to purchase some of their eagle feathers and other migratory bird parts. However, in trading with another group of Lakota Indians in South Dakota, he had overheard them saying that the Rapid City bunch of sellers had their suspicions relative to Bob's true identity. In short, they suspected he might be a 'fed.' And as was the practice of the outlaw Native Americans in those days, if they didn't like you, you got a good old-fashioned stomping! Bob went on to tell me that the products they had to sell were of such magnitude and volume that he could not pass up the chance at making a buy. To his way of thinking, that group of Indians were killing the hell out of the eagles and needed to be stopped and brought to justice.

That is where I came into the picture according to Bob. He wanted me to accompany him on the buy as a big money supplier and Indian artifact collector. But mostly he wanted my large carcass along to protect him in case things got out of hand. I agreed to meet him in Rapid City at the end of the ten-day period at his motel and then the two of us would make the buy. Then Bob would grab the quickest airplane out of Dodge while I was still around providing protection, and leave before the Indians caught him and stomped him into the prairie soils.

After that phone call, I went into my den and picked up my .44 magnum pistol and its undercover shoulder holster. Then careful not to let my Bride see me loading such a "hand cannon" into my gear, I buried the item and an extra box of cartridges in my clean clothing she had already laid out for my next trip. I usually only carried my .45 Colt Commander on routine investigations because it was so compact and easy to hide. But if things got "froggy" then I switched to my 6 ½-inch, .44 magnum "wheel gun" for the serious work. That plus I always shot Distinguished Expert with the big, heavy magnum. Additionally, if I 'ran dry' during a shootout, its massive 29-ounce steel frame made a great club in any subsequent required 'head knocking' action that followed.

After a great dinner, I kissed my Bride good bye, advised my two boys to mind their mom, and looked in on our just-adopted Vietnamese baby daughter, who was still so sick all she did was eat and sleep. Kissing her good bye, I covered her up once again so she would stay warm, aided by a quick prayer to the Old Boy upstairs that she would soon recover from her illness.

## DAY 1: THE CAMELBACK CRICKET...

Exiting my home in Bismarck, I loaded up my gear in the patrol car and headed north to J. Clark Salyer National Wildlife Refuge near the "booming megalopolis" of Upham, North Dakota. Once there, I would hopefully settle several battles between our refuge officers and several mean-assed North Dakota "dirt farmer" tillers of the land who had been illegally ditching some Service-administered lands.

Arriving that afternoon at the refuge and finding all its personnel out duck hunting, I returned to the one and only Upham Cafe. Still on one of my many famous diets over the years, I walked into the tiny backwoods, mom-and-pop, end-of-the-road in food quality, restaurant. Opening up the front door and entering, I was met with a head-high layer of intensely burned grease smoke from their griddle. Sitting down at a fly-covered greasy table I looked over the menu. Not seeing much of interest or what I considered safe to eat, I finally ordered a large bowl of their homemade vegetable soup (this was the only eatery for miles). I ordered soup because soup is like a hamburger. What damage can one do to either food item? I was soon to discover just how one of my two favorite food items could be 'altered' from 'class A' to questionable slop. When my soup arrived, I realized just how hungry I was and dug in. On the fourth spoonful, I noticed an item in the soup that I did not recognize. Closer examination revealed that it was part of a thorax from a camelback cricket! After paying for my unfinished bowl of soup, I headed for the refuge. There I dug into my survival box and had a nice dinner of sardines, cheese, crackers and a Diet Seven-Up. Sleeping overnight at the refuge on their office floor (the Division of Law Enforcement was broke; they didn't have the price of a room left in their budget), I met with Assistant Refuge Manager Jim Heinicke the next day. Once together, the two of us discussed several ditching violations on our wetland easements generated by two different farmers.

To you uninformed, the Service, under contract, paid farmers not to drain, burn or fill valuable wetlands used by migratory birds. Such restrictions were then covered under a federal wetland easement contract. Lying before me were two contracts that had been criminally violated by scrapers or plow furrows, ditching out and draining several wetlands. Since both farmers had threatened the refuge personnel who had initially investigated the violations by offering to break their legs if they ever returned, I "drew the short straw." That kept our refuge folks who had to live in the area out of harm's way with the local farmers and allowed the one doing the illegal ditching to vent his spleen on me. I didn't mind doing that kind of dirty work being the "heavy" though. The wetlands were needed for nesting and rest stops for migrating waterfowl and I liked those kinds of critters.

Besides the landowners had been paid not to drain, burn or fill those contracted wetlands and they had readily accepted the government monies to not do so. Then they turned right around, like they did with many other government contracts and programs of the time, and did as they damn well pleased...

As it turned out, the Service was the only government agency at that time which bothered to enforce the terms and conditions of their federal contracts. So as a result, the dirt farmers draining our administered lands many times became upset when we tried to enforce our contracts, while the rest of the government agencies just pissed backwards in the face of such political adversity...

## DAY 2: THE DRAINAGE DITCH, THE HOG FARMER AND ONE HELL OF A DINNER...

Well, being of German stock and as thick-headed as the rest of the draining dirt farmers in that area, I "went to war." Taking the needed aerial photos, wetland maps, copies of the contracts and other like information from the refuge officers, I ventured forth the next day to the farm of one Nels Horsten near the Town of Kramer, North Dakota. According to the information from the refuge's annual flyover of those administered wetland properties, Nels had drained several Type II wetlands under contract. Then he had hooked a deep plow furrow into a Type III wetland, effectively draining off the top two feet of the water in that basin. Since both actions were clear violations of the wetland easement contract, I drove onto the Horsten farmstead the next day.

Asking the lady of the farm where I might find Nels, she pointed to a pen where a rather large fellow was slopping the hogs. Walking over to the chap, I patiently waited until he had emptied his bucket of slop, turned and faced me for the first time.

"Good morning. My name is Terry Grosz. I am a Special Agent with the Fish and Wildlife Service and have come to talk to you about the three ditching violations you have on Section 16 of your land," I quietly said.

For the longest time Nels just looked at me, then he flung his slop bucket at me yelling, "I don't give a damn about you guys from the 'Wildlife' (that is what the locals called those of us from the Fish and Wildlife Service in those days). You get the hell off my land or I will feed your miserable ass to my pigs!"

My first thought was to put his just-thrown slop bucket where he would have a hard time in the future using it to slop any more hogs. But then another thought crossed my mind. Walking over to the thrown slop bucket that I had easily ducked, I stooped over and picked it up, all the while watching his actions carefully in case he "bull-rushed" me. Hefting the bucket, I headed for his barn where I had seen him emerging from earlier, carrying slop as I drove onto his property.

"Vot the damn hell," he bellowed as I quietly walked away with his slop bucket.

Walking up to his dairy barn without saying a

word, I saw other full slop buckets sitting along-side some fifty-gallon drums full of curdled dairy milk products. Sitting down my thrown empty bucket and picking up two full ones, I walked back to my now red-faced and thoroughly perplexed farmer. Since there were still about sixty unfed hogs in his pen hungrily looking for chow, I took a gamble. Walking past the perplexed farmer, I headed for the hog troughs in the pen full of loudly squealing, pushing hogs. Then I dumped one bucket of slop into one empty trough and drained the other bucket into another nearby empty trough.

Turning, I walked past my now really perplexed farmer and as I did said, "You going to give me a hand or not? There are still about 30 hogs that need slopping." Then I walked back to the dairy barn, dipped my slop buckets into a full fifty-gallon barrel, filled them, and headed back towards the pig pens. Once again, I walked back to the pig pens with my two buckets of slop and filled up a couple more empty troughs with the slop. Heading back to his dairy barn for more since not all the hogs had been fed, I passed Nels coming my way without a word with two more full buckets of slop for his hogs. For the next twenty minutes the two of us slopped his remaining hogs without a word being spoken by either of us.

Then when we finished that detail, I walked up to him and asked if he had a hay hook.

"Vot the damn hell you need a hay hook for?" he asked, "I haff no hay to bring in."

"You have a dead hog in the pen. It is not smart to leave him there for the other hogs to feed on. Since I don't enjoy carrying wet and stinking hogs in my arms and getting my shirt and pants dirty, I need a hay hook," I quietly said.

He just looked at me for a second and then said, "Inside the dairy barn, left-hand side on the wall." Then he just stood there looking at me wondering if he had a crazy government agent on his property looking for a hay hook.

Without a word, I passed my still-surprised farmer and walked over to his dairy barn. Walking inside, I spied a row of hay hooks hanging on some wooden pegs on the left side of the barn. Grabbing one, I walked back to the dead hog in the pen. Without a word, I sunk the hay hook into the open lower jaw of the hog. Lifting up the front portion of the hog, I headed for the gate dragging the dead hog (you see, I had once been a farm boy, too). Opening the gate, I continued dragging the dead hog out from the pen full of live hogs, then closed the gate behind me.

"Where do you want this damn thing?" I asked. Whereupon without any more words being spoken between the two of us, we loaded the dead hog into the back of his truck. Then we took it out to his farm dump and left it for the critters to enjoy.

After finishing up our chores and washing up with hardly a word being spoken between the two of us, Nels said, "Vee haff supper vaiting. Don't do me no good to keep that voman vaiting."

Helga, Nels' wife, put a super spread on the table that was fit for a king and a lowly government agent. We had fresh leg of roast hog, mashed potatoes with thick, spicy pork gravy, homemade dinner rolls fresh from the oven, fresh cooked green beans with loads of fatback floating therein, cabbage salad, and tankards of some of the best homemade buttermilk with gobs of real butterfat floating throughout that I ever drank. That meal was finished off with some of the finest homemade sweet rolls I ever ate as dessert. Man, those North Dakota farm gals sure could cook...

After that, the two of us sat down and quietly discussed the drainage business at hand. One week later, Nels had closed up his three illegal ditches and advised that would be the end of his drainage operations. Especially in light of the fact that I ate as much as he did and he couldn't afford to feed me now that his wife thought so fondly of "her own government man"...

By the way when I left later that evening, Helga gave me a breadbag full of wonderful homemade pork sandwiches and a jar of her homemade bread and butter pickles. Man, I thought, I hope the next "angry" farmer I had to meet was like Nels. As I was later to discover, the next "dirt farmer" with illegal ditches was a hoarder and a crackpot.

## DAY 3: THE LOON WHO HAD BEEN OUT IN THE PRAIRIE WIND TOO LONG...

The next morning leaving the refuge where I had once again slept on their office floor, I glanced upward through my windshield to take a gander at the day's weather. There were a few lenticular clouds signifying high winds aloft but other than that, a good day was in the offing. However, the air was cold and ice was beginning to form on all the waterways.

Heading east towards Willow City for my next wetland easement contact, I let my mind wander back through the pages of "high plains" history. The prairies in my "mental wanderings" were full of original perennial grasses and plants on top of good old dense prairie sod. The prairies were jug-full of numerous small bunches of bison eating their way across the lands, and off in a distance sat twenty or so Lakota warriors on a bison hunt. Man, it was great to be alive and a bit of a romantic dreamer...

A bit later, I returned from my dream world and got down to the business at hand as I approached my second 'mean-assed' dirt farmer's homestead. Turning down the typical long dirt road to his property, I let my practiced eyes drink in all the features into which I was now heading. There was a typical small, run-down farmhouse at the end of the road. In front of the house sat various pieces of farm equipment showing obvious signs of heavy use, wear and tear. No dogs ran out to bark at and greet me and that was good. I never did get used to tangling with the run-of-the-mill mean-assed farm dogs. There were two old run-down barns on the property and junk of every sort everywhere! By one of the barns, I observed a man working on the bucket of his front end loader. Figuring that may just be my man of the hour, I headed in that direction.

Pulling up alongside the front end loader, I got out and stretched my frame. I sure hated 'shoehorning' my more-than-ample frame into a mechanized roller skate of a sedan, I thought. My kingdom for that of a cab of a pickup, continued my thoughts, as I walked over to my farmer who had ignored my arrival and was still working on his front end loader.

"Good morning. Can you tell me where I might find Ernst Groober?" I asked.

"I be him. Why do you ask and what do you want?" he fired right back.

"Mr. Groober, my name is Terry Grosz and I am a Special Agent with the Fish and Wildlife Service," I said as I showed him my badge and credentials.

"I already paid my taxes and I ain't paying one thin dime more," he grunted at me as he went back to work on his front end loader as if I didn't exist and our business was finished.

"No, Mr. Groober. I did not say I was from the Internal Revenue Service but the Fish and Wildlife Service," I said.

Without turning around to speak he said, "I don't want anything you might be selling as well," he grumbled as he tightened a bolt on the bucket on his front end loader.

"Mr. Groober, I am here to talk to you about the 300-yard-long ditch you have on a Type III wetland on Section 20 of your land that is under a wetland easement contract," I continued.

Turning slowly, the man just looked at me for the longest time. It was then I got a really good look at my contact. He was filthy! His clothing was in rags and dirty as all get-out. In fact, it almost appeared he had wetted and defecated all over himself numerous times from the stained appearances on his clothing! And the sharp smell I was now smelling sure didn't come from his nearby pen of cattle or his wife's cooking (if he had a better half)...

"You 'Wildlife' basterds don't get the message do you? I thought I told your kind that if you ever came sniffing around here anymore, I would knock the devil out of you. What does it take to get you basterds to listen?" he growled.

"Just fill up that ditch on Section 20 with a 10% overfill to allow for settling back to the normal lay of the land and you will be rid of me," I shot right back.

"The hell you say!" growled my farmer menacingly as he walked out from his front end loader towards me with a large wrench in hand.

"Mr. Groober," I said, "I have identified myself to you as an officer of the federal government. The refuge staff from J. Clark Salyer National Wildlife Refuge, and now I, have asked you to fill in that ditch on Section 20 on your land which is under wetland easement contract. The very presence of that ditch is a violation of that contract and is a violation of federal law. And if you put it there or ordered it put there, then it becomes a criminal violation of 16 U.S.C. 668dd(c) of the National Wildlife Refuge System Act. In short, if you don't agree to fill that ditch, you are going to find yourself in federal court in Devils Lake."

"You gonna put me in jail after all the work I did in digging that ditch? What if I don't fill in that ditch?" he asked with a sneer.

"That is why I am here today, Sir. If you don't comply with my request to fill that ditch, I am prepared to either take you to jail in Devils Lake or issue you a citation for that violation of federal law," I said as I squared off with Groober in case he came at me with the large wrench he was holding in his hand. "Additionally, if I take you to jail, who's going to feed your cattle? The reason I ask is that the federal magistrate is not in town presently and you will have to sit in jail until he returns three days hence. And believe me, you will sit in jail until he returns," I said with a touch of finality in my voice.

When I brought up the cattle feeding situation, I thought I saw that issue of animal care fly across Groober's eyes for just an instant. I knew from talking to the refuge folks earlier that he was considered an odd duck and his neighbors did not have any use for him because of his always out-of-sorts attitude. Hoping I had finally gotten through to my irate and stinking farmer, I waited for his next move. To be quite frank, I sure as hell didn't want him in my patrol car if I had to place him under arrest and transport him to Devils Lake. I could just imagine how many days it would take to just get the stink out from the vehicle after I had placed him in jail.

"The government doesn't have to come in and feed my cattle if they take me off to jail?" Groober asked with a tad bit of concern now edging into his voice.

"Nope," I replied firmly.

"Well, I can't fill in that damn ditch right now. My front end loader is broken down and it will need some major work before I can use it to fill in that ditch. And by the time the loader is fixed, the ground will be frozen hard as a rock and then I can't fill the ditch," he said.

"Well, I will tell you what. If you promise to fill in that ditch with a 10% overfill to allow for settling within the next two weeks, I will just issue you a citation. You still have to pay a fine and if you don't fill that ditch, I will come back, arrest you and take you off to jail to await trial. Do we have an understanding on that issue?" I asked.

"Yeah, I guess so," he replied.

"You don't sound too convinced about my way of thinking. I meant every word I said. If you don't comply, I will be back, cattle or no cattle, and you will be going to jail," I said.

"OK, OK, I understand and will do as you say. If you are going to issue me a ticket, let us go inside. It is time for my stomach medicine so we can go inside and conduct our business while I take my medicine. Is that OK?" he asked.

"That will be fine with me," I said. Looking back on that decision now, I wished I had braved the cold and written him a citation on the hood of my car.

Following him into his house, it was all I could do to keep from gagging as I walked behind his dirty and unkempt body. Then once inside, I got a real shock. Entering his home, I discovered it was at least 80 humid degrees inside! Not from any source of mechanical heat but from all the rotting food stuffs and organic material in the house! The man was a "hoarder" and his home was clear full almost to the ceilings with refuse. Old clothing, magazines, garbage from his kitchen, trash, used furniture, and the like. You name it and that house was stuffed with everything rotting under the sun. Stepping over and on everything imaginable, I was aghast at all the maggots tumbling down the refuse from the rotten food items they were enjoying. And flies! The house was like someone had just dumped over a hive of bees there were so many flies on everything, and now, they were crawling all over me. Yuck!

Plowing our ways through the refuse into his kitchen, I was once again greeted with something only the devil could have created. The kitchen was a mess of maggots, dirty unwashed dishes, pots and pans, refuse clear to the ceiling and a fly-covered table surrounded with what used to be white wooden chairs! Suffice to say, that was the fastest citation for a wetland easement ditching violation I ever wrote!

Once back outside, I breathed in great gulps of cold, fresh air. While inside in order to not breathe in the foul, fetid sickly sweet warm air, I had only breathed in small puffs of air. However once outside, I about sucked in all the fresh air North Dakota produced that day. Then once back inside my vehicle, I could detect that same sickly sweet smell from inside his home that had now permeated my clothing. Heading back to the refuge, I showered and changed clothing. I also gave all the laughing refuge officers hell about not warning me about what I was about to get myself into with Mr. Groober... I also made sure my refuge lads clearly understood it would be a cold day in hell before I responded to any of their calls for assistance in the future (well, not really). Those harsh words put all my refuge lads on the floor in fits of more uncontrollable laughter over my most recent and evil-smelling experience...

Leaving my "gut-busted-from-laughing" refuge lads at the J. Clark Salyer National Wildlife Refuge, I headed southeast towards the small town of Anamoose. Earlier, one of the state fish and game lads advised me that the area's wetlands were chock-full of migrating redhead ducks. He also advised since the season was closed that year on taking redheads, many mistake birds were the consequence of those closure regulations. It seemed that warden's district of duck hunters couldn't tell a redhead from a canary (I was not sure my game warden could identify them either). Liking the reduced and in-trouble population-wise redhead duck species, I headed in the direction of the three small Antelope Lakes in his enforcement district. That area also included several historical redhead passes that I knew would be heavily utilized by the shooting locals.

To those of you of the non-duck hunting fraternity, a 'pass' is just that. It is nothing more than a commonly used pass frequented by various species of waterfowl to go from one area to another. And they usually present easy shooting and killing opportunities for those familiar with such historical flight areas.

## DAY 4: REDHEADS GALORE AND OUTLAWS, TOO, PLUS A FOUR-LEGGED SURPRISE!

Staying at a friend's home in Drake, North Dakota, near the Antelope Lakes, I was out early the following morning (again, no money in the budget to rent a motel room). One hour before daylight found me and the dog sitting off to one side of one of the better redhead and canvasback passes in the area. Their dog and I shared some of Helga's wonderful pork sandwiches from two days before and finished off her quart jar of bread and butter pickles. Shadow, my dog, loved the pickles as well and probably ate more than she should have. Later in the day when I needed her, it seemed she was always in the bushes taking a very liquid bread-and-butter-pickle dump! But that was OK because the hunting of humans was good. As is typical in North Dakota, there weren't many hunters out and about, but many of those that were, were in over their heads when it came to some duck identifications and counting, if you get my drift...

Staking out one group of hunters, I found that I was in the wrong place to really be able to count and identify all the birds they killed in the area of the redhead pass. That forced me to make a move by crawling several hundred yards to get into a better position. What a job that turned out to be. For any of my readers who have hunted in certain parts of North Dakota, you know of what I speak when I say that land is covered with many small cactus plants. Suffice to say in that long crawl, I picked up not only several knees full of needles but filled my hands full of cactus spines as well. But I finally arrived where I wanted to be and to my surprise, I couldn't find my three pass shooters! They had gone to ground and had just flat disappeared! I knew they hadn't spotted my graceful carcass as I sneaked up on them but sure as all the big cats of the world are disappearing, especially in Africa and Asia, my three lads were nowhere to be seen.

But I had been there before. That is, having had lads I was watching disappear in the blink of an eye like those big cats in a jungle as observed by Rudyard Kipling in his stories of old. To me, that usually meant they were up to no darn good. And sure as my Bride makes the world's best homemade sweet rolls (even better than Helga's), I spotted my three shooters hotfooting it across the prairie about 100 yards away heading for another good historical redhead pass. Taking off after them in a running crouch, I just happened to run through the area where they had disappeared from earlier. In so doing, I got one hell of a surprise. Lying on the ground in 'gay profusion' were the breasted-out carcasses of 23 redhead and seven canvasback ducks! All were illegal and from their body warmth, had obviously been taken by my three shooters before I could get into position to make 'drop' counts on them. Suffice to say, that sure built a huge bonfire under my last part over the fence! Hurriedly stuffing the carcasses into my game pocket in my hunting jacket and carrying the rest in my two hands, dog and I took off after my now furiously shooting lads in the next redhead pass over the hill. That particular pass was near the last Antelope Lake in the close string of those three bodies of water, so I figured I now had my lads finally cornered.

Finding I could not run as fast as I would have liked, I stopped on a small easily identifiable hilltop and off-loaded my 60 pounds of duck carcasses in a pile. There I figured I would return after taking care of my three shooters and if nothing else, would remove their wings and use them as evidence first, and then later in my waterfowl training classes at Glynco, Georgia. Lighter now, I really poured on the coal as dog and I headed for my group of shooters. Finally coming into view, I put a sneak on my three lads "that was as pretty as a picture" and found myself on them before "they could even pass a whiff of gas." Emerging from the cattails at the base of a wetland in the redhead pass, I caught my three chaps once again breasting out their recently killed illegal ducks. Only this time, there was no escaping the big fella carrying a badge who looked a lot like a young Robert Redford. Well, sort of...

An hour later, I had cited my three chaps from the small nearby town of Selz for jointly possessing 41 protected species of waterfowl, over-limits, and for possessing waterfowl in such a condition that species could not be determined. Talk about three sick dirt farmers who had earlier been out for a grand time. In fact, talk about predetermined violations. Each man had brought a small plastic container from home in which to place the breast meat from the illegal ducks they had collectively killed! How is that for "chutzpah?"

Walking back to my pile of waterfowl carcasses I had left on a small knoll after finishing up the paperwork with my three shooters, I noticed something moving where they were located in the tall grasses. Figuring I had a coyote who had found my duck carcasses and was making himself a meal, I drew my .45. Drew the weapon to scare him, not kill him, when I get closer, I thought. Besides, I liked coyotes. Liked them because they ate "range maggots" or sheep. Many times ate the very sheep that were generally destroying our western public lands through their overgrazing habits. Walking up onto my knoll, I was surprised to flare off a large golden eagle that was making fast work of my evidence carcasses. Surprised over being surprised, my eagle lofted off and sailed down into a nearby coulee. I had to give that bird his due though. When he left, he took a canvasback carcass with him. Nothing like taking the best-eating duck carcass with him in the process, I thought with a smile as I watched him flap his way out of sight.

Walking back to my parked vehicle carrying a load of duck breasts and breasted-out carcasses (minus the one the eagle had taken), I gladly dropped my load of parts at the rear of my vehicle. Sitting on my bumper to take a rest from carrying such a heavy load across the rolling, tallgrass prairie, I heard a vehicle engine coming my way on the same prairie trail on which my patrol vehicle and I now sat.

Sitting there, I soon spied a pickup slowly coming my way with a man standing up in the back with a scoped rifle in hand. Seeing me sitting there on the car's bumper, my rifleman in the back of the oncoming truck hurriedly pulled his rifle out of sight. Then spying all the duck carcasses lying on the ground before me in gay profusion, the vehicle operator continued coming towards me like I was a long-lost friend.

Stopping just feet from me, the driver of the pickup bailed out and whistling in wonder at my "kill" said, "Damn, Man. Don't you know the limit on ducks is only six per day and you can't have any of these kinds?"

Realizing he didn't "know me from Adam," I just let him carry on. Also realizing the lad in the back of the truck with the rifle was probably up to no damn good, I decided to continue my charade of "Joe the duck hunter and dummy outlaw."

"Well, you caught me. I hope the two of you aren't game wardens," I said.

My driver turned and said to his still-armed passenger in the back of the truck, "Shell, you hear that? This fellow hopes we aren't the game wardens because he has too many ducks and all of them appear to be the wrong 'flavor.'"

With that, the fellow in the back of the truck said, "Dang, Man, for a moment there we thought you were the game warden. But then we realized anyone driving such a 'purdy' yellow car like yours couldn't be a game warden. He would be the laughingstock of the town if he drove around in such an ugly colored thing."

I laughed with the two lads over the 'funny' about my baby diaper-colored vehicle as I rose to my feet. "I need to get these out of sight before anyone else comes along," I said as I opened up my dusty dirt-filled trunk and began lobbing the duck carcasses into it and out of sight.

"Hell, Man. Don't sweat the little things. Come see what we have in the back of our truck. We was coming here to gut and skin the damn thing out when we ran across you," said a man I was soon to come to know as Paul Drake.

Closing my trunk, I walked back to the rear of Paul's pickup for an invited look-see. Lying there in the bed of his truck was a freshly killed, not-yet-gutted illegal antelope! Man, talk about the luck of the Irish, I thought. I didn't even have to make this case, it just came right to me, I thought as I smiled inside over my good fortune. "Here, let me give you guys a hand and we can gut the damn thing right here. Besides, I hear tell that when you kill an antelope, you need to gut and skin it right away otherwise the meat will be strong tasting," said "the spider to the fly."

With that, Shell unloaded his rifle and laid it down in the back of Paul's truck. Then the three of us unloaded the critter, hauled it off a-ways into the prairie out of sight from the prairie trail and began gutting the animal. I was now silently laughing so hard inside that I found it hard to contain myself. Just imagine the story these poor lads now had to tell their friends once I sprang my trap. "Yeah, we drove right up to the game warden and had him help us gut out our illegal antelope," was probably how their story would go once they were back at work...

Once the antelope had been gutted and the two lads were wiping off their hands and knives on the prairie grasses, I walked back to my patrol car and opened up my trunk. "When you two finish cleaning up, I would like you to put your antelope into the trunk of my car if you would," I casually said.

"What!" said Shell. "This is our antelope and we intend to keep him. If you want one, go kill your own," he continued in a slightly pissed-off tone and tenor of voice.

"I don't think so," I said as I fished out my badge and walked back to my two lads standing by the gut pile. Now I was between them and Shell's rifle. Then I held up my badge and credentials for all to see saying, "Federal agent, Boys. The two of you are going to be cited into state court for possessing an antelope during the closed season and Shell's rifle will be seized."

Man, you talk about two surprised chaps. They just stood there looking like they were trying to decide if they should fight or flee.

Not wanting to give them time to come to some sort of stupid conclusion as to what to do next, I said, "Boys, I am between you, your rifle, and vehicle. And I am armed. If you decide to do something foolish, I would think it over. Think it over because to fight or run will just get you deeper into trouble with the law. Right now you are looking at a citation. If you decide to fight me or flee, you are looking at time in jail. Your call, Lads," I said as I parted my jacket so my two fellows could see the ever-ready .45 hanging in its holster on my belt.

Then before they could decide what to do, I quickly stepped sideways back to their truck, grabbed the keys from the ignition and removed Shell's rifle from the bed of the pickup. Then walking sideways keeping my chaps in view to the rear of my patrol car, I deposited Shell's rifle and Paul's keys into the open trunk and shut the lid.

That act resulted in the folding of the two lads like they had lost all their 'sand.' With that, I had the lads sit down where they stood by the gut pile and then hand me their driver's licenses. Transferring that information to several of my federal information pink slips, I also filled out two seizure tags. One for the antelope and the other for Shell's rifle (standard procedure when taking someone's property). Then I had the two men load the still-warm antelope into the trunk of my patrol car as well, after I had dug out Paul's truck keys and Shell's rifle.

Letting the lads know I was going to hand the information, rifle and antelope over to the local game warden so he could file on them in state court, I asked if they had any questions.

For the longest time neither man said anything. Then Paul said, "I suppose you must think we are pretty stupid?"

"Well, when it came to killing the antelope during the closed season, yeah, probably. Driving up to a game warden and showing him your illegal kill, maybe," I said with a smile.

"But you fooled us with all those dead ducks. I thought you were an outlaw just like us," said Paul.

"Well, I am kind of an outlaw in a way, I suppose," I said. "What I hunt are humans. And the government doesn't set any season or bag limit on what I catch. And I try to catch all I can. So I guess one can say, I am a little bit of an outlaw in that I always go for 'overlimits' on those I am pursuing." Later that afternoon, I met up once again with the local game warden for the area.

Standing together alongside the road, I thanked him for tipping me off on the illegal redhead and canvasback killing at the historical flight passes as I shared my "catch" information with him. He just more than smiled when I told him about the three farmers from Seitz that I had caught with the 40 illegally breasted-out ducks.

"Damn, Terry. I have been after those three devils ever since I took over this district. And you come in here and in just one day catch those three that I have been chasing for years," he said with a grin of amazement.

"Well, it doesn't make any difference as to who catches the bad guys, just so they are caught. Now I have a surprise for you," I said as I walked back to the trunk of my patrol car. Opening up the trunk, I reached in and dragged out the antelope, laying its carcass on the ground. "Turnabout is fair play. Since you helped me, how about you take this closed-season antelope case in return?" I asked with a smile. My game warden friend was once again amazed over his luck, especially when he discovered I had caught the Drake brothers with the antelope.

"Two more outlaws I have been chasing and never able to catch. How did you do it?" he asked.

"Piece of cake," I said. "They just drove up to me and surrendered," I continued with a mischievous smile as I handed him Shell's rifle as well.

Suffice to say after that day, any time that game warden had a lead on potential illegal duck killing, I got the information posthaste. Like I said, turnabout is fair play...

Moving further south ever-mindful of my meeting with Halsted six days hence at Rapid City, I stayed with another friend in the Jamestown area to maximize my patrol efforts. In so doing,

I made ready to work another area that needed my attention. Especially on some of its many redhead passes in that area as well.

## DAY 5:DUCKS, GEESE, SHARP-TAILED GROUSE AND FOUR "LAW DOGS"...

Before daylight the following morning found me further west in the Chase Lake area. Once again, I had fastened my miserable carcass into the area of a well-known historical redhead duck pass. Standing there in an old buffalo wallow, dog and I listened to all the aerial traffic as we had our usual breakfast in the pre-dawn darkness. A cold stick of salami for me and a can of Alpo dog food for my hard-working, four-legged partner. Chase Lake lying due west of my position was jug-full of ducks and every other kind of waterbird known to man. At least I have the makings of a good day, I thought as I munched on my cold salami. Dog in the meantime, her belly full, was now snoring away as she dreamed her way through other marshes full of ducks in her memory.

Below me I observed two vehicles pulling into the lake area proper. Then pre-dawn darkness followed as the vehicles' occupants departed and headed for their favorite shooting areas. Me,

I just kept to the high ground. There I figured once the shooting started, I could watch any action unfolding below me undetected with my binoculars. Then if any violations were detected, I could move to the sounds of their guns and the gunners. I didn't have long to wait. Once it became light enough to shoot, I discovered I had three guns just below my position hiding in the rushes at the lake's edge. Then out of sight behind some rolling hills further west, I had what sounded like a single gunner having a field day as well.

Fastening my attention on my three closest shooters knowing my best chances for an apprehension laid there, I watched and waited. That year the limit on Canada geese was two per person. I could clearly hear out on Chase Lake the 'worried discussions' of Canada geese concerned over the close-at-hand shooting on the south side of their lake. Shortly thereafter that 'excited goose talk' turned really nervous and then I heard flocks of geese taking to the air. Soon about 100 Canada geese could be seen approaching the south end of the lake as they climbed for altitude. However, they were not high enough when they crossed the south end of Chase Lake and seven geese fell to my three hidden gunners' shotguns! With that barrage of shots, the rest of the Canada geese on the lake arose into the air as they left for quieter places in which to rest and feed. Unfortunately, they, too, followed their brethren as they left the lake and flew south. Flew south right over my three shooters' guns and once again after the shooting was over, six more Canada geese fell to Mother Earth to migrate no more!

With those shots, dog and I were hotfooting it towards our three shooters since they had now killed seven Canada geese over the limit! However, as dog and I made our ways towards our shooters, they continued shooting at the hordes of ducks leaving the lake as well. Moving down on my three shooters, I saw at least eight redhead ducks fall to their guns as well. Now my three lads had real problems. They now had over-limits of Canada geese as well as eight closed-season redheads in their bags. Picking up our pace, dog and I hurried towards our shooters before they killed anything else sporting feathers that flew over their position.

Arriving on scene, dog and I discovered we had an additional problem with our shooters. Once intercepted by yours truly and his magnificent carcass, along with his wonder dog, "Gas Bag," who was always passing foul air in the worst places, we discovered we had other issues. After I had identified myself and upon asking for my shooters' identification, I discovered I had three Jamestown "Law Dogs" in tow! And suffice to say, North Dakota winters are cold, unlike those lads who were hot! "Hot" because they discovered they were going to receive federal citations for over-limits of Canada geese and closed-season redheads. "Hot" because they in turn had identified themselves to me as law enforcement officers and I was not going to show any of them "professional courtesy" and let them off with a warning. Suffice to say that "cool" North Dakota morning soon turned "warm" as I attended to the required paperwork. "Warm" in that I was soon advised that if ever caught in Jamestown breaking any law, they would see to it that I paid a heavy price for such indiscretions. It always amazed me that I discovered so many "badge carriers" who felt their badges gave them such powers and privilege over everyone else around them, especially when it came to breaking the law...

Gathering up my seized birds, dog and I headed back to our patrol vehicle. There I gutted all the birds so they wouldn't spoil. Dog, however, ate more than her fill of warm duck and goose guts as I continued the cleaning process. A move that six hours later I would come to regret. Especially when she was back in my car. It always amazed me how much gas that dog could pass when sleeping in the back seat of my patrol car... If I could have bottled it up and sold it, I could have made a fortune in the balloon business. That was until those balloons popped!

Finishing with the care of my evidence, I washed my hands and took a drink from a Diet Seven-Up in my ice chest. Looking back down on the lake area, I could see my three "crooked" Jamestown police officers had left the scene. However, my singleton or lone hunter still hunting and shooting on the other side of the rolling hills, was having the time of his life if the numbers of shots fired meant anything. In my past enforcement experiences, I had discovered that a singleton mucking around in the world of wildlife was always a good bet for a field inspection. It always amazed me how many times I discovered such "sports" hunting without a license, using unplugged shotguns, in possession of closed-season species or possessing over-limits. Finishing up my drink, dog and I started walking the half-mile or so across the prairie towards my shooter.

Finally arriving in the area where I had last heard shooting, I took a quick glass of the area with my binoculars. I saw nothing in that binocular sweep. Plopping my rather ample carcass down in the tall prairie grasses between the shooter's vehicle and the nearest wetland, I waited. Dog in the meantime, laid down, went to sleep and made ready for the 'balloon' business to follow. For the next hour, I neither saw nor heard anything except my dog passing goose-guts gas. *That is strange*, I thought. True, the singleton had been shooting for about two hours while I had earlier attended to my errant police officers and had plenty of time to limit out. But, where was my shooter now that I was on the "hunt?"

Then off in the distance, I saw about 100 sharp-tailed grouse on a far hillside on a 'rise.' That was followed by three quick shots and even more grouse rising into the air in alarm. There is my shooter, I happily thought. Getting Shadow to her feet, off we went towards the latest shooting which had to be my long-lost chap. Moving through several draws in the rolling prairie trying to stay out of sight, I finally topped a small rise. Sitting back down in the tall prairie grasses, I glassed the area once again. Then I saw a lone shooter heading down into a draw which held a small wetland. Moving in that general direction, I soon heard shooting once again but being out of position and he being in the wetland's tall rushes, I saw nothing. Saw nothing because he was now slowly walking and jump-shooting ducks in the wetland's tall marsh rushes.

Taking a "cut-off" position, I soon arrived, sat and waited. An hour later, I saw the head of a man slowly coming out from a long draw down below me near the wetland. He appeared to be carrying a load and would stop every so often and carefully look all around. Then he would continue his trek for another 100 or so yards. Then once again, he would stop and carefully look all around as if looking for "somebody." He was obviously carrying something but he was still in waist-deep prairie grasses in the bottom of a coulee and I couldn't see what he had. Then he disappeared in a long draw but when last seen, was still heading my way. For the longest time, I saw nothing. Then all of a sudden, he appeared out from the tall grasses and buffalo berries in a nearby draw coming right at me! Now I could see what he was carrying. He had a duck strap over his shoulder loaded with mallards, had a slug of hen mallards in his left hand, and was carrying a shotgun in his right. Once again, my man appeared to be very cautious and looking all around as if looking for something. Being well-hidden in the tall prairie grasses, I just held my position. Slowly my man kept coming and I just let him do so since he was heading right at my hidden position.

Keeping my binoculars on him, I saw the

complete surprise register on his face when he realized there was a man sitting in the brush watching him with binoculars from a mere few feet distance! Without missing a step, my man casually dropped all the hen mallards he had been carrying in his left hand. Pretending like he hadn't seen me, he now kept on coming like his walking around in the prairie was an everyday occurrence. Finally I put down my binoculars, stood up, and introduced myself to the man now not fifteen feet distant. I could see he was trying hard to show no surprise and appear normal but the look on his face betrayed his emotions. He was as nervous as a cat in a room full of pissed-off pit bulls and I knew it.

Letting him walk up to me, I asked for his shotgun and received the same. A quick check showed it to be unplugged (remember what I said about a singleton?). Then I asked to check his ducks on his duck strap. As he swung it off his shoulder, I noticed a suspicious bulge in the game pocket in the back of his hunting coat. However, I saved that bulge for last. Checking the ducks on his duck strap, I found that he had a nice limit of all drake mallards. Rising up to a standing position after checking out his ducks, I asked the man for his hunting license and duck stamp. Receiving those, I asked the man to stay put as I headed back on his trail to retrieve the hen mallards he had dropped.

"Those aren't my ducks," he blurted out.

Turning, I said, "Who said I was walking back here to pick up a mess of ducks?"

Realizing he had just screwed up, he said nothing more. Picking up seven hen mallards, I walked back to my chap. Laying the ducks down off to one side by Shadow who was quietly sitting by the trail, I took out my thermometer. Then I took the temperatures of all the ducks on his duck strap and those hens just picked up. All were within five degrees of each other.

"Why did you do that?" he asked.

"To add to my evidence just in case this goes to trial," I said. "I saw you drop those hens once you spotted me back on the trail. Then by taking all their body temperatures, I can show they were all killed about the same time in any court of law. With those facts, I can show a connection between you and the illegal ducks. Especially since there is not another hunter within miles of this place," I said. "Do you have any other birds on you?" I asked.

"You have seen all the birds I had," he said.

"What is that bulge in your game bag?" I asked.

For a moment he hesitated then catching himself, he said, "Those are just a change of clean clothes in case I fell in."

Reaching around behind him before he could change positions, I grabbed and lifted the 'bulge' in his game bag. It was heavy and warm! "I don't think this bulge is a batch of extra clothing," I said. "Please remove it from your hunting coat."

Once the 'bulge' had been removed from his hunting coat, I got one hell of a surprise. It was a collection of three plastic bread bags. Inside each bread bag were a collection of bloody chunks of breast meat! "What have we here?" I asked as if I didn't know.

For the longest time my chap said nothing. Then he said, "Breast meat from some birds."

"What kind of birds?" I asked.

Once again, he hesitated in answering. Then he said, "Ducks." Opening up one bread bag, I spilled the chunks of meat out on the ground. Duck breasts are dark in color and when breast-ed out, leave a rounded-out, not so thick, kind of a chunk of meat. In that bag were numerous chunks of like-in-kind meat. However there were other, somewhat smaller-sized chunks of dark meat that were deeper in the thickness of their cuts. "What do we have here?" I said as I pointed to the mystery meat.

"Grouse," finally came a hesitant reply.

I had suspected the same, especially after see-ing that rise of a huge flock of a 100 or so sharp-tailed grouse and hearing three shots from my shooter earlier in the day. Laying out the three bread bags of meat chunks, separating them and counting the same, I discovered their breasted portions of meat represented another 22 assorted ducks and 13 sharp-tailed grouse! My lad now had 35 ducks or parts thereof or 29 ducks over the limit, six female mallards over the limit (he was allowed one hen in the bag), nine sharp-tailed grouse over the limit, and use of an unplugged shotgun to take waterfowl! This "singleton" had a major problem...

Asking for his driver's license, I was handed an open wallet with a deputy sheriff's badge stuck to one side on the inside in plain view! Ignoring the badge, my fourth badge seen that day, I asked for and received his driver's license. Then I wrote out his citation for the gross over-limits of waterfowl and upland game birds plus the unplugged shotgun. At least the embarrassed deputy didn't ask for any professional courtesy (not that he would have gotten any). Then we parted company without another word being exchanged. Mainly because he was plenty em-barrassed over his actions and I was pissed at the stupid behavior of 'one' of my own kind.

Then I staggered back to my vehicle with my load of game birds. Once back at my vehicle, I gutted those birds not gutted and threw them into the trunk of my car. As for the duck and grouse breasts, they went into my ice chest for safekeeping. Then out came another soft drink as I stood there in the prairie winds contemplating the morning's events. Four lads checked and four lads, all law enforcement types, cited. Just goes to show one that the good, the bad, and the ugly come in all colors, flavors and kinds. Somehow, that just doesn't make it right though.

Sticking to my schedule, I headed back towards Jamestown, then turned south on highway 281 heading for South Dakota. There I planned on working the eastern part of that state on some known redhead and canvasback passes as well. Since I only had one Special Agent under my command in South Dakota and that was out west in Pierre, I figured I would give him a hand by working in the eastern portion of the state.

Thundering south with a fresh soda, an Italian cigar between my lips and a fresh chew of tobacco between my cheek and gum, I had time to think over my trip's events so far. In the short period of time in the field, I had investigated and made an Airborne Hunting Act case. Then I had settled several wetland easement investigations, bringing them to closure. That work was followed with several cases involving over-limits of ducks and geese, closed-season waterfowl species cases, breasted-out waterfowl cases, upland game cases, and even one involving an illegal antelope. Then it dawned on me. I was working on what I euphemistically called a "Dakota High Plains Special Agent Stew" in the making! When one threw into the "pot" all the goofy cases I had made so far, one truly had a "Dakota High Plains Stew" just as sure as God made little green apples... And I still had a few days to go in South Dakota before my work was done. I wondered what would be coming next in the way of 'ingredients' for my stew? I would soon discover that I wouldn't have long to wait once afield in South Dakota.

## DAY 6: FETISHES AND AN INDIAN LADY WHO ALMOST GOT ME KILLED...

After spending a night in a Motel 6 at my expense since my agency couldn't afford the price of a room and I needed a shower, I made ready for my next day's activities. Dog and I had our typical "game warden breakfast" in the warm motel room and then made ready for our trip south. Heading south on highway 281, I finally arrived at the Town of Aberdeen. Slowing down and looking for a gas station to take on a load of fuel and grab something more to eat since I was running out of salami, I happened to pass an antique shop on Main Street. In the front window of that shop was a homemade sign reading, "Indian fetishes for sale." That sure caught my eye in a heartbeat because many Indian fetishes came with migratory bird feathers attached. And many times, those migratory bird feathers attached to those fetishes were from protected bird species. Pulling over and parking, I rolled down the rear window so my dog would have some fresh air and then I walked back to the antique shop. Entering the shop, I was struck with the typical antique shop smells. Stale cigarette smoke, wood smoke from the potbellied stove merrily smoking away in the store's center, and the smell of things old. Always on the hunt for old wooden hand-carved duck and goose decoys, I wandered around like any other potential buyer. The owner of the store sat behind a huge old glass counter probably from an 1800's drugstore reading the local paper. She had quickly glanced up as I entered the store and since I was obviously browsing, quickly went back to reading her local newspaper. Not seeing any old duck decoys, I began looking for the fetishes as advertised in the front window. Sure as I make the world's finest mashed potatoes (loaded with spices, sour cream, cream cheese and butter), there sat the fetishes in the old glass case by the store's owner. A quick glance told me this lady had a problem. There were 15 fetishes made from odd-shaped, hand-painted small rocks each having migratory bird feathers attached to their sides. Closer examination revealed that the migratory bird tail and wing feathers were from Northern flickers, American robins, Northern kestrels and Northern shrikes. In short, all were illegal to sell under the Migratory Bird Treaty Act. And attached to each fetish were small price tags running from $10 to $15 each.

Seeing me standing there closely examining the fetishes, the owner put down her newspaper and with the 'scent of a sale' in the air, leaned over to where I stood.

"Beautiful, aren't they?" she asked.

"They sure are," I replied. "What do each of these fetishes represent?" I continued.

"Well, they can represent whatever you like but all of them will bring long life and good luck no matter what you do. I saw some like this for sale at an Indian Pow Wow in Bismarck at Fort Lincoln last summer and figured that would be a nice item for me to carry in this store. And since I have lots of birds in my backyard and at my feeders, it wasn't a problem getting bird feathers. I just had my son shoot some with his pellet gun and then I pulled their wing and tail feathers. Then a trip to several creeks in Paha Sapa, the sacred Black Hills to all you white folks who don't understand our culture, when I was there last summer and I acquired the rocks that I needed. Since I am part Indian, I can have and sell such items and they are in fact the real 'McCoy' when it comes to Native American fetishes."

"Are you sure they can be sold? Because I sure would like to have some for my kids as a reminder of their time in the Dakota's," I said like an uneducated lout.

"Absolutely. Like I said, I am part Native American and the law allows us Indians to possess and sell these feathers," she matter-of-factly advised.

Knowing what she was saying was a fabrication made up to enhance a sale, I asked, "Do you have any others like these? These might make great Christmas gifts for the rest of my family in California as well," I said.

"Well, I have some that are even more expensive in the back of the store and they run about $50 to $100 each," she advised.

"I don't care about the price. If you have the time, could you show me the others you have for sale," I innocently asked (said the spider to the fly).

"Sure," she said. "Follow me and I will show you what I have in stock."

Leading me into the back of her store by a workbench, she showed me the rest of her fetish collection. There just as big as you please sat six more fetishes on a table, all with eagle feathers attached! "What kind of feathers are those?" I asked like a dummy. I did so, so the seller would not have any defense in a court of law when it came to the bird's feather identification.

"Those are from eagles. We get eagles from the Fish and Wildlife Service from their facility in Pocatello, Idaho. Then I dismember the carcass they send me and re-use their parts on my special fetishes," she continued.

"And how much are all these six fetishes?" I asked.

"Well, because they are from eagles and white men are not supposed to have them, only just us Indians, I would have to charge you $100 for each of them," she said.

"You have a deal, young lady," I said like a very happy buyer. "If you would, would you carefully wrap them and all those in your counter in the front of the store. Then if you would make me out a bill of sale, I will give you a check for all of them."

"You from around here?" she asked cautiously.

"No. I am from Bismarck but I have good identification if that is what is bothering you," I said rather innocently hoping to cement a sale.

"OK, Mister. You have a deal. I will wrap these here in the back of the store so if any other customers come in they won't see the eagle feathers. Besides, if any of my Indian friends come into the store, they sometimes get mad if I sell eagle feathers to white men," she said.

With that, I walked back out into the main portion of the store and continued looking around for any other wildlife violations, while my young Indian lass wrapped up the eagle feather fetishes. Shortly thereafter, out she came with a small box all wrapped up in brown paper with my eagle fetishes. Then she began to carefully wrap up the 15 fetishes from under the glass counter so she wouldn't damage any of the feathers.

Finishing, she commenced writing up a bill of sale for all the fetishes. "Do you care if I just list 'fetishes' on the bill of sale and don't mention the bird feathers?" she asked.

"That will be fine with me," I said as I dug out my checkbook. Boy, talk about a happy look when she saw me dig out my checkbook...

When she finished, I said, "What is the damage, young lady?"

She took to figuring out what I owed her on

a piece of nearby paper and when finished said, "That will be $915. I will throw in the taxes myself since you made such a big purchase," she said with a smile that read 'greed' all over her face.

With that, I wrote out a check for $915 and handed it to her. I had deliberately left out the portion on the check as to who it was made out to. "Do you have a store stamp to put on who the check is made out to?" I asked.

"No, but I can write in the name of the store," she said. Then taking the check, she wrote out the store's name in the blank on the check (now I had a handwriting sample). "There you go," she said as she handed me the two boxes of fetishes.

"Dang," I said, "will you hand me that check back so I can enter it into my checkbook."

Without a thought given to my request, she handed me back my check and I made a big deal of entering the facts into my checkbook.

Then I pocketed the checkbook, check and all, and picked up the two boxes. Shifting them under my left arm, I took out my badge and showed my shocked shopkeeper who I truly was. She blankly looked at it as if to say, "What does that have to do with the high price of crowbars in Korea?"

Realizing that her eyes had glazed over from the huge sale just made, I said, "Young lady, my name is Terry Grosz. I am a Special Agent with the

Fish and Wildlife Service. For your information and from the rest of your earlier conversation, you already know that the sale of migratory bird feathers by anyone, Indian or white man in the United States is illegal. That also includes the sale of eagle feathers by Native Americans. If Native Americans lawfully acquire eagle feathers, you and yours can possess and pass them down from father to son. But they cannot be sold, offered for sale, traded, imported, or exported. That is the law of the land and like I said, your earlier conversation revealed to me your knowledge that such sales were illegal. As such, I am seizing these feathered products and when I return to my office in Bismarck, will initiate an Information (a federal complaint) in South Dakota against you for their illegal sale."

That poor gal was in shock, but just for a moment. Then she let me have it with the worst "sailor" language I have ever heard coming from a "lady." Boy, did she ever come unglued! For the next few minutes, she called me everything but a white man... After she had settled down, I got from my "sailor" those identifiers needed to file an Information in Rapid City, provided her a seized inventory document for the items seized, thanked her, and split.

Exiting her store, I stored the fetish seizures in

the trunk of my baby diaper yellow patrol car. Then without another thought, I headed for the nearest gas station to refuel for the rest of my trip to Watertown in the eastern part of South Dakota. Little did I realize at that moment, my actions in that antique store would lead me to almost getting 'scalped' a few days later!

Leaving Aberdeen, I headed east on highway 12 to Waubay National Wildlife Refuge. Once there, I off-loaded my wet seized items from my earlier cases into the refuge's evidence freezer in case I needed them later on for a trial. However, I kept the fetishes locked up in the trunk of my patrol car. That evening I slept in the refuge office on the floor because as said earlier, my law enforcement division was so broke my boss would not authorize any motel expenses! However, I still got my $12 per day per diem on which to live.

## DAY 7: ICE CHESTS MAKE GOOD PLACES TO LOOK...

Way before daylight the next morning, dog and I were already staked out in a complex of nearby wetlands close to Waubay Lake, waiting for the waterfowl shooting hours to begin (one-half-hour before sunrise). As usual, dog had her can of dog food and I feasted on a Coke and a stick of homemade German summer sausage purchased in a nearby mom-and-pop grocery store instead of my usual Italian salami. I did so because I had finally run out of my good Italian hard salami. Oh well, when in Rome do as the Romans do...

I had picked a good spot in which to start my day. The surrounding wetlands were full of just-waking-up ducks and geese. Standing there at the edge of the marsh, I could still smell the pungent odor of methane gas. Then a muskrat slowly patrolled by my inert carcass at the edge of the ice, while a noisy hen mallard let the rest of the world know she was in country. Man, it was great to be alive. Alive and out at the edge of the marsh away from my now loudly snoring Labrador some twenty feet distant in a clump of prairie grasses. Additionally, she was already converting the can of Alpo dog food into her form of Labrador methane as well...

Then off to my north, I heard the not-too-distant sounds of a softly popping shotgun. It was now shooting time and I wondered what my day as a hunter of humans would bring. I didn't have long to wait. South of me about 100 yards distant boomed two more shotguns. It was still too dark for me to see what my lads were shooting at (damn that legal half-hour before sunrise shooting allowance), but at least it gave me a possible, close-at-hand target for the day. Speaking Shadow's name softly to awaken her from her land of slumber, the two of us headed south towards the vicinity of our two close-at-hand shotgunners.

Finally locating my two lads partially hidden by a large muskrat house, dog and I dug into a nearby dense stand of rushes to watch and see what they were up to. Walking out from shore, the two shooters had broken the thin ice all the way. Then they had broken the ice all around the muskrat house so the ducks could see the open water. That was followed with a clumsy attempt at a "J" pattern of decoys in front of their muskrat house. Then as the waterfowl world arose for another day of trial and tribulation, numerous singles and pairs fell to the "looks" and clarion calls from the decoys and duck callers below. Soon, my lads had six ducks and one goose down for the count. However, four of the ducks they had down were closed-season canvasbacks!

Not wanting my two lads to kill every scarce canvasback in the flyway, dog and I headed out to apprise my shooters of their sad choice of ducks to kill that fine day. Splashing across a large body of water towards my two shooters breaking ice all the way, I saw them duck behind their muskrat house when they saw and heard me coming. *No problem*, I thought. I knew where they were located and it was just a matter of time before they were run to ground. Taking my time slogging across the open marsh to avoid breaking a sweat and cutting a hole in my waders from the shards of ice, I soon arrived at my two shooters' monster-sized muskrat house. Since I now couldn't see hide nor hair of them, I walked around behind the muskrat house and got the surprise of the day. There was no one there! My two chaps, suspecting I might be the law and that maybe they had the wrong flavor of ducks, had fled into the marsh's nearby rushes. However, they could run but not hide. The busted ice trail they had left was a dead giveaway as to their direction taken.

No big deal though, I thought. Plowing through such a partially frozen marsh, one has a tendency to leave muddy footprints in the normally clear water and a broken ice trail. And upon those markers I soon chose to follow. Realizing my "footprints and ice trail" was now beginning a long circle back towards the muskrat house in order to pick up their decoys or head for their vehicle, I cut across the marsh on an intercept course. Pouring on the steam to intercept my two shooters, I soon ran out of covering rushes. Then I just held my position in hiding where I could still intercept my lads if they chose to run. However, no such luck on the part of my two shooters. As it turned out, they were using my finger of rushes where I was hidden to now double back and collect their decoys. And in so doing and in short order, ran right into yours truly.

After "greetings" were made all around and my identification known, the three of us headed back to their decoys. Picking up their decoys, the three of us headed for my lads' vehicle. I carried their four illegal ducks and as my lads walked out ahead of me, I could see they were trying to carry on a hushed conversation. Clearly to my way of thinking something was rotten in the slaughterhouse but I couldn't put my finger on it. I knew my lads would not try and outrun me because I now carried their hunting and driver's licenses on my person. But what? As we got closer to their vehicle, I began to suspect why the worried conversations between my two lads. Parked at the end of a prairie trail was an old Dodge sedan. Off to one side of the vehicle stood a small tent and alongside that tent were two ice chests. Now to all you unwashed readers, an ice chest to a "tule creeper" like me is nothing but a damn plain invite for a look-see... Sure as God made my crawdad linguini some of the best and richest eating in the world (it's all that damn cream, garlic and butter I use), that had to be the bone of concern between my two lads. And once we arrived at their campsite, the first thing my two lads did was put the two ice chests into the backseat of their car as casually as they could as I wrote out their current day's citations on its hood. Filled out the citations for everything but the actual violation that is, as my lads tore down their camp and made ready to leave. Even though the ice chests had been loaded, I had not looked inside them and I damn sure meant to.

"You lads need to bring me those two ice chests so I can look inside," I casually mentioned. Man, you should have seen the looks of despair fly between the eyes of my two chaps. Then see the utter looks of plain damn seasick 'green' on their faces way out here on the wind-blown prairie as they placed the ice chests at my feet so I could look inside. I guess I could figure why the seasick look way out here on the prairie. Don't many folks consider the prairie just like a "sea of grass?"

Twenty-six ducks later, 14 of which were closed-season redheads and canvasbacks and the rest over-limits, along with nine closed-season pheasants, I finished up with my two lads' citations. Dang, I thought, it was amazing how many of these dirt farmers I had run across over the last few days that had trouble counting. It was obvious to me that Mrs. Wilson had done a poor job in her third-grade math class with a number of unable-to-count accurately, duck, goose, sharp-tailed grouse and pheasant-hunting folks I had run to ground these last few days.

I spent the rest of my day in the area checking all the local duck and goose hunters I could find. None had broken any laws and pleased over that, I took the long walk back to my patrol vehicle. By then, the South Dakota skies had turned leaden in color and the temperature now stood 'warmly' at the freezing mark. Winter was here for sure and soon to be in all its fury, I thought.

Heading south towards Brookings later in the day, I stopped at an all-you-can-eat place and about ran them out of business. It always amazed me how much of an appetite I had after working hard all day out in the Dakota bone-chilling cold of winter. Then I headed west on state highway 14 until I arrived in Pierre later in the evening. Once there, I stayed at the home of my agent in Pierre, Dave Fisher. Dave was a cool guy. He had been a B-24 bomber pilot during World War II and had been shot down over the oil fields in Romania. Trying to leave the stricken airplane after he had given the bail-out order, he discovered that one of his gunners in the excitement and danger of the moment had opened his parachute inside the airplane. As the plane spun crazily towards the ground, Dave tried to free his trapped gunner. Then realizing he had to bail out or be killed as the plane approached the earth, he reluctantly left his gunner and bailed out. He was only about 700 feet over the ground when his chute opened and he hit the ground with a heavy impact. In so doing, he screwed up his hips for the rest of his life. He was able to walk but he limped and did so in a lot of pain. But that pain was only equaled by the pain in his heart over the knowledge that he couldn't get his gunner out of the plane before it crashed. Dave spent about six months with the local partisans before being repatriated back to his own flying outfit. Then he returned to flying over Italy once again, causing hate and discontentment to those on the receiving end of his bombs.

Dave was a big old guy, about 6′ 4″ in height, and much loved by the South Dakota game wardens because of his easygoing ways and catch-dog attitude. He had come to the Service from the Iowa Game and Fish Department early on as a game warden. He then easily made the transition from Iowa game warden to that of a Special Agent for the Fish and Wildlife Service, much to the chagrin of all the bad guys he had tapped on their shoulders during their times walking evilly on the dark side of the world of wildlife.

Later that evening over a snapping turtle stew and fresh hot out-of-the-oven biscuits (he was also a great cook), Dave and I laid out our plans for the next few days. Pierre at that time had to be the Canada goose hunting capital of the country. The surrounding area was rich in high-energy goose food in the form of corn and every other seed grain in between. Then it was just a short flight for the tens of thousands of Canada geese from the grain fields back to the waters of Lake Oahe, or the open water of the Missouri down-river from the dam. Then came the dark part. Just about every farmer surrounding the lake and lower river had set up commercial goose shoot-ing clubs to take advantage of that bird resource and the money it brought into their coffers from sportsmen. Tax-free coffers I might add, in many instances. There was little way the IRS could keep track of the money the dirt farmers made off the goose-shooting public. And the moth-ers of those farmers didn't raise any 'turnips'... And if any of you folks reading these lines think ahead, one doesn't have a successful commercial goose-shooting club unless they have geese to shoot at. And many, not all mind you but many, saw to it that the 'shooting pot was sweetened' for the daily arrival of the thousands of hungry Canada geese for their rube shooters. How they 'sweetened the pot,' in violation of state and federal laws, was a closely kept and guarded family secret. A 'secret' to be discovered by the likes of Dave, his South Dakota counterparts, and now yours truly.

## DAY 8: A DETAIL GONE BAD ON MCDOWELL'S COMMERCIAL GOOSE CLUB...

The next morning, way before daylight, found me sitting in front of the Bryce McDowell ranch house and commercial goose-shooting club. Its parking lot was already full of numerous hopefuls all expecting a stellar Canada goose shoot. Me, since I was unknown in the area, had drawn the short straw so to speak. It was at least forty below and the wind was howling down from beyond the northern extremes for all its worth. That in turn with the wind chill, made it 2,000 degrees below zero! Coupled with those conditions was all of North Dakota's topsoil dirt in the air as well. It seemed with all those howling winds, all the poorly cared for lands in North Dakota were now being blown clear to Kansas. And let me tell you, all that cold and flying dirt sure played havoc with the mechanical workings on any shotgun used in the field...

Me, your classic Robert Redford type, was on an undercover assignment hoping to catch Bryce and his cronies breaking the law of the land when they shot too many geese or shot over a baited area. At least that was the plan. But if you readers remember back to your days in English literature in high school or college, "the best laid plans of mice and men often go astray"...

Just so my readers have a feeling for what the dickens was going on, here goes part one. The Bryce McDowell commercial shooting club had a bad reputation when it came to shooting geese illegally. His farm sat high up on the bluffs overlooking the Missouri River far below. Resting on a long bend of that river on the open water were at least 100,000 resting Canada geese. Up on the high bluffs above the river sat Bryce McDowell's commercial goose-shooting operation. His land had been planted to field corn which was still standing even at this late date in the year. The geese knew that corn was there and because of the energy-sapping cold during the South Dakota winters, had to have at that high-energy food source or die. So during days when they could fly (other than during blizzard weather), a steady stream of geese by the thousands would fly up off the Missouri River onto the McDowell farm to feed. In so doing, they had to pass numerous goose pits built into the sides of the bluffs on the river side facing the goose concentrations. To avoid using up too much energy, geese leaving the river would fly low up the bluffs and onto Bryce's cornfields to feed. Low meaning many times 25-30 yards off the ground as they struggled in flight up the sides of the bluffs. Ensconced in those bluffs would be dozens of paying gunners all hoping for a shot at a "goose dinner."

Part two of this killing field deals with Bryce and his outlaw bands—what he called "his money pit." Bryce was a reputed outlaw in and of himself. He would only take cash money from his rube shooters to avoid IRS entanglements ($20 per goose hunting trip for two birds), and freely allowed the illegal taking of geese over alleged baited areas in his unharvested cornfields bathed in overlimits as well. After Bryce got to know and trust me (I posed as a wealthy rice farmer from California looking for a farm to buy in South Dakota), he used to brag about what he made in his goose-killing operation as he tried to interest me in purchasing his land. He would show me rolls of $20 bills that would choke a saltwater crocodile and tell me he made over $100,000 a year in his shooting operation. In turn, Bryce would allow folks to goose-shoot over his lands and never claimed to Uncle Sam any more than $20,000 a year in profits because he only dealt in cash. Cash because the IRS could not keep track of his money-making operations that way. Therefore because the government never knew how much he took in, he allegedly could cheat them blind if he was of a mind...

Part three of this sordid mess was what Bryce called his 'Pit Captains.' Basically, trusted folks and fellow outlaws who managed the groups of rubes who were taken out on his lands for a good old-fashioned goose-killing venture. I never met any one of them who would rate any higher than a second-class maggot! For the most part, they were Bryce's drinking buddies and scum of the earth, looking to gain some local standing being a so-called "Pit Captain." Their job was to watch the hundreds of flocks of Canada geese pouring up the sides of the bluffs from inside a hidden dugout where they parked their vehicles underground. From their positions of hiding, they would watch for the larger species of geese flying up the cliffs from the Missouri River far below. When a flock of the larger species of Canada goose would fly over the pits governed by that "Pit Captain," he would yell out, "Take 'em." Whereupon the rubes in the adjacent pits would blaze away, crippling as many as they killed. The cripples would then turn and glide back down to the Missouri to more than likely die a slow death or be eaten by the ever-present coyotes or eagles along the river. Then his 'Pit Captains' would singularly run out and pick up the dead geese from that latest barrage of shooting to reduce the scare factor of other geese still coming their way. Then the event would be played out once again until everyone had two geese (not necessarily their own). Any over-limits or smaller species of Canada geese killed by mistake and not wanted would be hidden and later either sold or given to the Indians living nearby on the Crow Creek or Lower Brule Indian Reservations. Once everyone had their two geese, the "Pit Captains" would load everyone into the back of their hidden pickups and bolt for the farmhouse. There the shooters would be let out and another batch of shooters loaded into the pickups and once again that crew would head for the pit blinds on the edge of the bluffs. Then the selective killing of the largest species of Canada geese (there are many sub-species and sizes of Canada geese) would begin all over again.

Part four was the cause for another problem. Everyone would be jammed into a small shooting area out on the bluffs or in the cornfields. Then when the geese came over, everyone shot into the massed flocks of birds. No one ever knew which bird they killed. No fair chase here.

Later back at the clubhouse, the shooters would just pick two birds from the dead pile of Canada geese and that was that. There the "Pit Captains" would happily solicit tips for the "wonderful" hunt they had just provided to all the trigger-happy rubes holding their two geese probably shot by someone else...

Then came part five. If for some reason the geese became leery of the killing on the McDowell farm, Bryce, according to Dave, would sweeten up the pot. First, he would go out at night and harvest a set number of rows of his standing corn in order to give the geese a new look at the buffet below. Then he would fill in the goose pits previously used in his cornfield (yes, he killed the geese after they had braved the bluff shoots in his cornfield as well). Then to compensate for moving the available rows of harvested corn, he would dig new pit blinds in the area just freshly harvested. If that method didn't work, Dave advised, Bryce would just adjust the harvesters cutting the evening's corn, so that none went into the holding tank of the harvester. It would be just pooped out the back of the harvester and made available for the ultra-hungry geese. In short, not a normal agriculture harvest as prescribed by federal laws, but a violation of the federal Migratory Bird Treaty Act and associated South Dakota state hunting regulations.

Then came part six. If you had already killed your limit during the day and wanted to go again, Bryce didn't care. Just as long as you paid your $20, you could shoot at the geese all day long for all he cared. And the sad thing associated with that endeavor which I personally witnessed were the actions of the South Dakota Game Department at that time. They would have their biological types out there on-site checking the sex and age of the geese brought back to the ranch house. More than once I saw those folks confronted with shooters possessing over-limits of geese and they said nothing! They were there just to get the biological kill information and nothing else mattered...

So, it was into that fine mess of snakes I was dropped by my resident agent that cold damn morning so long ago. And I do mean cold. I was wearing a down jacket good for up to forty below plus a down vest underneath that. Hell, I still damn near froze to death! And that wasn't the funny part. Dave figured since no one knew me (I was a brand-new supervisor), he would use me in an undercover capacity to try and catch Bryce and some of his cronies doing something wrong. Well, I wasn't one to turn a blind eye to my wildlife in trouble, so I agreed. Hell, I was just like a dog hanging his head out the window on a moving farm pickup. I was ready to go... What a riot that turned out to be!

Dave had acquired a belted-battery device which I would wear. Looking back on it now, I looked like a Muslim suicide bomber from Iraq in that goofy, primitive device. It consisted of a heavy vest loaded with rows of old-style military batteries twice the size of normal batteries. A transmitting wire would then be run down my leg and theoretically, once I had the damaging goods on my bad guys, I would speak into my shirt mike and transmit the same to Dave. Dave in the meantime, a mile or so out of the way along with some extra help, would be listening to my transmissions. Then once I gave the signal, Dave and company would swoop in and gather up the bad guys. Then we would all go back to Pierre and have some more of Dave's wonderful snapping turtle stew. Yeah, right!

So there I sat in my unmarked patrol car in the McDowell's parking lot that fine damn winter, howling-cold morning so long ago. Then Bryce came out with his 'Pit Captains' and all 100 or so of us idiot shooters began loading up into various pickups to head out to the killing fields. That morning, I drew a 'Pit Captain' that was not only a maggot but an illiterate one as well. All of my fellow shooters and yours truly bailed into his pickup and lurched our ways across the cornfields and down the bluff to our shooting area. It was so damn cold, the snot froze on the side of my face! Then they drove the truck into the covered underground blind and all of us bailed out like a gaggle of dingbat guinea hens. Standing inside the vehicle's hiding place and in several adjacent high-walled pits, we were ordered to load our shotguns and wait for the signal. Then when it was given, all of us were to lurch out and blaze away at anything in the air overhead. That morning I carried my double-barreled 10 gauge with no shells. My job was not to kill but gather evidence for prosecution and that was what I would do. However, my damn heavy old-fashioned battery-packed vest kept slipping down and I was constantly lifting it back up with my free hand so no one would be any the wiser. I later found out the sewing on one shoulder strap on the military device had failed and one side kept slipping down. Finally, when no one was looking, I somehow hooked the side slipping down under my belt. That turned out to be a real mistake. However, as the events unfolded, I was able to quietly transmit the number of shots being fired, how many geese were killed, and if the Tit Captain' killed any to my agent listening in on the receiver.

However, trouble was now brewing. It was getting so damn cold that many of the actions on the rubes' shotguns were not working. The damn oil in their shotguns was freezing due to the intense winter cold and flying topsoil from North Dakota! Then there was the flying dirt kicked up by the howling winds off the barren bluff to our front as well. That, too, was flying into the guns' actions and forming an oil-and-dirt paste that fouled many a shotgun into just "one-shooters."

And then here came even more hungry geese flying low overhead to Bryce's cornfield killing zones above and behind us. After another barrage of shots, our illiterate "Pit Captain" retrieved one goose and I saw it was an adult and rare species for this part of the country. It was a Tule goose, one normally found in the Pacific Flyway to the west! My extensive background and training in waterfowl identification got the best of me as I saw the bird drop not twenty yards away. When it did, I yelled out in amazement, "Hey, a Tule goose!"

My stupid and now half-drunk "Pit Captain" took another swig on his almost-finished quart of peppermint schnapps (a common drink in the high plains during the winter months by many of the outdoor types), then said, "What the hell did you say?"

*Oh crap*, I said under my breath. I hope I haven't given my cover away. Looking over at my half-drunk "Pit Captain" I said, "Look at that goose! What kind is it? I don't think it is a Canada goose."

Picking up the rare Tule goose, the idiot-stick 'Pit Captain' said, "Hell, this is just a damn 'tool goose'."

Well, even if he couldn't properly pronounce the name, he did at least identify it correctly, I thought, as I now winced in pain over the extreme cold.

And that wasn't all. I could hear a soft crunching sound coming from my battery belt! Then I felt some burning oily liquid starting to trickle down my leg from the battery pack where I had tucked one corner of it into the belted area of my pants. Looking at my battery pack while pretending to attend to a call of nature, I saw some of the batteries were freezing and basically exploding open! Now I was in deep crap and I knew it. In fact, some of the liquid from the bursting batteries from our antique recording device was now solidifying on my skin and burning like hell! Fortunately, the group of shooters I was with finally limited out and all of us prepared to leave.

Rattling our way out from our bluff goose blind in the back of our pickup, we headed back for the farmhouse base of operations. Once there I was only too happy to grab my two geese (which I didn't shoot, remember, I took no shells), got into my car and left. Once out of the area, I headed for where I knew Dave would be located. Once there, I found him very worried. He could hear me talking about how cold it was and recorded my teeth chattering. He had received some of the information I had transmitted on that damn ancient Army surplus device but about a half-hour into its operation, it had quit transmitting. When it did that, he contemplated coming into the area with his state charges like the cavalry of old thinking I was in trouble. But he wisely held his ground and when I finally arrived back after the hunt, it was plain to see he was much relieved. You see Bryce, being a bit of a drinker, could be one wild and half-crazy hellion if he thought you were after him or challenging his authority. Dave fearing the worst when the device failed, figured he was going to have to shoot his way in and rescue me. Then when he discovered my batteries were bursting from the intense cold in the vest, he had a good laugh. I sure as hell didn't. The burst batteries required that I strip down out in that wind-howling, forty-below prairie and wipe myself off from all the battery acid that had run down my legs and into my groin area! Let me tell you, I was one very-cooked lobster-red and hide-burned individual!

Not having enough information to go in and make any arrests, we cancelled the operation and headed back to Dave's house. There I showered and rubbed Bag Balm (normally used for chapped and sore teats on milk cows) all over my reddened areas which were considerable.

Since the wind was now really howling and the temperature had dropped to 55 below zero wind chill, we cancelled any further operations for that day. Plus, I had to change out of my clothing which by now had almost disintegrated from all the battery acid. Into the trash my clothes went and I crawled into a new change. Thank God for that Bag Balm because I was one burned son-of-a-gun! However, a nice deep-fried walleye dinner that night with all the trimmings and a liberal application and self-oiling of peppermint schnapps, pretty well took my mind off my burning groin and my last part over the fence where the acid had drained all over as well! Damn, what one does for the critters. By the way, I still use that Bag Balm to this day when needed...

## DAY 9: GOOSE OVER-LIMITS AND A COYOTE "SCORES" AN AIRPLANE...

The next morning at "o-dark hundred" found Dave and me staked out on another commercial goose club on the bluffs above the Missouri. The winds had died down and it was now only a brisk 33 degrees below zero. Off in the distance, Dave and I could see five men setting out their goose decoys in a harvested cornfield, using the lights from their vehicles as an aid. Once the decoys were set, the vehicles left and parked about 100 yards distant in a row of cottonwood trees. Then Dave and I could see the lights from several flashlights working their ways back to the goose set. Once there, the men disappeared into their goose pits and then the blackness of predawn took over once again.

Once daylight arrived, Dave and I could see the men's heads above their goose pits looking all around for the anticipated flocks of Canada geese soon to come. They didn't have long to wait as a small family group of seven lesser Canada geese arrived, warily circled the set then finally convinced all was well, set their wings and floated in. Not a shot was fired as the men had since disappeared out of sight into their firing pits.

"I suspect they are waiting for the larger Canada geese to arrive. In the meantime, they will probably not shoot that smaller flock of geese and just let them act as live decoys," I said to Dave as I continued watching through my binoculars from the front seat of our hidden vehicle.

Dave just grunted his concurrence over my words and continued watching our five shooters as well. Soon, the cold, clear winter sky was filled with small skeins of geese moving every which way towards their favored feeding areas. Then we got action as a flock of 13 larger species of Canada geese warily circled our field of shooters several times. Satisfied everything was alright, especially after one of our earlier flock of geese rose up and flapped its wings, in they came like they were drawn in on a string. Just as our new flock of geese dropped their feet under cupped wings at the edge of the decoys and were most susceptible to the lethal shot streams, all hell broke loose. Our five shooters rose in unison and blazed away at our new arrivals. Ten of the 13 folded like car-hit chickens! But that wasn't the worst part. The seven lesser Canada geese that had arrived earlier, alarmed over the shooting, rose in unison into the air. In so doing, they flew right into the shot patterns being aimed at the flock of 13 and had five of their little group killed as well! Now our lads had 15 geese on the ground!

For the longest time, none of our shooters moved or showed themselves. It almost seemed as if they didn't want to be noticed. Finally, one of the five men casually got out of his blind and moved around in the decoys as if resetting the shell decoys. Then once again casually, after tending to a call of nature, dropped back into his goose pit as if nothing out of the ordinary had happened.

However, two sets of eyes had observed what had just occurred as they squinted behind their 7X50 sets of binoculars. Everything that had just happened smelled to high heaven to the badge carriers behind the binoculars. The casualness after the shoot, the one lad resetting the shell decoys as he casually slid the dead geese under them and out of sight like nothing had happened, and now the appearance of the man back in his goose pit, made the blood of those watching through their binoculars run a bit faster.

Moments later, I observed a larger flock of Canada geese heading their way over the cotton-wood treetops. Those geese, like the last, had obviously been shot at several or more times since hunting season had opened. They circled the field warily several times on slowed wing beats

showing obvious interest but great survival re-
straint. Then two geese peeled off from the flock
and dropped straight into the decoys. When they
landed, like most Canada geese, they just froze
and looked all around for any sign of danger.
Not finding any, they soon commenced looking
for corn kernels left over from the earlier harvesting process. That latter feeding movement was
all it took and the rest of the flock flew straight
towards the two geese happily munching away
on the ground. Once again, as the geese dropped
their feet and filled their wings with air to slow
their flight, our five lads rose from their goose
pits. There was instant pandemonium among the
flock, as they crashed into each other trying to
gain air speed and escape. However, their latent
air speed was no match for the flying lead pellets
coming at them at 1,200 feet per second! Nine
geese dropped like rocks and two crippled off,
just barely clearing the line of cottonwood trees
to their north. The rest of the survivors made
for the direction of Lake Oahe like there was
no tomorrow. Then one of the departing flock
peeled off and on slowed wings, returned to the
just-shot-over decoy set. Flying straight into the
decoys, that bird died in the air and fell the last
30 feet into the decoys, smashing one of the shell
goose decoys flatter than a flounder. It was al-

most a last act of defiance and one in which that poor goose had gotten even with the decoy for fooling it into the killing ranges of the shotguns below.

"That does it," I said. "They have a gross over-limit and need to go and see the judge to see if he agrees with us."

"Terry, let's see what they do with them. They have one hell of an over-limit and that won't be easy to hide. Let's just watch and see what they do with all those geese. After all, it's not like they will get away from us since we are holding the high ground," said Dave with a twinkle in his eyes and a grin on his face under a two-day-old growth of whiskers.

"Alright. But if they make ready to shoot another flock, we are going in and ruin their Christmas," I said. "Enough is enough..."

It was about then I saw the glint of a handi-talki antenna flashing in the sun's weakened winter rays above one of our shooters. As he was using the handi-talki, the rest of his partners were scurrying around the field filling their hands with dead Canada geese. Once any one of them picked up a handful of dead geese, he returned to their pit blinds and threw them down into the bottom of the hole. Twenty minutes later, here came a Dodge Ram Charger with two chaps inside. Our Dodge drove right out into the field and without even getting out from their rig, the men in the field opened up the back hatch to the vehicle. Into that vehicle hurriedly went what appeared to be two limits of geese. With that, the back hatch was closed and the Dodge sped rapidly from the field.

"Let's go get them, Dave," I hissed.

"No, I don't think so," said Dave. "I know both of those guys driving that Dodge and they have a commercial picking operation not far from here. They have always beaten me every way from Sunday and this is one time I am going to 'rock their boat for a change,'" he uttered as he watched our two chaps flee from the field. "I suspect those geese will be picked, cleaned and sold or given to hunter friends who have none," he continued. "Besides, our five gunners still have over-limits," he said as he lowered his binoculars and grinned a toothy grin over at me.

With that our five lads returned to ground and not a hide nor hair did we see of them for the next twenty minutes. Then here came our Dodge once again speeding down the field's dirt road and then bounced its way across the harvested cornfield.

"Now we can go and see what they have to say for themselves," said Dave as he started up his patrol car. A few minutes later we intercepted out two 'goose runners' and boy, were they surprised! Taking their illegal birds from them and holding their driver's licenses, we had them follow us back into the killing field. There we had a 'big party' with our pit gunners as well. Well, maybe not a happy party but a big one. Dave and I laid out what we had seen earlier and of course had documented. When confronted with the obvious, our local lads folded. For our field shooters, they were facing charges of taking and possessing over-limits of geese. For our goose runners, 18 U.S.C., Aiding and Abetting in the overages, would be hung on those chaps like a wet, goose-down comforter! Man, did they squall when I told them that I was Dave's supervisor and was contemplating seizing their vehicle as an instrument to the crime as well. Brother, that was all it took for their eyes to get as big as dinner plates! With those words of warning, they lost their mouthiness and the remainder of their "sand" ran down their legs like the battery acid had done mine a day earlier.

Dave and I cited the field lads and they were a sullen lot when told what the minimum bail in federal court amounted to for the error of their ways. Then Dave and I proceeded to the picking facility run by our two boys with their "goose-running Dodge." Once inside their facility, we soon uncovered the two limits of geese run earlier from the field, plus several more sets of illegally tagged geese. Before that little part of our day was done, Dave and I had more than paid for the gas and oil we had used to catch our little bunch of goose-killing outlaws who certainly needed the benefit of our attention.

Heading back to Dave's home in Pierre to get my clothes so I could meet up with Halsted the next day in Rapid City we sped along. I was about half-asleep in Dave's warm patrol car, when I noticed something odd out on the now-frozen Missouri River.

"Slow her down, Dave," I yelled. "Would you look at that?" I said. Dave pulled off the road as we both whipped out our binoculars and began watching a drama being played out over the now-frozen hard-as-a-witch's heart, Missouri River. Flying very slowly over a long stretch of the river was a green-and-yellow Super Cub. It wasn't more than ten feet over the river and was flying back and forth, up and down the river. At first I didn't see anything, figuring it was some-one airborne hunting coyotes crossing over the frozen river. But neither Dave nor I saw anything running. Then I finally spotted it. Lying near a small snowdrift on the Missouri lay a dark figure of a dead coyote. From the looks of it, our Super Cub pilot had caught a coyote who had been hunting wounded geese on the frozen river and gunned it down. Now from all apparent signs, our pilot was looking over the frozen river to see if he could safely land his airplane with his weight-distributing balloon tires and pick up the dead coyote, which would bring about $250 in those days' prices.

With that in mind, both Dave and I scanned the tail and wings when our pilot banked his aircraft, as we looked for any signs of aircraft identification. There was none because our pilot had illegally taped over his numbers so anyone on the ground watching him illegally aerially gun furbearers could not identify his aircraft and report him!

Finally our pilot with over-size tires flew low over the ice and briefly touched down. Then gunning his engine, he rose quickly back into the air, circled his landing spot and once again made a rolling, high-speed landing on the river. Once again the ice held. Then taking off and making another turn, our pilot very skillfully slowed his aircraft to almost a stall and softly landed it on the frozen river. Rolling to a stop, he sat there for a minute with his engine running as if ready to make a quick takeoff if the ice underneath his aircraft failed. It continued to hold once again. That time with the engine still idling, the pilot left his plane and sprinted for the dead coyote lying 35 yards off to his east. Grabbing up the coyote, our pilot ran his hands through the prime winter pelt as if happy with his kill.

Then Dave and I saw him staring hard back at his airplane in horror, then drop his coyote and make a mad dash for his aircraft. Swinging our glasses back to the airplane, I saw his still-whirling propeller strike the ice and splinter into a thousand pieces! The propeller had hit the ice because the weight of the aircraft's engine had overcome the strength of the still-fresh ice. And in so doing, it was now sinking nose first through the fresh ice into the Missouri river below! I could see our pilot running like a banshee for his rapidly sinking plane only to lose that race across the ice as had our coyote earlier. In seconds, the front of the plane under the weight of the engine, broke through the ice and began sinking out of sight.

That was the funniest thing I had seen in a long time. The tail with its taped-over numbers was sticking straight up in the air as the plane's wings spread out across the ice, held it briefly for a second from going under. Than all of a sudden, the current under the ice began to manifest itself.

In so doing, its drag on the airplane's submerged engine overcame the wing's resistance across the ice and finally pulled the airplane under. In the flash of an eye, the wings broke through the ice, were dragged slowly sideways and then it was gone! Nothing remained except a patch of dark water out in the ice in the center of the river. Talk about revenge of the coyote. I just broke out laughing as did Dave.

Then both Dave and I realized we had better hotfoot it down to the river to pick up our stranded, now-running-like-hell-for-the-shoreline, pilot. By the time we finally worked our way down to the river through the snow, our pilot, realizing he had been seen in all his stupid glory by a line of cars parked alongside the nearby highway, fled for the nearest dense stand of cattails bordering the eastern side of the Missouri. No matter how hard we looked, we never saw that pilot ever again. And neither did the local sheriff's office which had been called by another passing motorist from a close-at-hand farmhouse as well. I guess our pilot had correctly figured if his insurance company discovered he had been involved in an illegal act, namely a federal crime, they wouldn't cover the loss of his aircraft. And no newspapers ever reported the disappearance of our lad, so I guess we had a draw. Well, kind of a draw. One in which I like to compare to the old saying of, "lions one, Christians nothing."

We both searched until dark hoping we could snare this lad for a violation of the Airborne Hunting Act but we had no luck. Even after walking out on the ice and picking up that dead coyote, we still couldn't find our shooter. And after skinning out the coyote, we discovered it had indeed been shot down through the back like one shot from on high by a gunner in an airplane. Losing hope of ever finding our culprit and knowing I had to get on to Rapid City to meet Halsted and assist in his upcoming eagle-purchasing event, we left the area and the search in the hands of the local sheriff's office.

Back at Dave's home, I had a nice dinner of Northern pike fillets (without bones mind you) and all the trimmings. Boy, let me tell you, my man Fisher sure could cook. Then thanking Dave for the battery-burned hide, goose-field and picking-facility adventures, along with the sinking-plane funny, I split for Rapid City.

Heading south on state highway 83 then west onto Interstate 90, I headed for Rapid City. Little did I realize that buying a bunch of bird feather fetishes earlier on my trip would almost get me scalped by a bunch of suspicious Lakota Indians in Rapid City in the coming days.

## DAY 10: FOUR SUSPICIOUS GREEDY LAKOTA INDIANS, BUT THE "BUY" GOES DOWN...

Sliding into Rapid City later that evening, I located the motel in which Halsted was staying and rented a room next to his. That time I put it onto the government cuff since the Washington Office would be paying for it (Bob was an undercover agent assigned to the Washington Office). That evening Halsted and I formulated a plan in which we would work in tandem our Native Americans during the upcoming undercover purchase. It was decided by Halsted who was the senior officer present, that I would just stand menacingly behind him the whole time, providing 'top cover' in case the investigation went 'south.' Then Bob would conduct the business end of the purchase since he had purchased eagle feathers from this bunch several times previously. When it came time to transact the sale, I would provide the money since I was supposed to be the rich artifact collector and money man most interested in such a purchase.

The next morning as I got dressed, I put on my shoulder holster carrying my Smith and Wesson .44, 6 ½-inch magnum. Then I slipped two speed loaders full of .250-grain soft-point bullets into my right-hand vest pocket for quick and easy access. Then I belted on my Colt .45 ACP making sure it was one belt loop further back on my side so it could not be readily seen. Then into my large oversize fishing vest went two fully loaded magazines in case something 'stinky in the henhouse' came up and I "ran dry" with the .45. Finally, I slipped on my oversize fishing vest with all the pockets and such, as I looked approvingly at my Robert Redford, or should I say John Wayne, beautiful self in the mirror. Hell, you couldn't even tell I was packing so much artillery. Satisfied, I went to Bob's room and the two of us went out to breakfast.

As we had chow, I could see Bob's practiced eye looking me over very carefully for any "giveaway" flaws. Finally he said, "Aren't you packing? These four Indians are not only big men but mean sons-of-bitches. Especially if they have been drinking. Keep in mind I am getting older than glacial dirt and won't be much help if a fight breaks out. You will be pretty much on your own because I ain't packing. They always search me before any business is done, so I can't carry anything. But I plan on getting around having you searched by telling the men you are "not one of us." You know, one who likes other men. This group of Indians is a bit fidgety over someone they consider "special" to their cultural way of thinking. So, hopefully you are carrying and because of being "special" in their eyes, won't be touched or searched. If not, I suggest you go back to your room and put something on in case they get a bit "froggy."

"Bob," I said, "relax. I am carrying and if the shit hits the fan and I get any kind of a chance to go for my weapon, it is going to be a messy morning..."

"OK, partner, but don't let me down. I know you shoot Distinguished Expert but these can sure as all get-out be pretty evil-tempered men. You know, I would like to live long enough to retire." That day we spoke in Rapid City was in 1975. Bob peacefully died in 2010 at the age of 94, if I am not mistaken. It is sad for me to sit here, as I am writing these words, that most of you readers, except us old chaps, have never heard of Bob Halsted. A man who moved like a knife through many Native Americans who were selling out their culture for a few pieces of silver. Bob's work sure slowed down a lot of those walking on the dark side of Mother Nature in the Indian as well as the "white-eye" communities. Unfortunately, his work has gone basically unsung, as many more have emerged to sell off their culture to greedy white men once again. I still remember all the good work you did, Bob. However, my time and knowledge of your service is fast running out as I write these words as well...

Then Bob handed me a thick envelope. Picking it up off the table, I knew instantly it was clear full of money. "How much is in here?" I asked.

"There is $12,600 in that envelope. Enough for these chaps to try and take it and leave the two of us in a pool of blood," he said. It was then I fully realized just how scared Bob really was over the morning's coming events. It was also then that I began to get on my "game face"...

Precisely at ten o'clock, there was a knock at Bob's motel door. He looked at me from across the room where he had been reading the newspaper with a real funny look. The meeting had been set up for nine o'clock. Typically, our Indian chaps had not shown up until a full hour later... Nothing like running on 'Indian time,' I thought.

I got up and walked across the room, where I could stand in such a fashion whereby no one could get behind me. Bob went to the door and opened it. There in the doorway stood four HUGE Native Americans without a smile among any of them. Two of our chaps were carrying large, cheap paper-backed suitcases. And it was but an instant later that I recognized the smell of a lot of stale whiskey...

"Hello, Bull Bear," greeted Bob with a bit of a

wobble in the tone and tenor of his voice. "Come on in," he said. With that, the four men silently entered the motel room just about filling it. It was then that I was glad I had taken up a position earlier at the back of the room. The Indians looked all around and one I would later come to know as Robert Buffalo Calf, walked over to our bathroom, opened the door and looked in as if expecting someone to be hiding therein. I never said a thing, just stood there trying to look bigger than any of the room full of Indian chaps. That took some doing though. Two of the lads were not as tall as me but they sure as hell outweighed me, and I weighed in at a solid 320!

"I have to catch a plane pretty soon so I don't have a lot of time," said Halsted. "Did you fellas bring the stuff?" he asked.

Another fellow named David Small Bird, took one of the suitcases they had brought and placed it on Bob's motel bed. Then the last fellow, one David Franklin as I was soon to come to know very well, tossed his suitcase on the bed alongside the first suitcase. Then Robert Buffalo Calf, whom I took to be the leader of that motley crew, began opening up the suitcases. When he did, Bob surged through the four big-bodied Indians like the good undercover officer he was and without any thought of violence from the four men standing closely around him, began laying out the contents of the suitcases carefully on the bed. Me, I couldn't help but look intently at the artifacts being laid out on the bed as well. My *God, they were beautiful,* I thought. Then my law enforcement senses kicked in and I began listening to every word being said so I could testify in court as to who said what. And in the process make sure that no rights or agency policies were violated.

Then I noticed that David Franklin was sure taking a lot of interest in me... I could see peripherally that he was looking intently at me like a chicken would look at something small and moving on the ground in the chicken yard that might be good eating. Trying hard to keep some of my senses on him and the rest of my composure on the transaction, I redoubled my listening and observational efforts on the business at hand.

The first thing Halsted carefully removed from one of the suitcases was a golden eagle feather, magnificent, full double-train headdress. It was spectacular and had at least 20 juvenile golden eagle feathers on each side of the train, not to mention the 15 or so that made up the headdress portion. Next he carefully took out a single-train headdress that also had at least 20 golden eagle feathers in the train. Following those items came a dance bustle made up of eagle and red-tailed hawk feathers. That was followed with an eagle talon medicine necklace holding 32 hallux or thumb talons, made from 32 different eagles! Then came several eagle wing bone whistles, another smaller red-tailed hawk bustle, and several ceremonial bald eagle heads mounted on decorated sticks with feathers. Man, I was aghast at the cultural art work I was seeing! It was beautiful! But it was just like Bob had said. These guys were in the business of extinction, big time! The eagle items alone represented that at least 40 eagles had given their lives for such beautiful items!

Then Bob and Robert Buffalo Calf began to talk prices. I made sure I listened very carefully to the words being spoken from then on so as to avoid any entrapment issues being raised later on during trial. Here were the transactions and I wanted to make sure I was able to testify that the Indians, namely Robert Buffalo Calf, clearly offered to sell the items to Bob, a white man, who could not legally possess such items.

"We want $2,000 for the eagle talon medicine necklace because it came from 32 different eagles. We want $4,000 for the double-train war bonnet because Johnny Big Eagle took it from his dying father White Elk, and it has been in his family for a long time. We will take $1,500 for the single-train war bonnet because it was made recently from eagles the 'Wildlife Service' (the Fish and Wildlife Service) shipped us. It is new and not as rare since David Small Bird and his women made it for you for this sale today. We want $150 for each of the eagle wing bone whistles because they are very old. We would take $1,000 for the dance bustles and another $500 for the ceremonial eagle heads," said Buffalo Calf. Having had his say, he moved back out of the way so Bob could examine the items very carefully for the art work and values they represented.

After about five minutes of silence, as Bob looked things over carefully, he finally said, "As I figure it, you fellas want $9,300 for the lot. I think that is a little high because some of the work is not really museum quality. Since Terry (he used my real name since I did not have the time to develop a deep cover) here is buying these items for his private collection and I work for him as an evaluator, I would have to say the lot is solidly worth...$7,500." For my readers' sake, Service agents always tried to buy illegal items at a lower price than requested. That was not only good business but precluded any 'creation of a market' issues being raised by the defense team later on during trial.

Then looking at me, Bob said, "Terry, I think $7,500 is a good buy for all these items and they will go a long way towards completing your collection. But it has to be a cash offer and not a check or money order. These businessmen deal only in cash so the IRS doesn't get involved. And by you paying cash on the barrel head, that precludes any trail leading back to you on this illegal transaction."

Then Buffalo Calf popped up saying, "We need $9,300 for the lot or no deal!"

Bob looked at Buffalo Calf and then began fingering and examining the items offered for sale once again. After another long five minutes of silence, he turned to me saying, "Terry, I would say at the most $7,750 and not a penny more." Then Bob looked hard at Buffalo Calf saying, "Buffalo Calf, that is cash on the barrel head today if you accept this offer. And you know I know my artifacts and that is a fair price."

Buffalo Calf looked at his compatriots to see what they thought of the counteroffer and that was when the wheels came off the wagon!

"My sister in Aberdeen had a big guy like you come into her store and buy up all her bird fetishes. Then after he did, he told her he was a federal agent. You are a big guy and not one of us know you. What if you are that guy who ripped off my sister and are here now to rip us off?" menacingly growled a very-serious-looking David Franklin.

For a moment his accusations caught me off-guard and from the fleeting look I caught sliding across Halsted's face, it read pure fear! Realizing to say the wrong thing would end up in a free-for-all and possibly a killing as I saw all the Indians puffing up as they collectively looked suspiciously at me, I stared right back at each and every one of them. For the longest time nothing was said and the interior of the room seemed to be getting stifling hot. Shifting my weight slightly so I could throw a killing punch or draw and shoot if I had to, I quickly gathered up my thoughts.

"Just who the hell do you think you are? You don't even have an Indian name and as far as I am concerned, with a white man's name like that, I just realized something! I got fingered and busted by a man looking just like you in North Dakota when I tried to buy eagle-feathered items from a bunch of Indian dancers at Fort Lincoln. He was one of the men in the group I was dealing with and damned if I ain't sure it was you who squealed on me. That man who was your size and looks, shot his mouth off about a white man buying Indian items to a fed. That got me into a lot trouble and cost me $5,000 in fines. And come to think of it, the more I think about it he was a man built just like you! And if I am not mistaken, it sure could have been you!" I shot right back in an accusatory tone of voice.

Then turning, I said to Bob, "Let's get the hell out of here. I came here in good faith based on what you told me about these men and now I find myself in the company of a person who I am now sure turned me into the feds in North Dakota. As far as I am concerned, I have no money for these men because they all might be Indian Police!"

Then the room came alive with the other three Indians jumping on Franklin. It immediately became apparent that greed was now overriding their common sense and caution. Especially with this sale now up for grabs since the accusation about me had been made.

Soon Buffalo Calf got everyone cooled off and toned down. Me, however, I kept a sharp eye on my accuser in case he tried to make a run at me or worse.

"If you have the cash on you, we will take your offer of $7,750," said Buffalo Calf right out of the blue.

"Bob," I said, "I am not sure I want to deal with these men. I think I will take my money and move on if you don't mind. I just don't trust this one fellow and sure as hell don't want to be caught again by those damn feds."

Then Bob, getting over his earlier fright about being discovered and remembering why he was there, said, "Wait a minute, Terry. I have dealt with all of these men before and they are good for what they are offering. I think you need to reconsider because these really are top-drawer items they have for sale."

"Well, if that is the case, I will only pay the amount you originally specified these goods are worth and that was $7,500. I won't pay a penny more, especially after being insulted. If they are who the hell you say they are, they can take it or leave it," I grumbled like I was really hurt over being wrongfully accused.

"If you have the $7,500 on you right now, we will take it," said Buffalo Calf. It was at that moment I could tell he was fearing a sure sale was slowly slipping away and he wanted nothing of that... In short, his alligator greed was overloading his hummingbird cautionary hind end.

"What do you think, Bob? Should I trust these men or walk away from this deal before I have to pay another $5,000 in fines?" I asked.

"That is a good deal, Terry. Let's pay the men and get the hell out of here. Besides, I still have to catch my plane," he said.

With that, I reached into my vest and drew out my envelope of money. Once those four Indians saw the color of my green and the stack of $100 bills being counted out, their eyes glazed over and all their concerns and cautions were long forgotten. I counted out 75 $100 bills on the bed as Halsted took all the items and quietly placed them back into the suitcases. Then as a gesture of 'good will,' I tossed another $100 bill on the bed for the men's suitcases. That really clinched the deal and the Indians were now more than happy.

Gathering up their money, they hastily left Halsted's motel room much to Bob's great relief.

Then Bob looked over at me and said, "Where the hell did you come up with that goofy-assed story of being caught in Bismarck? If you had said anything else, they would have jumped us and I am afraid no matter how big you are, they would have stomped us into the carpet, taken our money and split!"

"I guess I couldn't think of anything else to say. I got accused once in California like that working salmon-selling California Indians undercover, and I used that same basic ruse. It worked there so I tried it here. Either way, the gods have smiled on us and we now have our case. I say let's wrap this one up, get you to the airport so you can get the hell out of Dodge, and I can get back to work."

One hour later, Bob thankfully boarded his airplane as I stood by making sure the Indians didn't come back and lift his scalp. He was more than glad to leave but before he did he said, "Any time you want to join us in Special Operations, I will put in a good word for you. You are pretty cool under fire and that is the kind of officer we need."

I just grinned. There was no way I would ever work for Special Operations. First, I was too big and did not fit into the crowd which is one of the first needs for the job. Second, I was never home now. Those poor guys in Special Operations were never home, period, because of the tremendous amount of illegal activity that required their kind of work at every step and turn! That breed of men we had in the Service in that capacity at that time, were very good at what they did. And that was hunting the most dangerous game without ever being armed for fear of discovery and seldom ever with any kind of backup. Men like Dave Kirkland, Nando Mauldin, Bob Halsted and others were nothing but 'Iron Men' and some of the Service's best officers ever!

As for our four chaps, they were later fully prosecuted for selling eagle and other migratory bird feathers contrary to federal law. Halsted Never came back to the Dakotas to work the

Native American feather trade again. He had enough close calls and I guess he loved keeping the full head of hair he possessed. Bob went on to complete a stellar career with the Service. There was one man who used up his life, that of his family, his heart and soul for the resources of this great land of ours. Today, as I said earlier, few of us old timers even remember his service or that he is gone...

The next day in a blizzard, I headed for home. On the way because of the slow and dangerous driving conditions, I had a chance to think over my latest adventure. I truly had put together a "Special Agent's Stew in the Northern Plains of the Dakotas." I had managed to catch a lot of bad guys who needed catching, made some wonderful cases and had weathered the great volumes of gas that only a Labrador dog can pass. Try that some time in a blizzard when one can't roll down the windows because it is 50 degrees below zero and your dog is in great form.

When I finally arrived back at my home in Bismarck, my wife and kids were happy to see their dad once again. Then my wife quickly marched me down to the laundry room and made me strip to my birthday suit because my clothing had picked up all of Shadow's great methane smells during the closed-window drive home during the blizzard. Ah, the adventurous life and times of a Special Agent as he made up another fine stew...

## WASHINGTON, D.C.,
## WITH ALL ITS WARTS AND HIGH JINKS

In the fall of 1976, I received a phone call from my Washington, D.C., Division of Law Enforcement Office. Picking up the phone, I was surprised to hear the voice of Bertram Falbaum, Deputy Chief of Administration, on the other end. Surprised because he seldom ever spoke to us "lower class" folks in the field. I think it was an attitude of 'looking down one's nose' issue at us lesser mortals grubbing around in the field, truth be known.

"Terry, we are soon going to advertise a new Desk Officer position here in the office and I would like you to apply. Chief Bavin has requested that I reach out to you and see if you might be interested in such a position. Before you say no, he feels your field experience in two separate regions would be a plus for our Investigations Division," continued the Deputy Chief.

Now as all my readers know, I am a country boy. And Washington, D.C., was as far away from the country as one with an ounce of horse sense and a viable heartbeat would ever want to go. Basically, that area in my low opinion, was nothing more than a swamp. However in those days, if one wanted to move any higher up in the organization, they had to have a minimum of two years' experience at that higher office level. I eventually wanted to hold one of the Nation's only 13 Special Agent in Charge field positions (now since reduced to eight). But I was just not sure at that time if I wanted to sacrifice my wonderful field career for one as a desk jockey for two years in a terrible city environment. More about that lousy environment later, as I gather up a head of steam for you unfortunate readers.

Then being a smart hind end since I had been caught off-guard by the Deputy Chief's phone call, I replied, "Make it a GS-13 and I will think about it." As it now stood, I was a perfectly happy GS-12 Senior Resident Agent in charge of North and South Dakota. I had the world and the bad guys by the throat with a downhill pull, to put it mildly. Hell, I even had some of the best Assistant United States Attorneys in the country named Dave Peterson and Lynn Crooks backing up my actions. Let me tell you, that is a wonderful plus when you have such good swords for Mother Nature like outstanding federal attorneys. Especially when one realizes I had just come from the area of the Ninth Circuit. You know, the one federal circuit that is the most overturned in the land by the Supreme Court for their lousy, fruit-loop liberal opinions. That shouldn't be that surprising though. After all, they are located in San Francisco... And knowing all the other Desk Officer positions were GS-12's in Washington, I felt my smart-assed remark would keep the D.C. mob at bay and get them off my back for a while. Besides, I hadn't caught everyone who needed catching in my two states of responsibility. So, I still had a ways to go and a strong desire to "get there."

The Deputy Chief just laughed at my less-than-civil or professional remarks, discussed a few other pressing issues relative to field operations, and then just as quickly, he was gone. Two weeks later, a new job 'Green Sheet' advertisement arrived at my home in Bismarck. Advertised therein was a new Desk Officer position in the Washington Office Division of Law Enforcement as a GS-13! Attached to that 'Green Sheet' advertisement was a cryptic note from Deputy Chief Falbaum. It read, "Your move."

Damn, I thought. I had shot my big mouth off and now they had called my bluff! Sitting down that evening with my always-supportive Bride, I explained the promotion potential and subsequent move issue fully with her. I eventually wanted to be a Special Agent in Charge but as I said earlier, I just wasn't too sure now was the right time. Bottom line, I still had a lot to learn before assuming a senior leadership position. Donna listened attentively then said, "You know I will go wherever you decide to go." With one of her beautiful blue-eyed looks and a great smile, she rose and went to attend to our newly adopted Vietnamese baby. That was it!

She had her say as my partner and was now off on another mother's errand. Damn, how I loved that women then and even more so today as these words are written some 37 years later, having just celebrated our 49th wedding anniversary! She really is some lady...

With my heart up in my throat, I applied for that new position some five days later. Then I was back to the wildlife wars in North and South Dakota. Some 30 days later a letter arrived from Washington, D.C. It began, "Congratulations"...

When Donna heard the news of my promotion, her beautiful blue eyes filled with tears. "Now what is the matter?" I asked, flabbergasted as a bug just caught in the mouth of a large bull frog. "I have just been promoted to a GS-13 (a big deal in those days in the Service's Division of Law Enforcement) and now you are bawling like a baby. What the hell is the matter?"

Her small frame quickly filled my arms and

then she blurted out, "Now the children and I will have you home and you will be ours for a change instead of belonging to everyone else!"

Thirty days later after a house hunting and buying trip, Donna, three kids, two dogs, and I were on our way to another adventure in the Nation's capital. And what an adventure that would soon turn out to be, warts, high-jinks and all...

Arriving days later in Fairfax, Virginia (Richard, my oldest son, got the flu en route and rode most of the way with his head in a garbage bag to catch the vomit), we finally met the moving van and began the job of settling into our newly purchased home. A home costing $65,000 more than the one we left in North Dakota (wart #1). It seemed every move we made we always bought high and because of the changing times, had to sell low. In our five government moves, that was always the case. We never made a dollar in any move! After a hectic few days getting unpacked and moving the furniture to where my Bride wanted it, I left for my first day on the job in the already-hated, noisy, polluted District of Columbia.

Right off the bat I got my first surprise. I had to take a bus to work (no surprise). However, it cost $1.40 each way (surprise).

A surprise especially since a GS-13 didn't really make that much money for the Washington, D.C., area, and the bus fare would be a large chunk of change out of our pockets each month that we naively had not planned for (wart #2). There is that dumb-as-a-stick country boy thing once again... Then if it was raining, that meant an hour-and-a-half trip each way to my office and home, because the idiots in the capital didn't know how to drive! I would later come to appreciate the hour-and-a-half ride each way when it only rained. Appreciated it because Ii took four-to-six hours each way by bus when it snowed! Sometimes 45 minutes travel per city block when the white stuff was on the ground because the locals couldn't drive if their lives depended on it (wart #3)!

Then I was exposed to my new office environment. Firstly, I had to wear a suit and tie. Have any of you ever seen a "water buffalo" wearing a suit and tie (wart #4)? Then I was introduced to my wonderful "office." A small eight-foot-square cubicle next to the windows facing Sixteenth and K Streets. I don't know what I had been expecting but something at least bigger than a bathroom stall in a Greyhound bus station would have been nice (wart #5). Nothing like the great outdoors I had come to love and appreciate prior to running headlong into this new assignment.

However, there was one bright spot. The Special

Agent in Charge of the Branch of Investigations was a man named Chris Vizis. He had been a former FBI agent and his reputation since coming to our organization was outstanding. He was the one reason I had decided to come into the Washington Office at that time. He was an outstanding criminal investigator and investigative report writer-two main weaknesses of mine. Being exposed to such a supervisor, I had hoped some of those traits would rub off on me.

Walking into his office that first day, I got the surprise of my life! Chris was cleaning out his desk and packing his things! Standing there in his office doorway, I couldn't believe my eyes.

"Chris, where the hell are you going?" I asked still in shock.

"I am leaving. I have had enough of this division's office politics, back stabbing and just plain lack of professionalism," he flatly stated. "Terry, I am sorry I will not be here for you. Especially knowing I was a main reason as to why you chose to come in here at this time. But I have had enough. You really need to watch out for all the office intrigue and just plain horse crap going on in here. Watch your back, my friend." And with that, he grabbed his personals and with a handshake, left (wart #6). Chris was later hired to be the city manager for Trinidad, Colorado.

However, he didn't last in that position very long before he left as well. The last time we talked, he mentioned something about the town being "run" by the Mafia...

Sitting back down at my desk in stunned silence over what I was experiencing, the phone rang. It was Deputy Chief Falbaum wanting to see me. Walking up the two flights of stairs to his office, I was now observing the 'heartbeat' of my division for the first time. First to greet my eyes were dozens of staff folks running around or typing furiously. Then one heard numerous phones ringing as couriers and other office types trekked by all around me. Finally I was ushered into the Deputy Chief's office by a cross between a hippie puke and a homeless person fresh off the street. There in the Deputy Chief's office, I was introduced to the proposed list of new agent hires for the Division of Law Enforcement for 1976. The tenth guy to be hired on that list was a pogue from the Department of Agriculture whom I personally knew. He was one of the most worthless human beings I had ever known and I immediately made my feelings known over his planned hire. Especially in light of the fact that the eleventh man on the list and one not proposed to be hired was an outstanding officer whom I knew. His name was Jim Klett and he was a North Dakota Game Warden who was excellent in every capacity as an officer. Raising hell about the selected Agriculture pogue and the non-selected Jim Klett, the Deputy Chief just smiled at his newest Desk Officer's reaction. Then I was curtly dismissed and returned to my lowly office hovel in the corner of the building overlooking K Street. There I was introduced to all the other Desk Officers on the floor with whom I would be working. They were a great lot and we eventually had a lot of fun and high jinks together to keep our sanity while so gainfully employed.

Then my phone rang once again. It was the Deputy Chief and he wanted to see me once again. Up the stairs I flew since I had early on recognized my status as a plebe. Well, maybe not quite as speedily as the Road Runner in the cartoons of the day. More like lumbered would be more descriptive regarding my velocity when moving up the stairs to his office. Once there, I was once again shown the new list of agent hires. Guess what? The Agriculture pogue was gone and my man Jim Klett was now the tenth and final man to be hired. Pleased over my small victory, I once again lumbered back to my warren on the fourth floor and quietly slipped back into my hovel. Then my memory returned to the words spoken by Chris, the boss who had just left. Especially those words about lack of professionalism. Then it dawned on me. I had looked at the new agent hire list and had grumbled over one of their selections. A hire that had been made after all the best and brightest had gone over numerous applications and made a selection based on that information. Then all that had been changed by a man from the field in his first day in the office grumbling about that selection list. As it turned out, Jim Klett became one of the top three investigators in the entire Division of Law Enforcement over the length of his career. He is now retired and living with his great little Bride in North Dakota.

Then my phone rang once more and this time it was Keith Parcher, Deputy Chief of Operations, and an old friend from my days as a U.S. Game Management Agent in northern California. The gist of that phone call was that I was now assigned as the new Endangered Species Desk Officer. That was just great! I didn't know 'sling from slang' when it came to knowing or understanding the Endangered Species Act of 1973 and its complex set of regulations! I was as dumb as a box of rocks when it came to that law since I had no experience in ever using it. Hell, I had hardly even read its regulations since my love and forte was as a migratory bird officer, not some 'ding-bat' Endangered Species Desk Officer... Typical government operation, sticking a round peg (well, a rather large-sized round peg) in a square hole (wart #7)!

Realizing I was as dumb as a stick when it came to that law, Migratory Bird Desk Officer and fellow Agent Dan Searcy took on my duties as well as his, while I tried to quickly learn my new role as the resident Endangered Species Desk Officer so I would sound like I knew something whenever called upon for answers. After a hard week learning everything I could about the Endangered Species Act of 1973, and the new associated Convention in Trade in Endangered Species of Wild Flora and Fauna (like an international Endangered Species Act), I was ready for business. Well almost! I routinely received between 30-35 phone calls a day about those two laws from every walk of life on the compass, be it agents, other federal agencies, the public, non-governmental agencies, foreign countries, or their embassies. In short, I was kept moving like a scalded cat because they were new laws and sometimes difficult to understand! On top of that, I had all the Congressional inquiries to respond to within 30 days relative to every law enforcement issue dealing with those two new massive and far-reaching laws. Then came all the meetings between our division, Congress, its staffs, other Service divisions, and numerous governmental agencies relative to those new laws. Then if that wasn't enough, I only got the toughest questions no one else could answer from field officers and the legal beagles on the "what and why for's" of those two sets of new laws. Zuhhhhh!

Additionally, I was also the Division of Law Enforcement's Foreign Liaison Officer. That meant I was also dealing with the U.S. embassies worldwide on issues relative to the two new laws and all the import/export problems relative thereto. Before long, I was running around like a four-legged paperhanger in a room full of razor blades, angry crocodiles (get the endangered species nexus?), and every other pissed-off importer and exporter of record with a greedy heartbeat...

Then at the end of the day with my head literally swimming from the work overload, I would get on my hot, crowded and humid (in the summer) bus for the usual hour or so ride home. Let me tell you, I earned every dime I was paid as the newest GS-13 Desk Officer. Then came the weekends where I was expected to work many of them in order to maintain my Administratively Uncontrollable Overtime pay (25% of my base pay which was added for my odd and unusual hours worked over a 40-hour week). Like I said earlier, I earned every dime I was ever paid during that period of my life (not that I hadn't when I was a field grade officer working the usual 80-hour weeks in a waterfowl-wetland easement district like I had just left)!

Then came the wonderful city life. Every time the government employees got a pay raise, everything sold to people in the District immediately went up in price (within a day or so) accordingly (wart #8)! So much for any help generated by a pay raise...

Walking the streets between meetings was also problematic. One had to be careful not to get mugged, robbed or killed by local predators, so we often went in groups to avoid such niceties of the Nation's capital. As for evening travel in the Nation's capital, you did so at your own peril in many parts of the city and heavily armed at that (wart #9)! In fact, that was the only place where I personally saw the spotlights of a police helicopter investigating ongoing criminal activity on the ground, get shot out three different times from gunfire from the darkened streets below. And that was in an area that has some of the strictest gun controls in the world. So much for the stupidity of gun control and the idiots who proclaim the values of such regulations...

Lastly, in my two-and-a-half years in Washington, I kept track of the police sirens and speeding cruisers going down K Street by my office window. I never counted less than six sirens going by daily in those two-and-a-half years (wart #10)! What a nice place to live...

Then came the unusual "right at home" in my office building. One day as I made my way up the steps between floors, I met two African-Americans coming down the stairs. Each man carried a new electric typewriter like they belonged somewhere within our office staffs. Then the law enforcement officer in me took over. I challenged the two lads as to why they were there carrying two brand-new typewriters down the stairs (normally they would have used the elevator). Both advised they were part of a typewriter repair and cleaning crew and were in the process of repairing the two machines they were carrying. Still not satisfied, I asked for some identification. That got me a thrown typewriter on the head, as my two crooks raced past me and down the stairs for freedom. Sorry, fellows, I played too much high school and college football as a defensive tackle. The tall skinny one who threw the typewriter at me, took a right cross to his jaw which I felt shattering under my fist like an egg! The other got horse collared as he sped by and I swung him upward so hard, he went over the stairwell rail and dropped onto the next flight of stairs below. Suffice to say, I created quite a stir when I came into the office on the next floor dragging two screaming-like-smashed cats, would-be thieves, one with a broken jaw and the other with a broken ankle. Nice place, our Nation's capital...

Then I had the usual office politics to contend with. It seemed some of the high chiefs had a following of ladies on the side if you get my drift (wart #11)... And if one didn't learn early on which side of 'the street to carefully walk on,' he got knots on the side of his head in short order. Fortunately, those of us agents on the fourth floor were all happily married as we just watched with shaking heads and clucking tongues the "frog amplexus" occurring on the supervisor-heavy sixth floor. I quickly learned a lesson in history of men in power from such activities. And to be quite frank, I saw the same behavior in every reach of government in which I had dealings. You know, how Rome must have looked like as it rotted from the inside...

Last but not least, we had the chance to observe the battles between the different divisions in the U.S. Fish and Wildlife Service for fiscal and political survival. It almost seemed that all the other divisions had nothing but disdain, disrespect (I could see why if office politics came into play), disinterest, or lack of understanding, on how the "tool" of law enforcement fit into the overall wildlife management scheme of things (wart #12). As a whole during those years, the Division of Law Enforcement enjoyed nothing realistic when it came to a financial base (wart #13). We always had the lowest budget of all the various Service divisions because we had few friends in our parent agency or Congress. Few friends, because law enforcement can embarrass many, especially if the wrong person, political or otherwise, is apprehended by agents from the division for stepping over the line of common sense or legality. Or because some of our leadership were such hind ends to other Service folks, that they came to hate the entire division because of the actions of a few. Suffice to say during those years and many after that I personally experienced, the Division of Law Enforcement learned how to very well "scrape crap with the chickens" in order to get their Congressionally mandated programs enforced at the field levels.

In fact as a Special Agent in Charge in later years, I can remember when Congress would finally get off their overpaid, dead hind ends and pass a budget by October 1 of each year. I can also remember only three years in my 28-year career as an agent in the U.S. Fish and Wildlife Service, when I had a budget that allowed us to do our job (wart #14)! I can also remember three years when congress passed the budget by October 1, that I was broke that very first day and could not even pay salaries of my staff already on the payroll because of insufficient funding (wart #15)! So in order to operate, I would spend an inordinate amount of time 'tin cupping' or begging for additional funds from other more flush divisions in order to just get the job done. If having any money to operate in a normal fashion was the ultimate indicator of being in favor, the Division of Law Enforcement had few friends both within and outside the agency.

I can still remember one sterling conversation in 1977 I had with the Assistant Director of Refuges and Law Enforcement. As such, he was responsible to see that the Division of Refuges AND the Division of Law Enforcement, both under his immediate command, had enough money in their respective budgets with which to operate. During that conversation, he mentioned to me that the refuge division had $130,000 in their budget that they just couldn't spend. And since it was almost the end of the fiscal year, Assistant Director David Olson was having fits trying to spend that amount of money so it wouldn't be returned to Congress. Since he had authority to spend that money and the Division of Law Enforcement was perennially underfunded, I mentioned law enforcement could sure use that money for new vehicle purchases. He just looked at me rather disdainfully saying, "I will have refuges paint their barns three more times in one year and use up that money before I would ever give it to the Division of Law Enforcement!" How is that when a senior supervisor carries such thoughts for a division under his responsibility? I rest my case when it came to many high-ranking members of the Fish and Wildlife Service really "looking out" for the law enforcement functions within the agency as mandated by Congress.

One weekend when I had a gut full of all the petty politics and office tomfoolery I could stand, I decided I would take my older son Richard squirrel hunting. We had a small piece of land on which we could hunt that was owned by our realtor, so we made ready for the big hunt and a father-and-son day afield. Little did I realize that outing would soon turn to some of the same crap I was trying to get away from...

Rising before dawn, Rich and I had a lumberjack breakfast and headed out to our small wood lot for some quiet father-and-son time squirrel hunting. Arriving at our chosen spot, we parked the car and the two of us walked out into the hardwood forest. It was a mild fall morning and as we walked, I rather enjoyed hearing the fall flights of ducks and geese flying overhead looking for a place to land and feed. Then I enjoyed the rustling of the oak and hickory leaves underfoot and the clean smell of the forest air. However, we could still hear the always-present faint incessant hum of nearby automobile traffic in the crowded Virginia countryside. Nor could we see the stars because of nearby cities' lights (wart #16). Finding a large oak tree center to our wood lot, we parked our hind ends in the leaves so we could watch the surrounding trees, and it soon got so quiet that we could hear our individual heartbeats. Damn, I thought, how great it is to be back in the outback in among what I so dearly missed and loved. Soon a rustling could be heard in the treetop "squirrel highways" as they came to life with our four-footed furry friends, soon to become a great dinner if our shooting was spot-on.

For the next hour, Rich and I shot a number of fat squirrels, as we had one hell of a good time together. However, off in a distance to our west, the incessant calling of Canada geese could be heard as they winged their way into a nearby cornfield only to be quickly followed by a slurry of shots from shotguns. This went on for over an hour and unless I missed my guess and from past law enforcement field experiences, my shotgunners had more than enough time to limit out. That is if they stayed within the limits of the land... And by now, my instincts were telling me my shooters in their little hidden spot had more than richly exceeded the limit on the numbers of geese that the law prescribed could be legally taken.

Now souring my time with Rich, I tried to retain any clues heard from my shooters as my young son, oblivious to his dad's discomfort at not being able to investigate the goose shooters, merrily popped away at the squirrels until he had his limit. With his last squirrel in the bag, Rich turned and excitedly said, "Dad, I beat you. You still have three to go. I can't believe I beat you," he excitedly said. I just forced a grin over his celebration and then said, "We need to go. We have more than enough squirrels for a great dinner and it will take me some time to gut, skin and get the hair off their meat for cooking. So we best be moving so we can get the other chores done Mom has for us to do before the day is through," I said quietly.

Walking out from our wood lot, I chanced seeing a muddy road leading past where we had parked, moving further to the west. Without saying a word, I put the car into gear and Rich and I moved down that same road.

"Where are we going, Dad?" asked Rich. "We need to be going back that way," he advised.

"Oh, we are just going to look over this wood lot and see if there is some better squirrel hunting down this way," I quietly said as my eyes scoured the area ahead for any signs of my goose shooters or any out-of-the-way secret goose shooting area.

Soon we came onto an out-of-the-way 60-acre cornfield hidden and surrounded by wood lots on four sides. A perfect killing field, I thought to myself, as my eyes continued moving for any signs of life from my earlier goose shooters. As suspected, they had finished up on what they had planned on doing and were now long gone. But I could see where they had parked in a copse of trees some 100 yards from a very cleverly camouflaged goose pit out in the middle of the freshly harvested cornfield. A goose pit area heavily surrounded with shell silhouettes and solid goose decoys. Off to one side of the cornfield were three giant shell decoys big enough to hide a man under. I had seen some of that type used in North and South Dakota along the Missouri River bluff to attract geese from long distances. In fact, the first time I had seen such large decoys, I figured they would scare off any goose with half a brain. Quite the contrary! They had acted very well as initial decoys seen by geese from great distances. Then as the geese got closer, they would spot the smaller, more life-like decoys and come into the feeding area without concern like long-lost feathered kin.

Stopping the car, I got out with instructions for Rich to remain with the car. I was still a credentialed officer and could trespass if I thought migratory birds were or had been taken over the area. Rich wasn't and I didn't want him to be part of any altercation between the landowner if discovered and myself. Walking out into the harvested cornfield, I saw about 150 decoys very well set out to bring in even the wariest honker. Then I spotted it! Yellow-painted Coke bottles scattered all around the decoys made to look like ears of corn (perfectly legal). Additionally, there were real ears of corn scattered all around as well (not so legal)! The whole damn area had been baited to the nines! Then spying a lot of footprints heading to or around the monster— sized goose decoys, I followed them in for a better look-see. Raising up one of the giant goose shell decoys, I discovered a ton of feathers and fresh blood spots underneath each of the three huge-sized decoys!

*Damn if they hadn't been using these shells as hiding places to stash their extra geese taken over the limit until they were ready to leave,* I thought.

Having seen enough, I hustled my tail end over to the camouflaged pit blind and looked in. It had a solid wooden floor, a heater, walls built into the dirt pit, and all were lined with full boxes of shotgun shells. *Man, I thought, these guys were sure loaded for bear and would return for another shoot if the loaded boxes of shells had anything to do with the high price of crowbars in Korea.* With that, I quickly lowered the spring-loaded camouflaged goose pit cover. Then I beat a hasty retreat back to my car and just as I started to get into the vehicle, I noticed a small suspicious-looking brush pile off to one side in the nearby wood lot. Stepping back out, I walked over to the out-of-place brush pile and lifted one end of it up. Underneath were three sacks of eared corn! Hot damn, I thought, this is their bait stash. Trotting back to my car so we could get the hell out of there before being discovered, Rich and I hurriedly left for home.

However, in the back of my mind raced a plan to come back and trap my goose shooters. Ones who had shot over a baited area that morning and more than likely had taken an over-limit of geese as well. That was if the numbers of shots they had fired that morning had anything to do with their kill ratio...

By the way for all my "eager-eater" readers out there, dismembered and browned fox and red squirrels, slow-cooked in an electric fry pan smothered with cream of chicken soup, is to die for. That is if you add a side dish of my world-famous mashed potatoes loaded with sour cream, cream cheese, a stick of butter, all spiced to taste with a fresh green salad on the side. Oh, you will need one of my Bride's famous homemade pies for a kicker as well. Do that and even God will come down and supper with you...

The following Saturday morning found my miserable carcass quietly standing alongside one of the giant goose shell decoys at the end of the previously discovered baited shooting area. The sun was a long way from being up, but I was practicing one of "Grosz's Rules:" Get into a work area long before shooting time to account for any mistakes on the part of the officer. Standing there shivering in the predawn dampness of a Virginia fall morning, I listened to the morning awaking around me. A red fox barked at me after almost stumbling into me as it scoured the baited field looking for any dead geese overlooked by the hunters. A great horned owl made his position known in a nearby stand of oaks as did another owl further down in the wood lot. Once again, the constant din of nearby highway traffic and city lights let one know man was obviously nearby and the stars would not be seen that morning as well. Overhead I could already hear numerous sets of whistling wings speeding by looking for a place to light their weary bodies and chow down.

Then at the far end of my wood lot danced a set of headlights from an automobile as it traveled slipping and sliding down the rough muddy road to the shooting area. It's coming my way, I thought, as I moved closer to the protective cover of the giant goose shell decoy. However, I was once again practicing one of "Grosz's Rules:" If the bad guys aren't looking for you, they won't see you. With that thought in mind, I continued standing in the pre-dawn darkness so I could clearly see what my guys were up to. Earlier in the morning before they had arrived, I had already collected a sample of the eared corn from the field in order to help prove my case in a court of law. Hiding the corn for later retrieval, I ambled off towards my eventual hiding place by the giant shell decoys. Then when my shooters approached (note I don't call them goose hunters because of the lack of fair chase since they would be shooting over a baited area), I began glassing my chaps as they disembarked from their automobile. There were four of them and after tending to a collective call of nature, they quickly headed for and entered their goose pit. Soon they were out of sight and then shortly thereafter I could faintly discern the smell of cigarette smoke. It always amazed me, even being of modern man, how many of our original senses, like smell, would return if one just paid attention to them.

With daylight now fast approaching, I crawled under my goose shell decoy and laid out in such a position where I could still clearly see my shooters' goose pit. I didn't have long to wait. Without a sound being made other than the wind whistling through their cupped wings, a dozen or so Canada geese quietly dropped right into the center of my shooters' decoys. Their sudden and unexpected quiet arrival even surprised my four shooters, so much so that they were unprepared for the shoot. Realizing the geese standing in their decoys were now very alert as they watched all around for any sign of danger, the men cleverly held still and did not fire. Soon another small flock spotted and soon joined the ones on the ground. Once again, our shooters held their fire. What the hell are they doing? I thought. Surely they could see the 20 or so geese feeding in their decoys. Then another flock of smaller Canada's noisily dropped into the decoys among the other geese. That triggered it, as my four goose shooters stood up in unison and shooting without plugged shotguns, fired 20 shots into the closely grouped geese! There was so much shot in the air, that I instinctively ducked my head as numerous spent shot whanged off my goose shell decoy. Looking up after the massive shoot, all I saw were wounded and flopping geese everywhere in the baited cornfield! My wonderful "sports" had ground-sluiced all the feeding geese and in so doing, had killed a mess-and-a-half of the unfortunate critters!

Lying still under my place of cover, I just watched with my notebook at my side. In it I had already recorded four shooters shooting 20 times at 8:03 in the morning. Then after things had quieted down and no outsider had come to investigate the barrage of shots fired earlier, my sports jumped up and left their hiding place. Soon all four of my shooters were scrambling around the harvested cornfield, running down the crippled and still-living geese, and in so doing, wringing their necks as they went. As near as I could count and record, my sports had killed 23 Canada geese—way over the daily bag limit!

Not breaking stride, my practiced outlaws with hands full of dead geese, collectively ran for the three giant goose shell decoys at the edge of the set. Two of the lads lifted up their goose shells other than mine and quickly tossed their geese underneath. Then my goose shell was abruptly lifted and I found myself buried in four handfuls of dead geese as I calmly lay there, in front of God and everybody.

"What the hell?" yelled the man closest to me, as he realized there was a human lying underneath his just-lifted giant goose shell looking right up at him!

"Holy shit!" yelled his partner as he also realized he had just thrown his geese onto a man lying underneath the shell decoy as well.

"Morning, Gentlemen," I said as I crawled out from underneath the giant shell decoy and reams of dead geese. Standing up on slightly numbed legs from lying so long on the cold ground and brushing myself casually off, I must have looked like a super-sized sandhill crane! "How is the hunt going or should I say how many geese do you four lads now have over the limit?" I quietly said.

"Where the hell did you come from?" bellowed a man built a lot like me only one hell of a lot richer as I would soon come to discover...

"I have been right here all along," I said as I pointed to the giant goose shell decoy.

About then the other two men came over to where the three of us now stood. *Good*, I thought. *That will make capturing this hunch one hell of a lot easier.*

"Who the hell are you?" bellowed out one of the fast-approaching men.

Waiting until all five of us were together at last, I said, "Gentlemen, my name is Terry Grosz. I am a Special Agent with the Fish and Wildlife Service."

"You got any identification?" said the man built like me but, of course, was not as pretty or beautiful-looking as I was...

"Sure do," I said as I held up my badge and credentials for all to see.

With that the fat one reached for my credentials as I took them quickly away from his outstretched hand.

"Gentlemen, I have identified myself as a federal agent and showed you my credentials. I am showing them one more time so you can all see them again but don't reach for them. If you do, you won't be able to do so because I never will surrender them to anyone except to those in my agency. You will just have to trust your eyes," I once again said quietly.

"Trust, hell! You are trespassing on my land and I want you off it right now," said the fat, wealthy-looking one.

"That I plan on doing just as soon as I write all of you up for taking an over-limit of geese, shooting over a baited field, and taking migratory game birds with shotguns capable of holding more than three shells," I shot right back as my eyes never left the fat one and his running motor mouth.

With the thunderclap of those words of observed violations, every one of my four shooters tempered right down. Then they just began looking at each other as if some one of the bunch had a better explanation of the morning's seamy events and if so, get to using it before this badge-carrying idiot gets up a head of steam...

"Gentlemen, I will need to see your hunting licenses, your driver's licenses, shotguns, and all of the geese if you don't mind," I quietly advised seeing I had made my control point earlier.

Soon the requested items were handed over. Upon inspection, I discovered that all the shotguns still remained unplugged! *Boy, these guys are sure rubes or felt they were really safe in what they were doing,* I thought. They hadn't even attempted to put their plugs back into their shotguns after blowing up the geese like real outlaws would have done. Then depositing the geese and shotguns in a pile, I took the lads through the field and showed them the ears of corn scattered about used as bait. By now, there was gloom and doom all around if the looks on their faces meant anything. Even more so when I removed a camera from my game bag and photographed all the eared corn, yellow-painted Coke bottles and my four rubes standing near at hand in the evidence...

Then the mouth that had flamed earlier asked if he could talk to me alone. That I consented to. Soon I discovered that he was a wealthy Virginia landowner and had brought three powerful Senate staffers 'out for a goose hunt they would long remember for a lifetime,' as he told it. Then he advised in a hushed voice that if I would let them go, he would talk to them about seeing that the Fish and Wildlife Service got some additional funding for that fiscal year (wart #17)!

Looking my Virginia planter right in the eyes, I advised what he had just requested constituted an attempt at bribing a federal officer. I also advised that if he persisted, I would file those charges against him as well. Those cold words brought my Virginian up short and with that he turned abruptly and rapidly waddled back to his hunting party in a huff.

With that bit of dirty business aside, I walked back to my chaps and commenced recording their information on federal citations for over-limits of geese, hunting over a baited area, and use of unplugged shotguns. Not much was said as I wrote out the paper on my four errant chaps.

Then the mouth spoke up once again asking if he could pay for all the citations. I advised I didn't care how they were paid because that was each man's responsibility. However, the matter had to be settled within 14 days as explained on the federal citation or a warrant would be issued. Then I commenced advising the landowner that his field would remain baited for up to a ten-day period after all the lure and attractant (corn) had been removed. With that, the men were given their citations, the geese were seized and I commenced gathering up another sample of eared corn as my four shooters loaded up and quietly left the field of battle. I later drove my previously hidden family vehicle back to the shooting area and picked up my load of geese and corn samples for a court of law if that later became necessary. Then I went over to the wood lot on which I could legally hunt and gutted out all the geese so they wouldn't spoil.

Later a local Special Agent from Virginia was contacted and all the evidence turned over to him to follow it through the courts since I was a Washington Office staff officer. I could have done the same but it was far simpler to just have a Virginia field agent take care of business and let me get back to the mounting 'boar's nest' of paper facing me at the Washington Office. As I later discovered, my Virginia planter had collected up all the citations and paid for them collectively to avoid embarrassing the Senate staffers any further.

That was fine with me just as long as the government and the dead geese had their say in that sorry matter.

Returning to the office the following Monday, I ran into Dan Searcy. He was a rather stout fellow and turned out to be a damn fine friend and excellent professional officer. But to avoid going nuts in the swirl of the Washington Office, Danny applied his talents as a bit of a prankster as well. Realizing that, it didn't take long for me to throw my dice in with Danny's and soon that sorry swirl of office politics became more livable. Bottom line, Danny and I were both pranksters who enjoyed creating a funny 'hoorah' at numerous others' expense. In fact, as I was soon to ascertain, the two of us working in tandem were some of the best when it came to creating intra-divisional office havoc. Man, you talk about chemistry...

Case in point. Danny was also responsible for running the mail room in the Washington Office in addition to his duties as a Desk Officer. Believe you me, that entity was a boar's nest of oddballs par excellence. The mail room staff were for the most part, a cast of characters not of this world. In short, they were a motley collection of loony-tunes, black female activists who didn't shave any part of their bodies or wash most of the time, poorly motivated expatriates from the District of Columbia scene, and a bit of everything else tossed into that mixture as added spice. On top of that, there was a mess of dog and cat infighting even among that odd cast of weird characters. Infighting that led Danny's natural inclination as a prankster to play funnies on them to no end.

One 'funny' in particular I remember fondly, even to this day (high jinks #1). Danny's head of the mail room was a tall, skinny, always gaudily and oddly dressed, 'hip-hop' character, who was one imperious sort of 'cheeky' fellow. He was always lording it over the others and creating mayhem to their extreme discomfort. And if that wasn't enough, he considered all the female creatures from his staff as part of his 'stable.' Danny, having a gut full of his idiot subordinate's high and low jinks, decided to play a trick on this imperious stiff in front of his help—a trick that was soon to bring that chap down to the level of the rest of us squalid and restive Washington Office unfortunates.

Arriving early into the office as the two of us normally did, we sat down over a hot cup of tea and laid out our devious plans. Danny decided he would put one of our portable radios into the safe after turning its volume up on high. After so doing, he closed the door to the mail room safe. Then my job was to lure the mail room chief into the sorting room holding the safe, under the pretense of asking him a lot of questions relative to his daily operation. Once I had the imperious mail room chief in the vicinity of the safe, Danny from another room would begin transmitting on a second handheld radio. In that transmission Dan would begin yelling for help.

Now picture this. Dan was a big guy over six feet in height and a tad over 200 pounds in weight. Inside the mail room was a small safe about three feet-by-three feet in size. Once Dan saw me take Chester into the mail room, he would give it a few moments and then begin his end of the act. Once up on step, Dan would begin yelling over the handi-talki at Chester to get him out from being inadvertently locked inside that small safe! With the trap set, I maneuvered Chester into his office holding the safe. There I began discussing his overall operation since I was a new Desk Officer and was trying to learn how his office worked.

Sitting down alongside his desk, I began asking operational questions. Chester in turn, from his always-looking-down from his perch and nose at the agents, responded curtly to my questions.

"Chester. Help me! I am trapped in the safe! Help me, help me!" said Danny's clearly muffled but distinct voice from the radio in the safe.

Man, you talk about an explosion! Chester came clear out of his chair and never stood up in the process! "Did youse hear dat? Did youse hear dat?" Chester yelled at me.

"Hear what?" I calmly replied.

"Did youse hear dat? Someone sounding like Danny was yelling for help and it was cornin' from dat safe in der corner!" Chester replied from behind two frightened eyeballs the size of basketballs, which were now pure white as the driven snow.

"Chester," I said, "what the hell are you talking about? I didn't hear anything. Sit your tail end back down in your chair. I still have a world of questions to ask you and haven't much time. I have a meeting to attend in twenty minutes, so we have to get 'crackin,'" I said.

"Chester! Chester!" said Dan, "I am trapped in the safe! Get me out before I suffocate," said Dan's voice clearly.

That time Chester vaulted out from his seat and ran from his office yelling something about "ghosts in his office" at the top of his voice. That was the last time either of us ever saw Chester The Molester, the mail room imperious one...

Chief Clark Bavin, suspecting Danny and I

were behind Chester's terrified exit, called the two of us into his office. There he asked for an explanation and if found guilty, a damn good ass "whuppin'" was to follow. But being good agents, we had alibis. Me, I was sitting there with Chester. So I obviously could not have had anything to do with his demise. As for Danny, he was clear across the office in the file room working on 100,000 misfiled index cards holding the names of Service-prosecuted chaps. Index cards misfiled because the person our Personnel Office sent over as a new hire did not know the alphabet (wart #18)! Now we had 100,000 misfiled index cards on previous outlaws the Service had arrested and prosecuted. It seemed the "heartbeat" our wonderful Personnel Office had sent over, had just filed the index cards wherever he wanted in the monster index card automated system, instead of alphabetically as needed for future retrieval!

So any time in the future a field agent queried our index files for a criminal name, he may or may not find the information he needed because the person's name had been misfiled by a "kumquat" (wart #19).

Or there was the time I went into the office of Special Agent Bud Lowery. Lowery worked in the Training Division for Special Agent in Charge

Richard "The Viper" Gordon—a new hire who came to our division from the Air Force after he had retired as a Lt. Colonel. He got his name "The Viper" because every time a complaint came in on one of our field agents, he wanted the officer automatically fired. It didn't matter if the officer was right or wrong, "The Viper" wanted him fired! As one can imagine, the agents soon came to hate and not trust the man, hence the name "The Viper." Anyway, I digress...

As I said earlier, walking in on Agent Bud Lowery, I saw he was sitting behind a large cardboard box holding a TV! "Where the hell is your desk?" I asked.

Bud just looked at me and somewhat ashamedly said, "Dick said I had not yet earned a real desk (Bud was a 12-year veteran). Therefore, I have to use this TV box until he feels I deserve a real desk." How demeaning! Suffice to say, that pissed me off big time!

Down to Dan's office I went and another set of high jinks was soon being plotted out by "Dan the Dumpster" and "The Grotz." For my slower readers, that was Dan and yours truly... The high jinks we plotted out this time was a PURE masterpiece. Dan and I decided we would get permission from Chief Bavin and spend the coming weekend in Delaware chasing illegal goose hunters. The Friday before we would come into the office wearing work clothing instead of suits and ties, since we would be working late shooters in the fields that evening. Then we made doubly sure everyone and their cocker spaniels knew the two of us were going to Delaware that evening to chase the goose hunters all around. Then come the end of the day, Dan and I retired to Blackie's, the nearest bar on K Street. There we stayed for Happy Hour drinks and dinner until around eight that evening. Satisfied no one was left in the law enforcement office, back we came with an evil glint in our eyes and elfin step. Once inside, we shoved Dick Gordon's desk off to one side in his spacious ornate office and then commenced filling it up to the ceiling with every loose box, bag and spare piece of office furniture we could find on the entire sixth floor reserved for all the "high chiefs" and their staffs. I am not kidding when I say we filled his office clear to the ceiling. I meant it! It looked like a storeroom or the home of a "hoarder" when we finished.

Then as a two-part finale, there was a small rug in Dick's office doorway. We laid the TV box with TV inside that Agent Lowery had been assigned to sit behind by his boss Dick Gordon, on the inside portion of that rug still in Dick's office. Then we closed the door over the rug. Lastly, we pulled that rug through the small space under the office door dragging that boxed TV right up to the inside of Dick's closed office door. Suffice to say, there wasn't enough space left inside Dick's office to put a mouse dropping! Then we split into the night like a couple of assassins... Oh, by the way — Dick was such a viper, we both realized that if we left a single fingerprint on anything, he would find out and have our last parts over the fence. So, Danny and I used gloves the whole time we moved all that crap into Dick Gordon's office to avoid being caught (high jinks #2).

Suffice to say, the illegal goose hunters in Delaware caught hell from two happy to be out of the Washington Office warriors those next two days. Then to close the ring on our alibis, those timely goose cases further strengthened our "we are innocent of all charges" stories.

Come Monday morning, Danny and I were right on schedule in the office all bright-eyed, bushy-tailed and of course innocent of all charges to be hurled by the afflicted that came our way. However, on the sixth floor where all the "super chiefs" resided, the evil brown smelly stuff was literally flying through the fan. Dick Gordon had returned to his office, only to discover that he could barely open up his office door. After finally smashing the TV box and TV all to hell, he was able to jam his now highly tightened arse into his trashed office. Man, you talk about pissed! He went straight to the Chief and raised holy hell. Soon the whole office was secretly laughing up their collective sleeves over Dick Gordon's office "surprise." Even Danny and I got into the act of coming by and looking at the trashed office, as innocents of course. Once there, we clucked our tongues in disbelief and gaggled around like the rest of the sixth floor office pogues eyeballing the evidence of "Dick's Dismay."

However, it wasn't long before Danny and I were considered prime suspects because it was commonly known neither of us had much respect for Dick and his high-handed office or personnel policies. Plus, regardless of our out-of-state goose detail, we being known as pranksters and all... Soon Chief Bavin called the two of us into his office. Mistake number one. A good officer always separates criminals so they can't get their stories to match and thereby jinx the investigation. However, Bavin brought the two of us together into his office during the investigation of Dick Gordon's trashed office for a good old-fashioned grilling. However, the grilling was halfhearted and perpetrated on two clever "criminals" who would only break if one tore out our fingernails or took away any hamburgers we might wish to eat... Hell, even the Chief knew Dan and I had taken the patrol car and left for Delaware to work goose hunters the Friday before. And we both had a handful of federal citations issued to numerous bad guys to show that is where the two of us had been the whole time. Fat chance in catching the two of us. Hell, we were slicker than good old-fashioned cow slobbers...

Then there was the now-forgotten-by-my-readers stage two of the high jinks, that had really pissed off Dick Gordon. Earlier when I had been working alligator poachers in Louisiana, I had discovered a great joke shop in the City of New Orleans. In that shop I had discovered a poster showing a naked man bent over with his head shoved up into a place where the sun never shone and it smelled badly. Realizing that poster was too good to pass up, I purchased the same for later and appropriate use. It now seemed that poster had somehow surfaced during the time Gordon's office had been trashed and was taped onto the inside back of his office door. Then when the office door had been finally opened, the poster had gone unnoticed being taped to the back of the door. After the office had finally been cleaned out by Gordon's subordinates some weeks later (he had left the clutter in his office so when he caught the bad guys he could have the culprits clean it up-never happened), he held a "still mad at the world" meeting. In that meeting he had summoned all his troops for a general ass-chewing figuring some of them were the culprits (poor guess because they were all too scared of him).

When all the troops were closely assembled in Dick's office, he slammed shut his office door with a bang. There in all its glory for all his staff to see was the wonderful poster describing Dick to a T! Man, you talk about creating an explosion! It got so bad that Gordon had the Secret Service come over and carefully remove the poster. Then he had them run a Ninhydrin Test, a test designed to lift fingerprints off rough surfaces on paper. Guess what? The beautiful Grotz (that is me) when he had purchased the poster item, had used gloves that time as well. Once again no fingerprints were found and the trail of the culprits had once again grown cold.

To all my readers, this story has never been told prior to this writing. I have chosen to do so at this time to show the great lengths some of us "unsavory" Desk Officer characters in the Washington Office went in order to maintain our sanity. It also served to bring smiles to those of us caged up in such a high-pressure environment once in a while for the good of the gang (high jinks #3).

I remember another set of high jinks that occurred on our floor that were once again pure gold. The Investigations Branch was located on the fourth floor of the building in which the Division of Law Enforcement called home. Since I would leave home at five in the morning to catch my bus, by the time I arrived at work I was usually the first agent in the office. However, there was usually another person in the office at that time as well. Miss Geneva Ables was our Investigations secretary. She was some type of lady. She was a little bit haughty and seemed in my eyes to always be a bit better than the rest of us hard-working slobs. And many times, instead of doing some of her administrative work assignments, she always found ways to dump those duties on us hard-working Desk Officers. And if you didn't do her work when she dumped it in your laps, many times your draft letters that needed to be finalized and sent to meet agency deadlines, would mysteriously "get lost or misplaced." Then the hellfire and brimstone would fall on the offending agent's head. In short, you didn't dare piss her off or else! In that captive atmosphere arose a small problem that soon mushroomed into a rather deep, evil-smelling mudhole.

Miss Ables would always wait for me to arrive

and fill up the hot water container so she could have her morning coffee. Even though she arrived 30 minutes before I did, she expected the higher-graded and -paid agents to fill up the water container every time so she could have her morning coffee. It seemed she liked having us agents wait on her hand and foot. As a result, the water issue soon blossomed into another set of high jinks in order to re-establish office "order."

Additionally, when I arrived, if the water container was not filled and sufficiently hot for her morning coffee, she would let me know of her displeasure in no uncertain terms. After about a year of such crap, I decided it wouldn't hurt her to fill the water container occasionally. Also, she never would pay her fair share for the coffee or tea used by everyone in the office. In fact, she used to tell me that since I made so much more money than she, it wouldn't hurt me to pay for her share as well. That did it!

One morning when I arrived, I was greeted by a very huffy Miss Ables. My bus had arrived later than usual because it had rained, and the other commuting idiots using the roadways simply could not drive for sour owl droppings. When I arrived, instead of having the coffee water quickly made ready for her morning drink, she jumped all over me like a chicken on a wingless bug! Not saying a word, I trudged down the hallway, filled the large water container and lumbered back to the Investigations office.

"About time," greeted my ears as I walked by her desk carrying the large water container. Well, a man does not get as big as me by standing last in line, if you readers get my drift... When I showed up the next day for work, I was carrying a small bundle in a brown paper bag. Doing as I was ordered by Miss Ables to hurry and get her coffee water ready, I obliged. However, that time after filling the water pot and plugging it in, I added a little something into the water...

Soon the water was boiling away and when she came into the room to fill up her coffee cup and use our agent-paid-for coffee, she said something under her breath that sounded like, "lazy whelp" to my kit fox ears (I will let you readers figure that one out). No problem, I thought, especially since I had been raised in a hardscrabble logging community back in the mountains of northern California, where the motto from the Good Book stated, "Do unto others before they do unto you." At least I think that is what the Good Book said...

Out she went back to her office to file her fingernails and enjoy her morning cup of coffee on the government tab (she was getting paid to be working that time of the morning). Getting

258 | TERRY GROSZ

up from my desk, I walked over to the happily boiling water pot, grabbed the unseen wire attached to the top of the inside of the pot and lifted out a small wire basket. Inside were four of the nicest bright-red, fully cooked, crawdads you ever saw! Walking back to my desk, I removed the crawdads from the wire basket and walking over to the desks of four haughty, Desk Officer-hating, female types who headed up our Intelligence Branch, dropped a boiled critter into each of their wastebaskets. I did so because those females were also very disrespectful of my fellow agents as well. And if anyone discovered the crawdads in their wastebaskets, maybe they would get blamed for this little "hoorah," since they always acted like the sun set on only them and the hell with the rest of us. That plus, after a while, the critters in their wastebaskets would begin to smell.

The root of their attitude problem was that they were lower-graded than the agents because their work activities were less complex. However, that stuck in all of their craws and they continually showed their disdain for the agents every chance they got. Hell, it wasn't our fault they were lower-graded. The Office of Personnel Management decided what grades folks were listed under as based on their assigned work complexity. That

didn't seem to matter much though as those ladies freely drank the coffee and tea purchased by
the agents as well.

As a sidebar, later in my stint in that office, we
had one of those wonderful, unvetted Intelligence
Analysts illegally steal a bunch of our intelligence
files and run off to marry a falconer we were currently working covertly in Illinois (wart #20). In
so doing, she blew the investigation wide open.
But once again I digress...

Then returning to my desk after 'the crawdad-
dropping incident,' I grabbed my tea cup, filled
it with bouillon cubes and headed for the water
pot.

"Ugggghhh!" screamed Miss Geneva. Innocent
little old me, upon hearing such squalling, never
flinched or lifted an eyebrow. I didn't know
how coffee tasted mixed with crawdad water
but didn't really care because I never drank the
damn stuff. Filling my cup with hot crawdad water, I headed back to my desk just as Miss Geneva
burst back into our section yelling at me and gagging all in the same breath.

"What is the matter, Miss Geneva?" I sweetly
asked. As my loyal and regular readers already
know, I was always sweet.

"My coffee tastes just awful," she replied
through a 'scrootched' up face (high jinks #4).

Taking a long sip of my beef bouillon drink, I actually kinda liked the crawdad mixture with my broth. "Mine seems to taste alright," I said trying hard not to laugh out loud.

"Well, mine tastes like crap. You get up off your dead hind end and go down and refill our coffee pot right now," she barked at me still trying hard not to gag.

"Yes, Ma'am," I sweetly replied (see how sweet I am). Getting up, I unplugged the water pot and headed for the men's bathroom where I could refill the pot and could have a good laugh away from our pissed-off office secretary. When the other agents came in, I informed them of Miss Geneva's "crawdad" wreck and we all had a good laugh. However, everyone's coffee sure tasted somewhat strange the rest of the day until the crawdad taste boiled away.

My little "teach Miss Ables a lesson" high jinks didn't end just there. The next day, I brought in a hot pepper! When the water pot had been refilled later on in the day by someone other than me (to throw anyone off the trail), I walked over and slipped in the hot pepper. As it cooked, the water got hotter and hotter if you get my drift (high jinks #5).

Now all the agents on the floor loved hot peppers. In fact, many times we would bring in hot peppers and share them around the lunch table as we agents ate lunch. So I just figured one little, teensy, itty-bitty hot pepper would not make a difference except among the women who hated us anyway. Well, I wasn't wrong. Soon all the women were complaining that their coffee water was burning their lips, and on it went into the afternoon when everyone finally went home. Then I carefully lifted out the hot pepper, ate it straight away and headed for my evening bus ride like nothing out of the ordinary had happened. After that episode, the ladies made sure they filled up the communal water pot and bought their own coffee and tea! Finally all was right once again on the Investigations floor...

Then the hated "Viper" put his foot into his mouth once again. One afternoon as the agents sat around our lunch table having lunch, there was a surprise treat. On the table sat a quart jar of home-canned peppers made by Danny's great little, southern belle wife, Linda. They were great! Hot but just right to experienced pepper eaters like the six of us sitting around the table. Sitting around our table having a good time, we were all of a sudden interrupted by one "Viper" bursting in on our lunch hour. Right off the bat he began grumbling about the quality of our work sent to the head office for their knowledge and review. Then he spied our quart of peppers sitting in the middle of the table. Without asking if he could have any, he helped himself saying, "I like hot peppers." It didn't end there. As he continued chewing all of us out, he kept at our jar of homemade peppers until he had wolfed down half of them! Then having had his say and fill of our peppers and ruining our lunch hour, the maggot slunk off like the low-life he was.

Looking over at Danny with "that" stealthy look, I said, "Is Linda still making and canning those hot peppers?"

"She sure is, and are you thinking what I am?" he said with a sly grin.

"Yep," I said. Then our conversation drifted off onto what poor manners Dick Gordon had in eating all our peppers. But the trap was now in the offing (high jinks #6).

The next day, Danny brought in two quart jars of peppers. Walking over to me he said, "This jar is full of the same strength peppers like we had yesterday. Inside the other jar are the hottest peppers Linda could buy at the farmer's market. Then she cooked and canned them in oil to make them even hotter. What do you think?" he said with a grin.

"I think I can handle the 'heavy lifting' if you are willing," I said with another grin.

That next lunch hour, all six of us agents sat around the dinner table like "knowing" foxes in the henhouse. And if one looked closely, we all had mouths full of sharp teeth! In front of each of us were three or four of Danny's mildly hot peppers on a plate for "show." Then sitting in the middle of the table, sat a half-empty jar of the world's hottest peppers. Hell, even God would not have tried any of those firebombs! The trap was now set. We had emptied out half of the hot peppers to make it look like all of us were eating that breed of killer. Then I called Gordon "The Viper" on the phone. When he answered I said, "Dick, bring your lunch and come on down to eat with us (said the spider to the fly). Danny brought in some more of those great homemade peppers and we would like to share them with you since you like hot peppers." I knew he had so few friends, that he would be more than anxious to join someone who showed any form of "love."

"I will be right down," he said.

Then the six of us put on the most innocent of faces that our mothers had ever given us. Soon we could hear the door slam in our office and here came Dick Gordon as if reeled in on a string. Sitting down at our table, he immediately grabbed a handful of those hot firebombs and never stopping from talking, shoveled two into his mouth as he undid the wrapping on his sandwich. Then all of a sudden, his eyes got bigger than garbage can lids and his face turned red as a fireball. Beginning to choke and not wanting to puke out the peppers in front of us lesser mortals, he took a quick bite of his sandwich and choked down the two half-eaten peppers. Meanwhile, all of us as if on cue, took up a pepper lying in front of us from the milder jar set aside and hidden earlier. Then we happily chomped down on them like they were nothing more than an M&M...

Then the unexpected happened. Dick Gordon jumped up from his chair knocking it back onto the floor. "I have to make a phone call," he "urrrrped" out. Then he was gone just as fast as he had arrived, leaving his partially eaten sandwich behind. A moment later as he headed up the stairs to his office floor, we could hear him puking like he was trying to cough up his liver! That horrible deep vomiting sound went on until he exited the stairway onto his floor and then silence. That silence reigned until all of us finally burst out laughing over our little "funny" as Agent Nando Mauldin used to always say. Like I said earlier, we all had a fine set of teeth. To our way of thinking, it couldn't have happened to a nicer guy.

Finishing our lunch hour, all of us returned to

our desks. Then my phone rang. Picking it up I could hardly make out what the caller was saying. "Looff whaff ooh didd tooo mee (Look what you did to me). I amph dying (I am dying)," came the stressed-out voice.

"Who is this?" I asked hardly understanding what was being said.

"Thiff if Gordon. Gettf yourr assph up here rightff nowff (This is Gordon. Get your ass up here right now)," said the voice.

Realizing the Special Agent in Charge was yelling at me, I split for his office. Arriving, I saw Dick with his face all screwed up in pain. "What is the matter, Chief?" I asked in an innocent and full-of-sweetness tone of voice.

"Ittiff tthose pepppers. They arrefe killing meeff (It's those peppers. They are killing me)," he mumbled through now rapidly swelling lips (he now looked like Angelina Jolie). "Loof at meee (Look at me)," he said. Then opening up his mouth, all I could see were water blisters all over his lips, gums and tongue where those fireball peppers had ripped him a new one.

Trying hard not to laugh (tells you how much he was hated), I said. "How the hell did that happen? You saw us eating the same peppers (remember those we had taken from the normal jar and had placed in front of us, like we were eating from the "purdy fire jar" in the middle of the table?)."

"I donff noof buff look atf mee (I don't know but look at me)," he replied in pain.

"Gee, Dick, I don't know what to say. You saw us eating the same peppers. Maybe the particular acids in your mouth reacted with those peppers in a bad sort of way," I continued with a spider-like smile.

With that he waved me from his office and I returned to my floor. There I duly reported to the other officers that it would be a long time before Dick helped himself to our peppers during lunch in the future. Come to think of it, he never came back down to dine with us during lunch on our brand of hot peppers. In fact, he took sick leave for the next three days...

As mentioned earlier, one of my duties in the Washington Office was that of Foreign Liaison Officer. Within that responsibility were expectations that I would assist foreign governments working their way through our Endangered Species Act of 1973, Convention in Trade in Endangered Species of Flora and Fauna (CITES), and associated import/export regulations. Additionally, I was responsible for updating what was called the LAW-17 file. That file held all the foreign nations' of the world official import and export laws. Those LAW-17 regulations were in turn utilized by our port of entry officers, our field contingent of Special Agents, U.S. Attorneys, and our Special Operations Branch of undercover officers when it came to understanding the legalities of foreign laws. In short, when such an international import case presented itself, it was my duty to provide that jurisdiction's prosecuting U.S. Attorney's Office with updated foreign country regulations so a Lacey Act, Endangered Species Act or CITES case, based on foreign law, could be proven beyond a reasonable doubt.

In order to maintain a current LAW-17 file, I was required to keep in constant contact with foreign countries through our embassies located in each country to obtain their newest sets of import/export laws. Then with those current laws in hand, our U.S. Attorneys could prosecute smugglers and importers/exporters of record for bringing illegal wildlife/plants and/or their parts or products into the United States contrary to each foreign country's specific laws.

During my time in that position, the United States was having one hell of a problem with all sorts of illegal wildlife/plant imports entering this country. It seemed every African country and the associated importers of record were making money off the illegal imports being entered into the United States by the bucketsful. In short, it was all I could do just keeping the LAW-17 file current and complete because of the number of countries in the world community. Plus a lot of countries were so backward or with only fragile government infrastructures in place, that many times I had to go to great lengths just to acquire copies of their laws of the land.

Then a great opportunity arose. There was to be a multi-country conference held in the upcoming future in Gaborone, Republic of Botswana, Africa. Realizing I had a 'bird nest' on the ground, I began the lengthy process through the State Department of arranging a trip to Africa to attend that conference. My plans were to go to that conference, and set up training sessions with each country's export representative or officials on understanding U.S. laws. I also planned on having current copies of our import/export regulations shipped earlier to hand out to each country's representative as well as our code books of regulations. Additionally, I planned on meeting those African countries' import/export representatives, establish a personal contact and make sure I had their correct names, mailing addresses, fax numbers, copies of their authorized signatures on export documents, and phone numbers. Then in turn, I was going to have each country's representative train me on their export laws and provide me a certified copy (one signed by that country's import/export official representative) to be returned to our now-updated LAW-17 file.

Pretty damn simple, I thought. Good enough reason for the trip—a very efficient way to utilize my time and resources, and using $7,000 of the taxpayers' dollars could accomplish something that had never ever been done before.

Now before I get into the "coming political muck-to-be," my readers must realize that politics are the bane of any wildlife law enforcement agency. As such, politics come at you from every quarter and degree of the compass of reality with their varying levels of intensity. In those arenas, one will discover politics will morph from the very subtle to that which will destroy one's career in a heartbeat. Yet bad politics are an evil stain that every officer must carry every day they are afoot or on horseback in the field or office arena.

In today's world of wildlife, politics are translated from the 'dark side' into insufficient budgets, poor equipment, tainted hiring practices, poor supervision, dangerous assignments, questionable supervisory practices, legal insufficiency, officer shortages, and the like, including direct interference from the highest levels of government as you shall soon experience. Yet it is something borne every day by the wildlife officer like the chew of tobacco he or she carries in their shirt pockets. But, once again, I digress...

My readers must bear in mind once again that if our government prosecutors can't prove a criminal case beyond a reasonable doubt, we have no case. A current certified copy of the country of origin's exports laws goes a long way towards making sure we proved our criminal illegal import cases beyond a reasonable doubt. Sounds logical and I shouldn't have any problem with my agency's justification to the State Department for such a trip, doesn't it? Wrong!

I submitted my very-detailed justification to all the required agency heads and had it approved straight away as a very good mission in light of all the illegal imports coming into the U.S. from Africa. Then the State Department reviewed my proposal (required since foreign travel was involved), and heartily approved the mission as a damn good idea and one designed to get the most bang for our buck. With all the approvals in motion, I began the process of acquiring all the needed passports and visas for the countries visited. That done, I also began the lengthy process of procuring all the necessary shots against the various diseases in that part of Africa to which I would be exposed as I worked. As an aside, believe me, some of those shots hurt like hell. In fact, I can remember getting several in my last part over the fence, that formed large painful lumps that made sitting difficult for the next few days...

Then I was ready. Well, not quite, because a little favor performed for the U.S. Customs Service caused the wheels to come off my wagon. It seemed Secretary of State William Rogers had just landed on the local airbase reserved for Washington dignitaries. And as luck would have it, in the ensuing required Customs inspections, some things were discovered by the inspectors that didn't look right. Since I had a close working relationship with the local Customs folks (now ICE), they gave me a heads-up. It seemed the Customs lads didn't want to crap in their mess kit over the Secretary of State's possible illegal imports, so they punted to the Fish and Wildlife Service, namely yours truly. Not being very bright, I accepted their squeal for assistance since I was the alleged expert and soon found myself examining some imported items from the Secretary of State's personal luggage. In that inspection, I discovered 34 ivory and elephant hair bracelets made from Asian elephants. Asian elephants that just so happened to be an endangered species under our Endangered Species Act and an Appendix I species under CITES. An endangered species which carried an import restriction except for propagation or research purposes under both sets of federal laws.

The Secretary's staff left behind to take care of the grunt work claimed the Secretary had imported them as personal gifts and therefore had an exemption from the Endangered Species Act and CITES. Being schooled in both sets of statutes, I knew better. I found that 34 sets of endangered ivory bracelets far exceeded what I called personal use, and were outside the permitted exemptions which required a permit for either research or propagation purposes. Realizing I had an import violation, since the importations were not for research or propagation and lacking any such permits, yours truly seized the items on the spot. Can all of my readers see where this one is going in light of my little tirade above on politics? Suffice to say, that seizure created a ruckus by someone (read Secretary of State) who felt he was above the laws of the land. Boy, does that sound like many of our Members of Congress of 2009-2010 before many of those soft-headed fools got voted out of office by the American people? However, to avoid the public relations fiasco, the items were quietly forfeited to the Smithsonian Museum as is required under a quaint federal law and that was that. Well, not quite... As I alluded to earlier, that thing called politics can be a multi-headed snake.

Three days before I was set to go to Africa, I came to work at my usual time. The entire office staff had a cake for me and a party planned when I arrived. Not understanding what the hell was happening, I inquired as to why there was a cake for yours truly. I was promptly told by a smiling Agent Dan Searcy that I was 'now sick' and the cake was for me to get well fast. What the hell, I thought. I not only was built like a horse but was as healthy as one as well. Especially now with all the painful shots I had received in my hind end for the upcoming trip to Africa. There was great laughter in the group and a still-quizzical look on my face when my friend Agent Dan Searcy stepped forward once again. With a flourish, he handed me an official State Department outgoing telegram addressed to Mr. E.T. Matenge, Director of Wildlife, National Parks and Tourism, Gaborone, Botswana, Africa. Scanning the telegram, I recognized my conference contact for the teaching and training trip to Africa three days hence. Then I read on. The telegram apologetically went on to inform Mr. Matenge that the unexpected had occurred and I had become seriously ill. It further stated that it was unlikely I would be able to heal up in less than a few weeks and as a consequence, would not be able to attend the important upcoming wildlife conference! You know, the one I had received all the previous clearances from the State Department in which to go. The one I had suffered through ten thousand painful shots in my hind end. The one in which I had made contact with numerous African countries and made arrangements for copies of their import/export laws to be on hand when I arrived. The one in which I had sent numerous copies of our sets of regulations for the African countries and my foreign counterparts to use as well. You know, the one set up by the IUCN (International Union for the Conservation of Nature and Natural Resources) so a major effort could be made to encourage other nations to join CITES. You know, the one...

The telegram had been signed by one Lynn A. Greenwalt, Director, U.S. Fish and Wildlife Service. As I later discovered, Secretary of State William Rogers had subsequently discovered my already-approved trip to Africa. You know, the one being taken by the "Nit" that had seized his precious 34 endangered elephant ivory bracelets to be used for personal uses contrary to two United States laws!

Being familiar with State Department tele-grams and State Department "speak," as well as having inside contacts, I soon discovered another thing. The Secretary of State didn't have the guts to sign the telegram himself but had our Agency Director sign it. And in so doing, had Cyrus Vance, Under Secretary of State, do the dirty work in cancelling my trip and being the one to advise my agency so the Secretary wouldn't soil his hands or be identified as the avenging point of the blade by the ivory bracelet-seizing "Nit."

Suffice to say, I was pissed. Hell, I am still pissed some 36 years later as these words are written! Being the German hardhead I was, I went downstairs back to my office after the party was over. There I penned another trip request, only this time for countries in Asia. Asia being the next worst place for illegal imports coming into the ports of entry in the United States. That re-quest basically went through the same channels, was approved and when the Secretary of State was out of his office, was signed and off I went. It was that trip that damn near got me killed by Burmese insurgents, plowed over by a typhoon, and almost torn apart by about 300 mad import/exporters of record in the City of Manila. But to find out about that series of wild adventures, the readers will have to read Chapter 8, "Asia, 1978," in my book titled, *Defending Our Wildlife Heritage.*

However, surprise, surprise. Missing that trip to Africa got me quickly assigned to another adventure in which swinging oars, flying fists, being the target of flying beer cans filled with cement and lobbed firebombs, under the guise of saving a river full of hard-pressed king salmon (now just about all gone, as these words are written), was unexpectedly dropped into my lap.

Sitting at my desk one morning, Agent Joseph Hopkins, the new Deputy Chief of Operations, came into my office and advised me to clear off my desk and make ready for another emergency field assignment. He further advised that an order from the Office of the

Secretary of the Interior, Cecil Andrus, had just been issued for the Klamath River in northern California, prohibiting the local Native Americans from running their gill nets for salmon. The regulatory closure had been effected upon the recommendation of the Service's Fisheries Division because king and silver salmon numbers on that river were approaching a critical biological threshold. Any further fishing by the Indians with their extremely effective gill nets would drop the migrating salmon numbers below the safe escapement threshold and would be deleterious to future salmon runs. It seemed the agents in Region 1 (California, Oregon, Washington and

Nevada) had been trying to hold the line under command of that order from the Office of the Secretary, and prohibit the Native Americans from illegally running their gill nets. They started their closure patrols July 13, and by August 1, were running into stiff resistance from the Indians who outnumbered the agents by 20 to one! On the evenings of August 28 and 30, agents patrolling the Klamath River were overwhelmed by about 200 mad Indians manning 20 boats. Service boats were put out of commission with damaged motors and destroyed steering mechanisms. During those battles, Indians clubbed agents with oars and fists, hammered them with beer cans filled with cement, and threw firebombs at the agents' boats. Several arrests were subsequently made by those agents who were quickly driven off the river by the superior force of Native Americans who were using extreme violence in defying the non-fishing order.

At that point, the Washington Office Division of Law Enforcement took over the operational command of the situation on the Klamath River. Then a call went out for the toughest agents within the division to muster up and head for northern California and the Klamath River operation to enforce the Secretary's order. With that explanation, Deputy Chief Hopkins ordered me to close up shop, head home, gather up my gear, and fly out to the Klamath River area posthaste.

There I was met with 35 determined agents, many of them exceeding six feet in height and weighing over 200 pounds! Additionally, the agent core was augmented with a number of burly National Park Service Police sent by the Office of the Secretary to augment the scant Service agent corps.

One day after my arrival, an evening meeting was held by all the officers present. There I was surprised by being called out from the crowd of officers and placed in charge of the river operations. Taken completely by surprise at my new assignment, I said, "Who?" when my name was called out by Joe Hopkins, Deputy Chief of Operations. This was just after he had relieved Keith Parcher, the Region 1 Special Agent in Charge, of duty and placed overall command of the operation into the hands of the Washington Office! Then he appointed Special Agent in Charge Jack Downs (from another Region and fully experienced) in command of all the operations on the ground, with the press, on meetings with the various Indian groups, local law enforcement detailed to the operation, and the Indian Police who controlled the land side of the operation, especially when we brought in Indian prisoners from the Klamath River.

"Terry, I just appointed you as my chief of operations on the river," said Joe. "Get your tail end up here and start putting together a plan so we can take back the river from the Indians and enforce the Secretary's order providing for salmon escapement," he ordered.

To give my readers a flavor for what had just happened, I was a GS-13 Washington Office Staff Desk Officer. I had just been placed in charge of the Klamath River portion of the operations over the Special Agent in Charge of that region who was a Line officer (not a staff officer pogue like me). One who held a GS-14 rating with a hell of a lot more field experience than I had. Suffice to say, that order created a lot of hard feelings from the Special Agent in Charge and his Assistant Special Agent in Charge from that Region, who were both relieved from supervisory duty now that the Washington Office was in charge. The Assistant Special Agent in Charge was later assigned to boat duty because he still was a resource-loving agent and to his and my way of thinking, still a valuable agent. He served under me well in that capacity throughout my stay. May God rest his soul after his recent passing.

However, no matter how one looked at it, I was in an awkward position to say the least. But, orders were orders.

I only had the time it took to move to the front

of the room to get my head on straight and produce a battle-ready plan that would support the
Secretary's order. Talk about going from just being there to help in any way I could, to sticking
my head into a "gill net." However, that was then,
and away I went. In front of a group of many of
my peers and one hell of a lot of superiors, I took
stock of the toughness of the present officer core.
Those who were physically challenged or just
not ready for the rough and tumble, for whatever reason, I quickly relegated to other duties.
Then a question-and-answer assessment of our
operational needs was made. I questioned those
officers as to their thoughts on equipment needs
and selection of boat crews.

Once done, I realized I needed to arrange the
boat crews so they could defend themselves and
work as a team. I also needed to make arrangements for not only the crew on a boat so it was
not overloaded, but left room for prisoner transport as well. Then I was apprised of our physical
boat situation. It seemed of the six boats we had
started out with, four had been badly damaged
by the Indians during earlier battles. In those
battles, the Indians had used a clever strategy of
ramming our boat engines with their boats until
we no longer had power sources. Something
that could be very dangerous in light of the fact

that we were on the mighty Klamath River near the mouth with the dangerous and rough ocean currents a few hundred yards distant! It quickly became evident that we needed more boats and stronger ones. The next day, an order went out for bigger boats like large-hulled Boston Whalers and 30' Boice-built, inboard-outboard jet-drive jet boats. As they were being purchased from local contractors or assembled from throughout nearby Service entities, several Service aircraft were sent down to Sacramento to the police supply houses. After talking it over with the battle-experienced boat crews that evening, I had ordered a wealth of new equipment for our agents. In short, we were now SWAT agents and needed the same kind of gear used by such personnel if we were to safely gain control of the river and avoid any further injuries to our officers. When the Service aircraft returned, they were stuffed with extra ammunition, night clubs, Mace, night vision gear, bulletproof vests, net hooks, and the like.

Then I was faced with a potentially very dangerous turn of events. It seemed a contingent of Indians from the American Indian Movement had come into town. According to local law enforcement, they had arrived to show their California brethren how to manipulate the press and the agents in the field, so the media would support the cause and in the process, embarrass the agents. It was also rumored that they were in town to show the local Indians how to fight back against what they called "government oppression." Now I had gone against the AIM members in North Dakota during one of the Sioux uprisings against the government while stationed there as a Senior Resident Agent for the Service. Several of my close U.S. Marshal friends had been killed by AIM members and I knew firsthand they could prove deadly if we didn't get the upper hand and fast. As it turned out, our preparations against such a movement and their meddling proved out to be well-planned and they ended up becoming an insignificant problem. Meanwhile, the Indians fished undisturbed with their gill nets, celebrated their victory over the Service on a nearby sandy spit, and the salmon died struggling to breathe in the Indian gill nets by the hundreds!

By day three after my arrival we were ready. We had amassed six powerful boats, the smallest being an 18'-Chrysler inboard/outboard. In that armada were two brand-new powerful, very fast metal-hulled jet boats as well. All the boats were staffed with the meanest agent teams I could muster. Then I made one of the new jet boats my command platform. Advising the nearby Coast

Guard to man a rescue boat at the dangerous mouth of the Klamath River, expecting there would be Native American 'floaters' after the battle, we sprang into action.

That evening as we gathered our forces along a set of Coast Guard docks prior to setting out, one of the agents asked me what they were to do with the Native American women.

"What do you mean?" I asked.

"Well," said Agent Tom Wharton, "they fight just like the men. They are just as vicious when they are throwing cement-filled beer cans, rocks shot from slingshots, manning gaff hooks, throwing fire bombs, and the like. Plus they are putting juveniles and pregnant women at the point of battle so the media can show our brutality when we tangle over their uses of the illegal gill nets (an American Indian Movement tactic)."

Thinking for a moment as the sea of officers' faces examined mine, I said, "In the case any of you run across Native American women or juveniles in the fight, they are to be treated just like the adult men. And if that means you bust them in the chops when they try and assault us, so be it!" With those words, you could see a swell of new commitment running across the men's faces (there were no women on this law enforcement detail). Those words of instruction bode poorly for the Indians running illegal gill nets, but happiness incarnate for the migrating salmon.

With those words, I laid out the evening's plan. I selected two 18' boats with six men each to head down the river by themselves towards the mouth of the Klamath River where most of the gill nets were currently being run in violation of the Secretary of the Interior's order. Those officers looked at me questioningly knowing they would be outnumbered 20 to one if they sallied forth with only two boats and 12 men. Then I advised the remainder of our four even-stouter boats filled with approximately 35 Service agents and Park Police officers (and damn good ones I might add) would quietly follow the "bait" boats. Then the men got big grins on their faces. It quickly became apparent what my plan was. I would send in two boats acting as bait to suck in those Native Americans illegally running their gill nets. And typically they would come if they had those kinds of odd on their side. Then when the boatloads of Indians swarmed those 12 men in their two boats, my four-boat flotilla of heavies would join the battle in a surprise move and attack from behind. A classic pincher move, I thought. It was a good thing I was of German stock and had read all of General Rommel's books on the principles of desert warfare...

Sure as a spring-run king salmon makes for fine eating, my little trap worked slicker than cow slobbers. The Indians on the nearby sandy spit saw the arrival of the two lightly manned Service boats. The alarm went out and soon the Klamath River filled with boatloads of mad-as-hornets Indians like the Persian fleet of old when they took on the Greeks hundreds of years ago. Shortly thereafter, 20 boatloads of Indians swarmed my 12 agents in their two boats. Within moments, they were hurling rocks, beer cans filled with cement, exposing their side arms the Indian men carried menacingly, smashing down on the agents with their boat oars, and the like. Only this time, the newer and tougher agents were freshly armed with night sticks and had on bulletproof vests and were able to fight back like wildcats.

As the Indians closed ranks around my two boats, the rest of us came charging downriver from out of the dark and smashed right into the Indian's small 10-14' wooden boats. In an instant, there were swimming Indians, floating gas cans, oars, life preservers, and swamped, busted-in-half wooden boats floating away from the battle scene! As expected, the Indians quickly broke ranks in the face of force and fled like the cowards they were. But not before leaving 12 of their kind back in our boats in handcuffs for assaulting federal officers and resisting arrest! Then as the first two boats took the prisoners back to the Coast Guard landing to be incarcerated by a tough squad of Indian Police, the remaining boats began gathering up and pulling the numerous gill nets being run in the river. When that happened, a great howl went up from the nearby sandy spit occupied by the Indians standing by several big bonfires. That was followed by two rifle shots fired at our boats from the darkness from the opposite side of the river! In an instant, those agents cradling AR-15's and M-16's turned their eyes towards the north bank of the Klamath looking for their assailants. It soon became apparent that our shooters had fled the scene before they invited a hail of bullets coming their way. Meanwhile, the Coast Guard boat sent downriver earlier, began gathering up the floating Indians from sunken boats, floating gas cans, oars, and life preservers that had floated away from the battle scene.

When the two boats returned which had been ferrying prisoners, they were loaded up with seized gill nets and those were taken back to the boat docks as evidence as well. Then as a typical north coast fine foggy mist and slight rain rolled into the mouth of the Klamath, a new scene greeted 'its' eyes. Six Service patrol boats were stretched across the area above the mouth of the Klamath River normally occupied with illegal Indian gill nets.

As the officers stood there in their respective boats watching the Indians on the far banks of the Klamath raising hell over being displaced, the river offered up its own surprise. Soon salmon could be heard bumping into our boats as the run of migrating fish pushed their way upstream. With the rains came increased water volume and the higher dissolved oxygen levels favored by migrating salmon. Both were welcome news to the salmon and boy, here they came. Soon the river occupied by boatloads of heavily armed, raincoated agents came alive with jumping salmon. In fact several times agents had to remove salmon which had just jumped into their boats and throw them back overboard so they could continue their upstream migration.

Then in the bow of my jet boat named the "Gorilla Boat" operated by Agent "Doc" McCallum, so named because of all the massively sized officers it contained, spoke up Agent Tom Riley. Tom was an ex-linebacker for the old Baltimore Colts and built like a brick outhouse. In his arms he cradled an M-16. Tom was so massive and muscular, the M-16 looked like a toy squirt gun. Then as more salmon jumped all around us migrating upstream unhindered by gill nets, Tom said, "I would stand out here all night, naked, just to make sure this miracle of life could occur." Tom was from Texas where they had just enough water to drink, much less any for salmon in which to run. His words pretty much echoed the sentiment of all of us gathered on the Klamath River that rainy night so long ago.

Once we had the Indians' attention at the mouth of the river, I set into motion Plan B. There I had not only men and boats on the river all night at the mouth but had boats with operators using night vision gear, running upriver as well. For that operation, we assembled six-man teams and used one of the powerful jet boats. With that plan in motion, 43 miles of the Klamath River upstream was run every night to prevent a repeat of the nightmare downstream for the salmon. Dangerous as all get-out but necessary because gill nets are run at night when the salmon cannot see the nets' fine mesh. That type of patrol was run by my very best boat operators and when I was there (remember I was leaving shortly for a trip to Asia), we only had one accident. And that was a bad one. It seemed in one of the darkest river canyons, the night vision oscillator went out unexpectedly and before the boat operator could shut down the engine, they slammed into a sandbar going over 40 miles per hour! All of the men were injured in various degrees, suffering broken bones and terrible bruises to almost-crippling, career-ending injuries when they went from 43 miles per hour to zero in a second against an unmovable sandbar! But the patrols continued and the salmon escapement mounted day by day.

Before it was all over, 115 Indians had been arrested for violating the Secretary's Closure Order. Additionally, another 14 were arrested for assaulting federal officers and 5 more were charged for resisting or impeding federal officers with force or violence! Seventy-five illegal gill nets were seized, along with 247 salmon, 2 sturgeon, guns, boats, motors, fence posts (used as weapons against the officers), and the like.

In the end, 115,000 king salmon escaped up river to spawn. As a result of that escapement, three years later the Klamath River sported four times its normal escapement of salmon. Four years later three times the normal escapement of fish ran the river. Five years later, the Klamath River sported two times the normal escapement! All of that was done without any loss of life. Just another example of how one of Aldo Leopold's five wildlife management tools, namely law enforcement, can help the world of wildlife in the management scheme of things. A lesson that was soon lost on the Service, if budgets for the Division of Law Enforcement, as these words are written, are any measure of lessons learned...

Shortly thereafter, I left my post as river operations officer and headed for an 80-day stint in Asia. There I ran into a host of adventures and damn near got killed (that story can be found in my book titled, Defending Our Wildlife Heritage). However, like many of my Washington Office experiences, I learned a great deal and more importantly, matured as an officer as a result.

Now I know I have grumbled incessantly about being cooped up in the stifling atmosphere of the Washington Office. However, not all that went on in that office was a negative experience. The Washington Office experience served many of us well who went onto bigger and better things. As a result of my two-and-a-half years in that office, I learned many things. I learned where all the Service skeletons in the closet were located. That information was invaluable to know for obvious reasons... Then when the serious fights took place, those in opposition knew of my service. As a result, many times I was able to slip by the "horse pucky" and help those who worked for me as well as many of the critters I had sworn to protect. I learned all of our laws more thoroughly and became sort of an expert in the endangered species side of things in the Service. As such, I never had to hold a candle to anyone here or abroad when it came to fighting for, or the normal infighting involving such matters.

In one major area, I learned how to take care of my subordinate officers and staffs and came to care greatly for all those within the ranks who served the critters and me. I clearly related those actions to my times and experiences in the Washington Office. For it was there I saw how some power-hungry folks, not knowing or really caring, stomped on those less-fortunate souls to better themselves or their standing in the eyes of their superiors. Seeing that kind of destructive behavior and having come from a hardscrabble early childhood background myself, I learned and vowed never to do any of that to my fellow human beings. And I never did! In fact, I can count at least 70 people I positively helped throughout my career, who as a result of my assistance and the power of the office I held, went on to bigger and better things in their lives. And for many, I changed their lives absolutely for the better.

Through identification frustrations for our field officers when it came to endangered species parts and products, I initiated the idea of our own forensics laboratory modeled after that of the fabled FBI laboratory. Today that wildlife forensics laboratory is a crown jewel, and is pure poison to those who challenge the Service in the federal courts of the land over forensic matters within criminal investigations.

Through other 'needs-knowledge' gained while in the Washington Office, I was later able to use that wisdom gained to create the National Wildlife Property Repository and the National Eagle Repository, both of which are one-of-a-kind and continue to this day serving the Native American people and the needs of the Service.

Additionally as the Division of Law Enforcement's Foreign Liaison Officer, I was able to travel abroad and use the power and resources of that office to initiate that division's successful programs in the years following, whereby investigative work, training and foreign liaison duties were performed by other Service agents for the good of the peoples and critters of the world.

Without a doubt, my Washington Office experience with all its warts and high jinks was some of the better years of my life. They were hard years but they were some of my best. In fact, those experiences still follow and support me to this day in my work as a multi-published author with movie credits gained from my true life adventure writings, and acting as a technical consultant for some in the media world. Knowing what I now know, I would do the same once again if I had the chance. Just imagine with all my experience at this point in my life, what high jinks I could come up with during this day and age if still assigned to the Washington Office and was in the traces. Then again...

# CHAPTER SIX

## NORTH BY NORTHWEST

BY 1979, after two-and-a-half years as the Endangered Species Desk Officer in the Washington Office, I almost had enough of that scene. One fine spring morning, the whole issue of working as a Desk Officer in that office came to a head. Since I was fairly well-known and trusted throughout the Service, I had a number of field officers who also wanted to move up in the ranks, call me on the "Q.T." They called and asked what it was really like working in the Washington Office. Never being raised a liar by my folks, I gave the inquiring officers the "straight skinny." I told them it was expensive to live in the area surrounding the District. That they would have to give up their patrol vehicles when they came into D.C., and would lose a great deal of their freedom of choice because of the very nature of the work performed at that level. I also advised they would be looking at riding on expensive public transportation or carpooling on a daily basis. Or having to pay really stiff parking fees if they decided to use their own vehicles going to and from work. Then I gave all of them the straight skinny on what it was like to work in the Washington Office dutywise. During that specific point of the conversation, I advised the work was intense, complex at times and many times somewhat political. They were also advised that production was always framed by deadlines, the changes they would be able to make would be small, and the leadership they would be working under throughout the highest levels in the agency could sometimes be problematic. I also advised that they would be working in a bubble and if they didn't perform in the eyes of the field officers and their supervisors, their chances for promotion out of the Washington Office would be difficult. In short, such an assignment could be a dead end careerwise if they chose to drag their hind ends and not give the job their all. I also informed all those callers that the work they performed in that capacity would be some of the best they would find themselves ever doing and the learning potential was "purdy" vertical. The lads were also advised that the field experience they brought

with them was invaluable in the decision making process, and as a result, "mountains could
be moved." Lastly, I made sure all were told that
when they left the Washington Office having
done well, they would have no trouble running a
region when they made Assistant Special Agent
in Charge (#2 in command) or Special Agent in
Charge (#1 in command) of a law enforcement
district.

Well, suffice to say, that kind of plain talk
didn't sit well with Chief Clark Bavin. So much
so that one day I discovered my rather large carcass sitting in his office or the 'principal's office'
as I liked to call it. It seemed what I had been
advising those calling in to see if they wanted to
make the leap and join the ranks of those in the
Washington Office, had reached his ears. And
that information reaching his ears was not what
he wanted to hear coming from his 'troops.'

Now to you unwashed readers, holding a field
position in the days of old as a Special Agent
for the Fish and Wildlife Service was plain and
simple a dream come true. In that arena, you
were your own boss, you caught as many bad
guys a day as you wanted, you lived in the great
out-of-doors surrounded by Mother Nature,
and enjoyed freedom in its purest form. Most
importantly, as a Special Agent you could make
a difference all by yourself if you worked hard

and used your head. In short, to trade all that for a stint into the Washington Office was like going from a dream into a nightmare for many of those less-disciplined or -focused folks in the field. And to do so on an uneducated and unfocused whim could cost one his career. As such, many better officers chose to end their careers right there in those field positions, as opposed to moving up in the organization through the dreaded Washington Office 'two-year experience requirement.' An experience requirement that for many was a disappointment or, if they were not circumspect, simply dashed their careers on the rocks. And numerous times in the process cost one his wife and family through a divorce because of the pressures that position held.

It was simply because of those concerns on the part of the Chief; i.e., the best officers stayed in the field, that he took offense to me telling the truth when field officers asked what it was truly like living and working in Washington. Chief Bavin figured that if I told the field agents what it was really like to work in Washington, D.C., many would never take the plunge. Then as a result, his office would never be worth the powder to blow it to hell. Especially if he only drew in the dregs of those field-grade officers; i.e., those officers who could not cut it working in the field or those who just wanted to move up in the organization regardless of the fact that they were dumber than a box of sandstone rocks.

To my way of thinking, he had a point but not one to be made at the expense or sacrifice of an agent and possibly his family. Having come up through the organization, I had the opportunity to observe just how the process worked. I also discovered that regardless of the work issues at the D.C. level, really good officers who wanted to make a change did so and took the plunge. They in turn, for the most part, developed and matured into the leaders the Service and the world of wildlife really needed.

At the end of that day's stint in the principal's office, I was simply told by Chief Bavin to lie to those requesting information about working in the Washington, D.C., Law Enforcement Office! I was further told that I could only tell those inquiring officers the good things and, "Let them find out the hard way about what it really is like working in here."

Then I was curtly dismissed and headed back to my hovel on the fourth floor. Once there, I called Special Agent in Charge Bob Hodgins in the Minneapolis Office. During that conversation, I asked if his recently advertised position for Assistant Special Agent in Charge was still open for applications. He replied that it was to close that very day but if I decided to toss my hat into the ring, he would hold it open for another ten days. Two days later I applied! Thirty days later, I was informed I had been selected as the new Assistant Special Agent in Charge in Minneapolis, Minnesota.

Thirty days later, my Bride, three children and two dogs began our move to the Region 3 law enforcement office in Minneapolis, Minnesota. Donna, my Bride of many years, bawled all the way to 'the land of eternal cold in the northwest.' She had good reason. First, she loved living in Virginia. Second, she hated the thought of living in cold and wintry Minnesota. Third, the moving van arrived three days later than the moving contract stipulated at our home in Fairfax, Virginia. As a result, we had to pay a $150 per day non-move penalty even though it was not our fault! And the government did not reimburse anyone for such a problem! Fourth, it rained the whole day when we were finally able to move from Virginia, soaking all our furniture being moved in the process. In fact, wetting everything so badly, Donna had to stay in the moving van so she could wipe off all the rain and re-polish our wood furniture as it was loaded. Fifth, our moving crew was comprised of what I shall say was mostly "local talent." Twice I caught several of those "folks" going through our things and helping themselves to what caught their eyes.

Having enough of that foolishness, I got out my handgun, placed it in my belt and told the moving crew's local talent that the next one I caught stealing our possessions (jewelry, etc.) would end up having a bad day. That pretty well broke them of sucking eggs. Well, sort of, as you shall see later. Sixth, all our furniture had to be loaded at our home and then off-loaded in a warehouse in Fairfax, Virginia, because a long haul mover was not yet available. While there in the warehouse, a chain saw, two large boxes of tools, five boxes of reloading equipment, and all six of my fishing rods and reels disappeared! Obviously "our" moving crew found more "unguarded eggs to suck." Seventh, our household goods were reloaded on a long haul truck that went to New York to pick up another part of a load of household goods from another family. From there, the moving van was to go to Minnesota. While in New York, that long haul trucker was fired for theft and all our household items were once again off-loaded in another warehouse. Eighth, those goods were later reloaded on another long haul van bound for Minnesota. Ninth, that van broke down on the freeway in Illinois during a terrible snowstorm. Tenth, our household goods were reloaded alongside that freeway during that snowstorm into another long haul van. Then that van proceeded to Minneapolis, Minnesota. Since we lived in Prior Lake, just outside Minneapolis, the van driver called me for directions to our new home.

Oh, before I forget, my patient Bride, three kids and two dogs had been living for two weeks in sleeping bags and eating out of an electric frying pan at our new home until our household goods finally arrived. When the goods finally arrived, we could hardly wait for them to be off-loaded so we could get back to living like human beings and not cave rats.

Eleventh (when will this ever stop?), as our van driver disembarked in Prior Lake, he walked over to me and said, "I cannot unload the van. The damage to your goods is so great because it has been loaded and unloaded so many times by folks who didn't give a damn, that there is a lot of damage. Also, loading your goods into another truck in a snowstorm alongside a highway in Illinois didn't do any good for the condition of your items either. Since the damage is so great, I have asked for a company insurance adjuster to be at my side as the truck is unloaded. Hope you don't mind waiting until he gets here."

When the adjuster finally arrived (remember, this is February in Minnesota), the van doors were opened. Brother, what a mess! Everything had gotten wet and had frozen harder than a cannonball. Boy, you talk about a mad and upset

Bride! And when I said everything had gotten wet and had frozen, I meant it. We had a beautiful, antique solid-mahogany piano that belonged to my mother. When she had given the family prize to us, it was in spotless condition. As it was unloaded from the van, I hardly recognized it. It was busted into many pieces and its sounding board area had filled with water and was frozen solid! In short, it was destroyed! All our clothing and dry goods had gotten soaked and were now frozen into blocks of mess and that included all our furniture cushions. Fortunately, I had transported all of my Bride's houseplants and my firearms in our personal vehicle. Otherwise, they would have been stolen or destroyed as well. Our soft furniture had all gotten wet and frozen solid as nearby Prior Lake! And in so doing, had split all the fabric and busted the wooden supports holding the fabric! In short, we had an appraised $19,000 in damages and that was in 1979 dollars! I had wisely insured all our items but even then, that did not cover all the damage. In short, we received only $7,000 for our losses and the government did not pay for any additional losses above and beyond that figure. Now I could see why my Bride had bawled all the way from Virginia to the lovely State of Minnesota.

The only thing that seemed to help her get over the above disaster was the fact we were moving into a new home that had been built by and for the developer. And once again true to form, we had to sell low in Virginia and buy high in Minnesota. Then a short-time later, we got a taste of Minnesota's very high state income taxes. Brother, it seemed every way I turned, someone was taking a bite out of my last part over the fence on this new adventure...

Then came my first day on the job as the brand-spanking-new Assistant Special Agent in Charge for Region 3's Division of Law Enforcement. The first person I met in my new office was the Special Agent in Charge, Bob Hodgins. Bob was a World War Two veteran tailgunner on a B-17 and damn lucky to be alive, if one knows their history as it relates to tailgunner survival rates on our bombers over Europe. He had come to the Service as an outside hire after a very successful career as the Director of South Dakota Game, Fish and Parks. As I was soon to learn, Bob was one of the finest human beings I came to know. Additionally, he was one of the finest managers I ever met and a genius when it came to his profession and how to deal with people! Bob died a few years later from a rare form of cancer, and the critters and I lost one of the most outstanding human beings we ever had the pleasure to know. The more I look around, it seems the good Lord always takes the very best and leaves all of us old toads behind.

Like my mother used to say, "God doesn't want just all weeds in His garden."

But for now, and after a genuine 'glad to have you on board' meeting, we promptly got down to the Region's law enforcement business at hand. First, he assigned me the responsibilities as the region's operations officer and general manager of all the office staff. Then he began advising me on the region in general. I was to be the second in command of the States of Minnesota, Wisconsin, Iowa, Michigan, Illinois, Indiana, Ohio, and Missouri. An area that held about 65,000,000 people during that day with all its resource-related warts and human-caused, resource-related problems.

We had a crew of basically younger, inexperienced Special Agents, because Bob could not afford the most senior agents commanding higher salaries (this was before per agent funding spread the division's wealth around more evenly). Our equipment was very old and run-down, and we basically had little or no budget. *What the hell else was new?* I thought. After all, this was the Service's bastard cat Division of Law Enforcement, and perennial underfunding was the agency name of the game.

To Bob's way of thinking, we had several problem states within our command. He considered the Illinois governmental agencies, including the United States Attorney's Office, crooked and corrupt from stem to stern. His evaluation of the State of Michigan was that it was run by strong states' rights leadership that considered itself top of the line in the world of wildlife law enforcement. As such, their leadership brooked little or no interference from us Teds' on all matters related to the world of wildlife, since they felt they had a "lock" on that position. Gee, the last perfect "person" I was aware of, we had hung on a board some 2,000 years earlier. Since then, most of us were less than perfect...

The States of Minnesota, Wisconsin, Illinois, Indiana, Michigan, and Ohio were bordered by the Great Lakes. Therein lay some of the greatest wildlife crimes from wildlife smuggling to and from Canada, to a multitude of illegal commercial fishing sins. That area had deeply rooted Mafia and Native American problems, especially in the fisheries illegal take and crooked commercial fishing industry. And to his way of thinking, Illinois was nationally known for its widespread migratory game bird illegal baiting and over-limit violations. As Bob droned on about the pluses and negatives of the district, it was not hard to visualize we had a mad bison by the tail as he whirled around, and there was only one hold shorter, to illuminate an earlier story in this book...

Then Bob advised we had one of the best Regional Directors in the Nation in Harvey Nelson. Bob considered him one of the Nation's best waterfowl biologists and senior Service managers bar none. However, his opinion of the Deputy Regional Director, Art Hughlett, was one, shall we say, far less enthusiastic and supportive. To Bob's way of thinking, Hughlett did not see the real need or value in having a Division of Law Enforcement in the overall wildlife management scheme of things. That was other than to require Bob give him a free ride to the Regional Office and back home each day and night in a government patrol vehicle...

Then he got into the history of the other supporting divisions in the Region 3 office. It seemed we had great up-front 'show' of support, but little rubber-that-met-the-road from the other divisions. As for the Personnel Division, they proved to be mostly a roadblock on everything he tried to do as a division leader. Or should I say, the Chief of the Personnel Division. In short, he considered it his God-given duty to keep the Division of Law Enforcement personnel as low-graded as he possibly could. According to Bob, it was like all the money used by the Law Enforcement Division came from the Division of Personnel chief's OWN pocket.

As for his field-grade law enforcement officers,

with three exceptions, we had some of the finest and hardest working officers in the Nation. They were mostly young but when it came to coming up with ideas on how to catch the bad guys, they were top-notch. In fact, I was pleasantly surprised to learn that my new region had more covert operations ongoing, 16 to be exact upon my arrival, than the rest of the Division of Law Enforcement carried in the entire Nation! Officers like Rick Leach, Jeff Lang, Karen Halpin, Neill Hartman, Ed Nichols, Don Berger, Dick Dickinson, Ed Spoon, Larry Keck, and many others, although young to the trade, were as aggressive and industrious as a hatful of killin' mink in a henhouse! In fact, once that group of officers got on your trail, the only way to get them off was to shoot them!

Then came Bob's appraisal of our Regional Office staff. He considered all of them some of the finest workers he had ever known. However, he had three females in the office who did not care for each other and sometimes those feelings erupted into open hostilities. "'Big T,' that is why I have left you in charge of the office staff. I can't keep those three from fighting, so you can give it a whirl. I also expect you to take charge of 'Secretaries' Day' when it comes around. I don't go for that kind of pressed-upon-us crap from our Human Resources section, so as of now, you are in charge of those kinds of shindigs. Try not to screw it up, so the folks in Human Resources find a reason to take a 'shot' at us for not complying with all this political correctness 'horse crap.'"

As it turned out, Secretaries' Day was one of the first things I screwed up. Bob was in Washington and I was now in charge. Wouldn't you know it, here came 'Secretaries' Week.' Now in those days and later on in my career, I never could understand how a supervisor could just take his secretary out to lunch and not the rest of the hard-working staff. So, I improvised. Come the fateful day, I took the whole staff out to lunch! Included in that bunch was Bob's wife, Dollie and my Bride, Donna. If I remember correctly, we went to one of the finest Mexican restaurants in the Twin Cities area. And wanting to do it up right since the boss had left me in charge, I ordered several pitchers of Margaritas for the ladies. And when those were gone, I ordered several more. If I remember correctly, those ladies were sure thirsty that fine day... Nine pitchers of Margaritas later, I sent Dollie and Donna wobbling back home and returned my staff of ladies to duty back in the Regional Office. If I also remember correctly, the luncheon bill came to over $200 and that was back in 1979! Boy, I sure didn't have any rumbles between my three ladies who had difficulty getting along the rest of that day! In fact, if I remember correctly, my entire office staff were as quiet as a bunch of mice pissing on a ball of cotton for the rest of that day. Suffice to say, when Bob returned and him being a bit of a pinch penny, his eyebrows went up clear into his hairline when I told him what his share of the Secretaries' luncheon was going to cost him ($100)! But I must have done OK, because when the next special day rolled around the following year, I was allowed to repeat my stellar performance with all the ladies as if it was the right thing to do.

I soon settled into the law enforcement management routine and much to my surprise, I seldom ever had a weekend to myself or with my family. My agents were so damn busy, it seemed every weekend I was on a takedown somewhere in the district, throwing a mess of meatheads into the local bucket. In fact, I remember one taxidermist's takedown in Chicago very clearly. Typically for our region, we had a zillion arrest/search warrants to serve and could only muster about ten agents to serve all that paper. So each team would grab a warrant, serve it, put our chaps in jail, and return to the Chicago office to write out our case reports for that fellow just incarcerated. Then we would grab another unserved warrant, head out, serve it, and return to the office after booking our chap to fill out our case reports on the most recent batch of chaps once again. One of those times, when all ten of us were sitting around a table furiously writing out our case reports so we could serve the remainder of the arrest warrants, one of the agents said, "Damn, I am missing my little girl's third birthday today." It got really quiet in the room and then another agent said, "I am missing my fourth wedding anniversary today." Before it was all over, all ten officers admitted to missing an important family event that particular day! Like I said earlier, my group of officers more than "paid for the gas and oil" when they were out and about. And with that example and many other like ones to follow, you can see why I was never home on the weekends. In fact, I used to limit my two fast-growing sons to one weekend a year for a hunting or fishing trip because I just didn't have the time to do otherwise!

After about a year-and-a-half of such a hectic schedule, I had just about extinguished both of my candles that I was burning at both ends. I decided it was time for me to do about a week of a regional "walk-about" by myself. A walk-about to see if I could get back into the groove of things as a line officer when it came to catching "wildlife varmints" in the field on my own. With that idea in mind, I went into Bob's office and tiredly dropped my carcass into one of his chairs.

"What's up, 'Big T'?" he asked, followed by one of his famous smiles.

"I think I need some time off from this office routine of 12-hour days, seven days a week. I would like to take a couple days off and take my boys squirrel hunting. Then I would like to take a week and just mosey around in Minnesota looking for some bad guys and see if I can't ruin their day. You know, just do some old-fashioned field law enforcement. Think I can do that without screwing up what plans you might have for me in the hopper?" I asked.

"Can you hold off for a week?" he asked. "We have a pissin' contest between two agents that I need to settle and I would like to have you along so you can see how to handle such an event. After that, I can cover for you and you can run. You deserve it as near as I figure it. The two of us have been running pretty hard trying to meet all our agents' demands, so I don't see why I can't hold down the fort so you can go out and sow some hate and discontent. Then it will be my turn to take some time off and do some crappie fishing with my girls. But as I said, we first have to fly to Chicago and settle a pissing contest between Glenn Orton and Rick Leach. With that in mind, I have had Miss Barb (Bob's wonderful super little secretary) make our airplane reservations. After that, unless something catches us from behind, you can take off and see what kind of trouble you can create."

Well, that was fine and good but the following week flew apart at the seams. First out of the gate, I had a major pissing contest between the three ladies in my office that I had mentioned earlier. That took two days getting things back on even keel and gathering up all the feathers plucked from all their hides by each other, which were scattered all over the office floor.

Then came our regional pilot, Agent Skip Lacey, and a surprise hatful of hornets. We had a pilot who I was somewhat wary about. He was what I euphemistically called our "bomber pilot." In short, he wasn't a good enough pilot to be a "fighter pilot," so the Service made him a "bomber pilot." Bottom line, I had my suspicions having flown several types of aircraft since my high school days and having flown with him as well. I quickly came to the conclusion that he was not what I would call a "comfortable in his own hide while flying" pilot. Then to top it off, he had been doing a lot of flying below 500 feet or what we called "hazardous duty flying." When Service pilots did that, they were entitled to Hazard Duty Flight Pay. Since our budget was lower than a snake's belly in the Marianas Trench, I had ordered Skip early on to stop flying so low and just make due at higher altitudes. Well, Agent Lacey thought he knew better and disobeyed my direct orders. Weeks later, Bob and I received a rather lengthy bill from Agent Lacey totaling $6,800 for Hazard Duty Flight Pay!

Suffice to say, Bob and I tangled over that one big time! He immediately ordered me to ground Skip and put our regional airplane up for transfer and use in any other region who wanted it! Suffice to say, Bob blew every fuse he had in his tool box! Man, it took all I had to calm Bob down. He was furious to say the least and extremely pissed, if all his hollering heard clear down the Regional Office hallways meant anything! Knowing Agent Leach really needed the airplane's services in his designated part of Illinois, I finally got Bob calmed down but just barely. If I am not mistaken, we had to pay Skip Lacey the $6,800 Hazard Duty Flight Pay since he had previously flown the hours and submitted his bill. However, that cost him dearly. As a trade-off, he was restricted as our pilot and could only fly when I ordered it and you can bet it wasn't done below 500 feet unless he wanted to be fired! And the rest of the agents in the region were grounded for a bit as far as any duty that required per diem expenditures as a result of Skip's large bill. Suffice to say, I would imagine Lacey got an earful from his grounded, not-traveling-very-much, hard-working, fellow agents.

Then if that wasn't enough hell-raising during the week, Agent Dickinson called me. "What is the old man's blood pressure like this week?" he asked.

Knowing Dickinson could be somewhat of a "flyer" sometimes, I asked, "What the hell did you do now, Dick?" I asked.

Ignoring my question, Dick persisted. "How is the old man's temper this week, Boss? Is he OK to deal with or is he on a short fuse?" he asked.

"Well, if I were you, I sure as hell wouldn't do anything to further piss him off," I said. "Now, what the hell did you do?"

"Well, I was helping the local police department on a drug case (we had no authority as Service officers to enforce such laws, but Dick was an ex-cop and just couldn't leave a running bad guy alone) and during a chase gone bad, I slid my patrol car across a parking lot and whanged it up a bit," he quietly admitted.

Now in those days, we could not afford new patrol vehicles due to having a non-existent budget. Figure this one out. Our agents were on call 24 hours a day and drove special emergency-type vehicles purchased from the lowest bidder, of course. However, since the Service saw little need to adequately fund our law enforcement operation in Region 3, we had to drive our sedans beyond the agency-prescribed 60,000-mile limit for emergency vehicles. Way beyond that mileage limit, truth be known. How about 100-140,000 total miles beyond the safe limit! In fact, several times when the refuge division turned in their worn-out, non-emergency type vehicles sometimes at the 80-100,000 mile range, we would trade our old worn-out ones for theirs. That is, we would trade in one of ours with 100-200,000 miles for one of theirs with only 80-100,000 miles on their turn-ins. That should give you readers some idea as to how precious a new vehicle was to our agents in Region 3. Now back to Agent Dickinson and his little "boo-boo."

Agent Dickinson had just received one of the very few new vehicles in our fleet. Now, as I was later to discover as a result of losing it on a high-speed chase while helping the local police department, it was junk! Try telling that to Bob after Agent Lacey just drained what little budget we had against orders with his low flying! Suffice to say, my week had been filled with wildness and now Bob and I were heading to Chicago to settle a pissing contest between Agents Leach and Orton!

Arriving in Chicago, Bob and I were met at the airport by Agents Glenn Orton and Rick Leach. Glenn was our Senior Resident Agent in the northern part of the State of Illinois and Rick commanded the southern portion. Therein laid part of the rub. Glenn also had the responsibility for the port of entry in Chicago, in addition to supervising the northern portion of Illinois. To Rick's way of thinking, Orton had all he could do in just managing the Port of Chicago and its import/export activity. That plus there was a general ideology pissin' contest between both men. Glenn was laid back in what he did and Rick was a fireball. Like oil and water, the two did not mix for sour owl droppings...

Back at the Chicago office, both men laid out their concerns and management plans for their respective districts. Additionally, Agent Leach recommended that Orton just manage the Chicago port of entry and that he, Leach, manage the rest of the State of Illinois. To my way of thinking, what Leach proposed made sense. He was a hard-charger and thrived on complex and extensive field work. Orton on the other hand, was more laid back in his approach in what he did. And to my way of thinking, the Service would be best served if he was just relegated to supervision over the extensive port duties. However, Bob was the boss and it was up to him as to what he did since it was his district. True to Bob's management style, he listened to both sides of the proposal and asked numerous questions. Then he advised both feuding officers that he would need some time to think it over. Then hav-

ing done so would give them his response to the
issue. With that, off we flew back to Minneapolis
since we had a mountain of unfinished administrative work staring both of us in the eye.

However, once back on the plane Bob said,
"'Big T,' what are your thoughts?"

"If it was up to me, I would keep Orton at the
port of entry and give Leach the rest of the state
to run. Leach has the support of all the agents
in Illinois, and Orton has no support from his
troops because of his erratic management style.
Leach on the other hand is equally at home
working either overtly or covertly and does one
hell of a job in each arena. Orton on the other
hand is more of an old-fashioned duck cop and
one-dimensional in his management style. And
after a year of reviewing his caseload such as it
is, I would consider him a marginal field-grade
officer at best. Additionally, the agents in the
state would follow Leach down the barrel of a lit
cannon if asked. So, that is the direction and why
I would lean the way I have suggested."

Bob true to his management style, said nothing. Then leaning back in his seat on the airplane,
he fell asleep on the way home. One day later,
Bob called me into his office. "'Big T,' I am going
to call Orton and tell him I am reassigning him as
my Chicago port of entry supervisor. Then I am
going to call Rick and advise him that he is now

in charge of the rest of the State of Illinois. Any concerns or questions?" he asked.

"No, Sir," I replied.

"Good. Now get your tail end back to your desk and rewrite both of those position descriptions to reflect what we discussed here this day so Personnel can sign off on our infrastructure change. When you get that done, rattle my cage and the two of us can go out to lunch and for once, I am buying," he said with a fatherly grin. Like I said, Bob was the best supervisor I ever had. I doubt I will ever see another man quite like him in my lifetime. At the time these words are written some 31 years later, that prophecy has come to pass. I have yet to find another man quite like Bob...

With the bulk of our supervisory problems addressed in the office, I prepared for my period of "relaxation" as I contemplated rattling out and about in the district, hunting my fellow man. However, all was not to be just fun and games. I had planned on meeting the District Director of Customs at our Canadian border port of Noyes, Minnesota, and several of our refuge managers at their respective refuges as well.

The following Monday after spending the weekend with my sons duck and squirrel hunting, I was up and off around five in the morning. I had a fair piece to go and wanted to get there before dark, so maybe I could get in a little work other than driving all day to the far-flung northwestern portion of Minnesota.

Heading out on Interstate 10, I headed northwesterly until I arrived at Moorhead, Minnesota. Then I headed north on state highway 75, taking it clear to our port of entry at Noyes, Minnesota. Fortunately, I had a clear but cold winter day of travel and made good time without being tangled up in some bad weather, as Minnesota is known for at that time of the year.

## A GIFT FROM THE U.S. CUSTOMS SERVICE

Arriving at the Customs port of entry, I parked my patrol car in the parking lot and entered the Customs facility. There I was directed to the Director's Office and after introducing myself, the two of us had a great discussion on what the Service did, its responsibilities and how the two agencies could work more closely together. As the two of us sat there talking, a Customs Inspector quietly knocked on the Director's door.

"Yes, what is it?" the Director asked of his Inspector.

"Boss, we have a large RV full of American duck hunters coming home after a week of hunting near Ste. Anne in Manitoba. In inspecting their vehicle, we discovered a lot of illegal Cuban

cigars and several cases of undeclared rye whiskey. We also discovered a lot of ducks, geese and
other game birds in the trailer they were hauling.
Additionally we found a pile of freshly killed
ducks and geese hidden in the RV's shower.
Since wildlife is not our area of expertise and we
have a Fish and Wildlife officer here who is up
on that stuff, we thought he might like to take a
look and see if they have over-limits or not."

Man, with those "golden" words ringing in my
ears, I was up and out of my chair in a heartbeat.
"Care if I accompany you while you are checking out these hunters? I always wanted to do this
and now that I have you here, we can make it a
duo, if you don't mind?" the District Director of
Customs asked.

"Hell, yes. I would love to have you along to
see one of the things we need to do a better job at
and that is American sportsmen border control,"
I said.

With that, out the door we went like a couple
of school kids heading for a bag of free candy!
Stepping outside, I observed a huge, very expensive bus-like motor home pulling a trailer de-
tained in 'secondary.' Secondary being the place
out of the way of the other foot and vehicular
traffic, where Customs Inspectors could give an
entering vehicle a closer once-over without de-
laying other returnees to the U.S. Outside the bus

stood six unshaven men all shaking in the cold afternoon temperatures like a mess of dogs passing peach pits, after leaving the warm interior of their vehicle. On top of the bus was a large carrier and behind the RV stood an enclosed trailer with its backdoor open. Off to one side on an inspection table were what appeared to be four seized cases of Canadian rye whiskey and three boxes of illegal-to-import in those days, expensive Montecristo, platinum series, Cuban cigars.

Walking over to the mess of men standing by the RV, I said, "Evening, Gentlemen. My name is Terry Grosz and I am a Special Agent with the U.S. Fish and Wildlife Service. I am told by my Customs counterparts that you lads have a mess of game birds that you took earlier in Canada. They have requested that I inspect those game birds to ascertain if they are legal." Then holding up my badge and credentials for all the men to see (the Division of Law Enforcement gave up wearing uniforms in 1973, in exchange for just civilian attire fit for the conditions at hand). Then I said, "These are my credentials of office, Gentlemen. It is under that authority that I will inspect the birds you fellows brought down from Canada as requested by my Customs counterparts (note how official I sounded in front of the Director of Customs...). If all of you will dig out your Canadian hunting licenses and driver's licenses,

I will be back shortly to check the same." Then I beat a hasty retreat back to my patrol car and dug out my briefcase full of Canadian Hunting Regulations for each province. Grabbing up the Manitoba regulations, I hustled back to my Customs Inspectors and sportsmen. Walking back to the trailer being pulled by the motor home, I hurriedly scanned the duck and goose regulations. Then turning the corner so I could look into the trailer, I was amazed! There in piles on the floor and hanging by their necks on duck straps from hooks on the interior walls of the trailer, hung hordes of feathered birds by the bucket load. Scads of ducks, geese, cranes and grouse greeted my eyes! Seeing I had a potential hatful of violations, I once again, only this time more carefully, reread the current Manitoba hunting regulations. With the details of those regulations swirling around in my head, I began counting the numbers and identifying the various species of game birds. Since Manitoba allowed a mountain of birds in one's possession limits in those days, it took some time for me to make an accurate species count. Finishing, I was surprised there were exactly six possession limits of every type of game bird allowed! Man, talk about being disappointed. Then the sun finally shone on the manure pile behind the barn. When the Customs Inspector had interrupted my discussion with the Director, he had mentioned a pile of freshly killed ducks and geese hidden in the shower of the motor home! Leaving all the birds stacked on the pavement outside the trailer, I headed for the door of the motor home. Walking up to the men still shivering outside the motor home in the dusk of the day, I said, "Alright, Gentlemen. Now I need to see some hunting licenses." With a bit of grumbling from my chilling-out, in-bad-need-of-a-bath sportsmen, they dutifully handed over all the required licenses. All checked out and were legal for the limits of birds back in the trailer.

"Now, Gentlemen. Do you folks have any more game of any kind to declare?" I asked.

A tall skinny man who reported to be the motor home's owner stepped forward saying, "Not a stitch of anything. These Customs people pretty well stripped us of whatever else we brought back from Canada. What you see is what we have, Officer."

"What about the extra birds the Customs Inspector discovered that you folks stashed in the stall of your shower?" I asked. Man, it got so quiet that you could have heard a sharp-tailed grouse passing gas from a hundred yards away, and that was in the blowing evening winds of northern Minnesota! "Well?" I continued after a long awkward moment of silence. "Well?" I said for a second time, thinking the grouse-passing-gas noise in the prairie wind had drowned out my earlier question directed at my shivering lads...

Then the tall skinny motor home owner growled out, "I already told you. Them Customs guys fair cleaned us out. What you see is all that we have."

Turning, I said, "Which of you Customs Inspectors saw birds in a pile in the bottom of the shower in the motor home?" I asked. A stocky Customs Inspector raised his hand like a school kid would do in the classroom when he had to go to the bathroom. "Show me," I said (I had planned on shaking down the entire RV anyway but for show; now I chose to make everything look very formal and serious). Whereupon the Customs Inspector entered the motor home as I closely followed. However just before I did, I stopped and turned. Pointing to a smallish Customs Inspector standing there watching this whole affair, I said, "Would you use that ladder on the back of the motor home, crawl up on top of the roof, go forward and check and see if they are carrying anything in that car top carrier?" Then I followed my Customs Inspector into the bathroom of the motor home. Sure as God made Minnesota winters something to respect, there laid a pile of gutted ducks and geese in broken-bodied profusion! I am sure the mallard ducks and white-fronted geese (the best eating) were not happy with their situation, but that sure brought a great big smile to my lovely, Robert Redford-looking mug upon observing such a "find."

Twenty-eight mallard ducks and eleven white-fronted geese later, I had cleaned out the shower. Then I tore apart that motor home for any other contraband. In the process, discovering five more boxes of Montecristo Cuban cigars hidden at the foot of one of the beds under the covers and buried by dirty laundry. Boy, that was all that it took as several burly Customs Inspectors now fairly tore that new motor home a "hind end" looking for any more hidden contraband they had overlooked during the first go around!

Stepping back out of the motor home so my Customs counterparts could do their thing, I noticed my smallish Customs Inspector standing on top of the motor home beckoning towards me. "Yeah, what have you discovered?" I asked knowing such a top carrier on the roof of the RV was a usual hiding place for contraband by a bunch of outlaws.

"There are two untagged, gutted mule deer and one antelope up here!" he said.

*Hot dog!* I thought. "Hand them down over the side of the motor home," I yelled back with a grin.

Well, suffice to say, our six chaps from the Twin

Cities had a "come-to-Jesus" meeting with the Customs lads. Smuggling was one of the charges laid on the motor home operator if I remember correctly. As for me, I had a patrol car clear full, and I do mean clear full, of a number of God's critters. That evening, I stayed at a motel in Pembina and the next morning early, I headed for Agassiz National Wildlife Refuge. There I off-loaded my game birds, mule deer and antelope (Lacey Act violations). The migratory and endemic game birds went into the refuge evidence freezers to be held until the cases cleared the federal dockets in the Twin Cities. The two mule deer and antelope went into one of the refuge's unheated sheds. After a call to the United States Attorney's office in Minneapolis, I got permission to release the big game animals after taking their pictures and removing their tails (physical evidence). Then I had the refuge manager call a local church and release the three big game animal carcasses to several needy families at the local pastor's direction. After getting the opportunity to meet all the refuge staff and get a tour of the area, I headed almost due south to the Fergus Falls area. That area between Fergus Falls and Detroit Lakes was chock-full of dozens of small lakes, redhead and canvasback passes, wetlands and every other place in between that water-loving shore birds, ducks and geese called home. And since hunting season was in full swing, I figured I would check out that area for those who had problems in counting what they had killed...

Heading south, I had to smile. I had come north to meet our Customs counterparts and introduce myself. I also wanted to emphasize to them the need for them to be on the lookout for wildlife violations coming into the United States from Canada. As it turned out, they had given me a super investigative gift involving waterfowl, small game and big game violations. And in so doing with such a spectacular catch, it had whetted their appetites to be more on the ball when such loads of wildlife crossed their port of entry in the future. During the rest of that waterfowl season before everything froze up in Canada, we received several calls from our new friends at Customs at the Noyes port of entry. As it turned out, some pretty good cases were made as those walking on the dark side suddenly had the 'light of the law shined in their eyes' by the now-alerted Customs lads.

## FROSTBITE FALLS

In the area of Fergus Falls, Minnesota, or "Frostbite Falls" as my younger son Christopher called it after a particularly cold and challenging hunting trip, lie hundreds of small lakes, beautiful wetlands and adjoining "birdy" uplands. In short, nothing but a wonderful haven for waterfowl migrating through in the spring, nesting in the spring and summer, and migrating back through the area to their wintering grounds in the fall.

Early one morning after a good night's rest, I rose before daylight and headed out. Moving northwest from Fergus Falls, I soon arrived at my twin destinations for the day's work at Ottertail and Rush Lakes. I had been told by the refuge biologists at Tamarac National Wildlife Refuge that both lakes were loaded with migrating redhead and canvasback ducks. Knowing such species in early fall plumage would be hard to identify by most sportsmen and since there was a bag restriction on redheads and canvasbacks, I headed in that direction. To those readers who are not duck hunters, identification of such species for many sportsmen is difficult. Especially when the birds are migrating south and in various stages of their plumages. Identification becomes doubly important when states regulate the numbers of species that can be lawfully taken. And sad to say, even with the dead bird in hand, many waterfowl hunters have difficulty in the bird identification of many of our waterfowl species. So it was on that fine day I headed in that direction, knowing someone out there hunting ducks would more than likely screw up. Little did I realize at that time just how "stinky" they and yours truly would screw up...

Hiking into the north end of Rush Lake in the cold pre-dawn darkness, I managed to surprise a striped skunk enjoying his breakfast of a previously killed and lost duck. Realizing at the last moment I was within "gun range" and the skunk was pissed over his breakfast interruption, I got both barrels! Damn, you talk about smelling putrid, I was now no wonderfully smelling flower like I usually was. Then when the skunk reloaded, raised his tail and stomped his front feet once again before I could escape, I figured I was about to get a double dose of Ma Nature's perfume. However, it is amazing what a quick draw and a 230-grain .45 ACP bullet can do to a previously dining skunk at six feet! Suffice to say, the skunk got off one shot and I got off two! Returning to my vehicle, I stripped down in the freezing morning temperature to my most beautiful bare-naked self. It was at that very moment that a nearby pack of coyotes began howling and I am pretty sure there was some laughter therein as well. To preclude any further funnies on their parts or embarrassment on mine, I dove into my emergency reserve of clothing for just such occasions for a fresh, better-smelling change. Standing there shivering, I finally got dressed and then I sacked my "skunked" clothing into a plastic bag and left it on the roof of my patrol car to air out. Resuming my trek back to the lake, I was more careful that time so I wouldn't get surprised once again by another "Quick Draw McGraw," straight-shooting skunk.

Ahead of me on my trek, I stumbled across three duck hunters putting out their decoys. This was as good a place as any, I thought, as I burrowed into a thick stand of brush for the cover it offered. From my vantage point, I could plainly see my three shooters as they finished putting out their decoy spread. Then they moved back to shore and hurriedly constructed a duck blind from rushes, brush and great handfuls of tall prairie grasses. After that activity, they settled down out of sight to await legal shooting hours. Moments later, I could detect the faint smell of nicotine as they smoked cigarettes and could overhear the quiet sounds of their conversations. All of that took place over the faint Ma Nature's perfumed aroma I was now experiencing as I ate my stick of dry hard Italian salami for breakfast.

Out on the lake, I could hear the faint whistling of Arctic swans, the croaking of coots, the harsh call of a nearby hen mallard, and the nervous calls of several flocks of awakening Canada geese. It was sure great to be alive, I thought. That was except for that ever-so-faint smell of "pissed-off skunk."

Then I became aware that the air was filling full of whistling wings of hundreds of ducks as they left the lake for their nearby hoped-for harvested wheat field "breakfast tables." Then I heard several "swish-plunks" when a mess of ducks sailed unobserved into the water near my three lads' decoys. The ducks had come unexpectedly from the north and before my lads had realized it, they were happily swimming in their decoys. Caught by surprise in the pre-dawn light, I observed my three shooters hunker down even further in their blind. *What the heck are they doing?* I thought. They had at least ten ducks happily swimming around in their decoys well within gun range. Why the dickens did they not shoot? crossed my mind. Then it became painfully obvious. There were seven Canada geese winging their way just feet above the water, heading right for their decoys as well! The geese made not a sound as they honed in on my three lads' decoys as if pulled in on a string. Then swinging into an opening of the "J" pattern in my lads' floating decoys, the geese dropped their feet and quickly slowed, with their large wings making many loud swishing sounds amidst much happy honking back and forth. That happy scene quickly ended as my three chaps arose as one and unloosed a hail of deadly lead into the just-landing geese and now rising-in-panic flock of ducks (use of lead shot was legal in those days). It was as if all hell had broken out among the lifeless sirens called decoys and the soon not-to-be-living groups of waterfowl. It rained ducks and geese so rapidly that I never had the opportunity to even make an accurate count of their kills!

As nine shotgun blasts echoed out across the lake, my three lads stood there in unison mesmerized by the killing effect they had just introduced into the quiet end of their little inlet. Then the three quickly hunkered back down into their blind. The abrupt roaring of their collective shotguns had aroused all the other already nervous birds sitting at their end of the lake. Soon the air was once again filled with birds of all sorts and colors zipping to and fro.

Since my three lads appeared to be under their daily bag limits of ducks and geese, I also held my position. It was still too dark (damn that half-hour shooting time allowed before sunrise) for me to make out the species of ducks they had just killed. So I just swore under my breath and kept watching with my binoculars for the next action coming their way. Then a small flock of mallards and pintail zipped by the decoy set and then changing their minds mid-flight, jacked themselves abruptly around. With their air brakes on and feet down, my flock landed among the decoys and dead birds en masse. Realizing something was bad wrong, they hurriedly sprang back up into the air. However, this time my lads were ready and it fairly rained ducks from the close-at-hand frantic birds. It was then that I made my move. As my three chaps hurriedly reloaded their shotguns, I made my always-gorgeous appearance. They were so engrossed over their reloading drills that not a lad saw or heard me approaching until it was too late.

"Morning, Gentlemen," I said. "Federal agent. I would like to take a look at your hunting licenses, duck stamps and shotguns," I said as I continued walking into their midst. Man, you talk about three lads jumping like bugs on a hot rock. My guys went so abruptly skyward in surprise that it was like a returning prairie dog when it realized there was a black-footed ferret down in its burrow looking right at him!

"Who the hell are you?" said a heavyset man as he picked up his hat which had fallen off his head when he jumped sideways at being so surprised over my unexpected arrival.

"My name is Terry Grosz and I am a Special Agent for the Fish and Wildlife Service," I replied.

"What the hell is that? I ain't never heard of such a person. Are you a game warden?" he continued as his partners looked on from their frozen-faced positions.

"Well, kind of. I am like a federal game warden," I replied as I showed the three men my badge and credentials.

All three men squinted at my badge and credentials and then apparently satisfied, laid down their shotguns and began digging for their wallets. As they did, I replaced my badge back into my shirt pocket and picked up the shotgun of the man who was closest to me. Unloading the shotgun, I quickly checked it with my plug checker (a small ten-inch-long rubber fuel line improvised so I could check the capacity of one's tubular magazine, because the rubber would not scratch the metal surface while so doing). I found his gun legally plugged. Reloading his shotgun, I repeated the process with the remaining two chaps. By then my lads had their hunting licenses out and they were checked and found to be legal.

Then I waded out into their decoys and picked up the seven dead Canada geese and threw all of them up onto the shoreline near their duck blind. Following that I gathered up 15 assorted ducks and tossed them shoreward as well. Satisfied I had retrieved all the dead birds, I headed onto shore. There I knelt down and began identifying the ducks.

"You boys know what you have here in the way of ducks?" I asked.

With that question, my three lads gathered around and began identifying the ducks. Between the three of them, they identified three drake mallards and two drake pintails. That left a mess of brown and gray ducks unidentified.

"What have we got here?" I asked pointing to a mess of female or hen mallards and pintail.

"Them is brown ducks," said a smallish man who appeared to have no front teeth and those he did have, were heavily tobacco-stained.

"Anyone else want to give it a try on the identification of this pile of ducks?" I asked. I had no takers. "Well, Gentlemen, these two are hen mallards and these three are hen pintail. Fortunately for you lads, the three of you are not over the limit on hens." Then I reached over and placed six all-grayish ducks in a pile. "Anyone want to guess what we have here?" I said looking up at the three men.

"Them is gray ducks," said the man with no front teeth as his other partners nodded affirmatively in concert at their partner's identification.

"What we have here are female and juvenile redheads," I said as I looked back up at the men. All three men just looked quizzically at me as if I had just identified some form of unheard-of pudding. "Do any of you lads know if these are legal or not?" I asked. Now the three men figured they had stepped into it and not a man moved nor did the beat of their hearts, I suspected. Not getting any response, I advised the men the so-called "gray ducks" were closed-season species and could not be taken that year. Once again with that news, not a move or heartbeat was evident among my chaps. "Well, Gents, these six are illegal and cannot be taken or possessed. Since I am not sure which of you chaps shot these ducks, I will have to issue all of you a citation for taking a closed-season species. Especially since all of you shot into them at the same time. Normally in this situation, since I am not sure who killed what, all of you would get a citation for the whole mess. However, since you lads are above the boards on the rest of the hunt, I am just going to issue each of you a citation for taking two redheads apiece which means a reduced fine. Any questions?" I continued.

"Officer, I am not trying to be a smart-ass but are you sure on the identification of those six ducks?" asked the third man who up until then, had not said a word.

"Yes, I am," I replied. "I have a Bachelors and Masters degrees in Wildlife Management and have taught waterfowl identification to state and federal officers for numerous years. All six of these birds are redheads and as I said earlier, are protected from being taken this year in Minnesota," I said.

Looking up at my chaps, I saw resignation in their eyes as to the correct identification. Then

I dug out my citation book from my game bag and requested all of their driver's licenses. As my three chaps were digging out those licenses, I became even more aware of the intense shooting taking place on Ottertail Lake lying just to the southwest of where I currently was located. That intensity and frequency of shooting told me I needed to be there as well and as such, I sped through the three citations' issuance with my chaps. Taking time to explain the citations to my three chaps, I could see all of them now wrinkling up their noses at how I smelled as I stood close to each man explaining the citation and their responsibilities.

"Sorry about the smell, Men. I ran into a skunk in the dark this morning and he didn't miss," I said with a weak grin.

With those words, my three chaps broke out into polite laughter and then unable to control their mirth, just roared! Finally one of the lads said, "We are sorry, Officer, but none of us were going to say how badly you smelled for fear of getting even more tickets than we are getting. But now that you have identified the smell to us, we thought it was funny. This makes our tickets easier to swallow, knowing that we got 'peed' on by a game warden and the game warden got 'peed' on by a skunk." Then there was more laughter. I sure as hell didn't think it was so funny but if they did, I guess I had to laugh as well over my misfortune. However, you should have seen the 'the guy doing the peeing,' after the .45 cleaned the skunk's clock from stem to stern, I thought to myself with a smile.

In the meantime, my heavy shooting continued over on Ottertail Lake. Hurriedly writing out a receipt for my duck seizures, I quickly left my chaps and headed back to my patrol car. There I put my "skunky" clothing into the trunk and beat a hasty retreat over to Ottertail Lake. However, by the time I arrived and staked out the area looking for my shooters, the shooting had ceased. I neither saw nor heard any more shooting after that. It was then I realized my shooters had probably had an early morning shoot, then quickly left to get to work. Well, that was OK with me, I thought. Since my shooters had such a good shoot, I surmised, they would return another time since they had not been bothered. If that was the case and I was still in country, guess who would be lurking in the bushes waiting the next time they showed their faces? And hopefully it wouldn't be another damn straight-shooting skunk...

Getting back into my car, I began driving all around the area where I had heard such volleys of shooting earlier. In such roving, I finally ran across an old house trailer parked alongside a wooden outhouse. Its overall sad state of affairs told me the trailer was probably being passed off as a duck club where one could get out of weather if need be. Further exploration revealed there were fresh vehicle tracks in the area next to the trailer, with a smoldering fifty-gallon barrel smoking out back. In short, that lent credence to someone's recent use and subsequent departure. Walking around behind the house trailer, I discovered an area near the lakeshore where numerous ducks had been recently picked, cleaned and gutted. I also discovered another skunk helping himself to the duck leavings and I sure as hell gave that one a wide berth. However, during that inspection and surprising the skunk, I discovered dozens of fresh, illegal redhead and canvasback wings scattered about in the cleaning area! Further examination revealed there were 47 separate fresh wings representing one hell of an illegal redhead and canvasback massacre! Well, that was OK to my way of thinking, now that I was onto their little illegal operation. Since my shooters had such good shooting in their little inlet and had not been bothered by the law, I was sure they would be back. And now when they came back, they may have one hell of a duck shoot but there would be another surprise waiting in the bushes for them as well. Further investigation discovered a well-used trail leading out through some marshy low ground, heading directly to a rather-substantial duck blind in a small bay on the south end of the lake near their trailer.

I suspected that was where my shooters had hunted previously but now I had a problem. There was hardly a close-in place in which I could hide my rather amply sized carcass without the chance of being discovered in the process. The landowner had hayed or plowed most of the ground in the immediate area and for the life of me, I couldn't find any close-in place where I could hide, then quickly ambush my shooters when they came back from the blind.

Finally I located a rock pile that I could hide behind where I could make good use of my spotting scope. It was not in the best location but it was one in which I would not be discovered. However, its location was a good quarter-mile away and up on a slight hillside overlooking my shooters' blind. Then to make matters worse, the ground had been plowed all around, leaving hardly a living plant behind which I could hide as I sneaked out of the area during broad daylight when ready to spring my trap! But I had been in worse situations during my career and just figured if they weren't looking for me as I sneaked in or crawled away, they wouldn't see me.

Then when I needed to spring my anticipated trap, I could exit the area by crawling through the plowed area like a Marine on maneuvers. Then when out of sight of my shooters, I would run to my patrol car and drive back to my suspects' house trailer. There I would have my chaps in the vise grip of the "German Tank Driver" (that's me)! Little did I realize, as so often happens, the best-laid plans of mice and rather large-but-beautiful game warden types, could go so far astray and end up in the "crapper" in the truest sense of the word.

For the next two mornings way before daylight, I was at my rock pile overlooking my suspect duck blind. However, no one showed up either day. Knowing those chaps still needed some of my tender loving care, I remained in the immediate area working other duck hunters. Then on the third day, my duck hunters arrived just before daylight. I was sure glad to see them drive up to the trailer in two vehicles after my several-day stakeout.

For those readers who have never been on a stakeout in the fall in western Minnesota, it gets kind of cold and lonely. Then as you are standing there in the pre-dawn dark, cold and shivering, your mind begins playing tricks on you. Tricks that lead you to begin doubting yourself. Especially those kind of doubts that say maybe you had been seen by your shooters and they were laying low for a while. But today had the makings of a good day, as I soon saw several flashlights coming out from the house trailer moving in the general direction of my suspect duck blind. Then through my binoculars I could see my chaps setting out their decoys around their blind. Finally, with the decoy work done, the lights went out and quiet reigned once again over the land. That is except for my clattering teeth in the cold Minnesota morning chill on a lonely hillside overlooking the southern portion of Ottertail Lake.

Come dawn, I could hear shotguns going off around me as other duck hunters began their duck and goose hunts. However, my guys just sat tight and never fired a shot as numerous shooting opportunities presented themselves in and around their blind. Then a flock of mallards numbering about 50, came flying off the lake and sailed right into my four shooters' decoys. That did it! Up from their seats came my four shooters like a carp after a gob of worms and the cold morning air was rent with the sounds of 12 shots being quickly fired into the surprised flock of mallards. Even from where I sat behind my rock pile, I could see my shooters were having a deadly effect on the once-large flock of mallards. In fact it "rained" mallards so fast that I lost count after 18 had being killed!

Then for the longest time after my shooters

had fired their killing volleys into the ducks, my lads disappeared back into their blind and no one moved. After a wait of about twenty minutes, two of my four shooters finally arose and walked out into the decoys. There they picked up 21 mallards and walked back to their duck blind like nothing out of the ordinary had happened. Once there, I could see the other two members helping by stuffing some of the retrieved ducks into the two men's game bags in their duck coats. Then with two duck straps full of nice fat mallards (excellent eating by the way), my two shooters slowly trudged back to their house trailer. However, I noticed they stopped every so often and just looked all around as if looking for a game warden to pop up out from nowhere and put the grab on them. They finally arrived at the house trailer and disappeared behind it where I could not see what they were doing. Meanwhile, my other two shooters back at the blind never fired a shot even though they had many opportunities to kill numerous ducks flying into and out from their decoys.

Twenty-one minutes later, here came my two duck shooters from the house trailer and neither was carrying a duck strap. And the big bulges in the game bags in the back of their hunting coats had disappeared as well. My two chaps quickly covered the ground back to their duck blind and once inside, disappeared out of sight. Then for the longest time, nothing happened as skeins of birds flew about Ottertail Lake looking for a good place to 'breakfast.' Once again, I noticed several small flocks of what appeared to be diving ducks sailing into the four men's decoys. However, the shooters never stirred but remained out of sight in their blind. Then I saw why. The little inlet in which they had set up their blind and decoys was jug-full of emergent and submergent vegetation. To you non-duck hunters, that spelled out a ready and waiting buffet for any hungry member of the Anatidae or waterfowl family. Before long, the little inlet was soon full of happily bobbing and diving ducks feasting on the vegetation buffet.

In fact, several times I observed wigeon (medium-sized duck) sitting on top of the water as if patiently waiting for something from below to occur. Then when a redhead or canvasback emerged from the depths carrying a mouthful of vegetation from the inlet floor, the wigeon hastily grabbed onto what that duck was carrying and quickly moved off to enjoy the stolen harvest from the hard-working diving ducks. For my non-waterfowl hunting readers, the wigeon is not as good a diver as is the redhead or canvasback. In short, it is easier for a diving duck like the redhead or canvasback to get the best goodies from the floor of the bay more so than a puddle duck like the wigeon. Then when the diver surfaced, the ever-watchful wigeon would see the duck coming up from the bottom, time its position and then grab the mouthful of food from the diver just as it emerged from the waters below.

Fascinated by the behavior of the feeding waterfowl, I was surprised by the quick action of my four duck shooters standing up in the field of view of my binoculars, rapidly blazing away at the numerous ducks happily feeding in their decoys! Suffice to say, that created quite a deadly picture in my binoculars. I observed in my optics my four shooters being jostled from the recoil of their rapidly firing shotguns. Then I observed many small puffs of smoke from the muzzles of their shotguns and the subsequent splashing of shot patterns over the water. In that splashing of the shot patterns hitting the water, I could see ducks being rolled by the lethal shot streams, as well as others springing into the air in alarm. Then many of those that had sprung into the air in alarm began falling back into the water as they lethally crossed into one of the spraying shot patterns. In a moment, it was all over once again.

Just like when my four "shooters" had sluiced the mallards earlier, my lads disappeared back down into their blind and never moved a muscle for about twenty minutes. It was like they realized the destruction they had caused and were now hiding to avoid any game warden's "late viewing eyes" trying to locate the place where such shooting had just occurred. By now having worked waterfowl sportsmen for many years, I began to see a recognized pattern being shown by the real killers or game hogs in front of me, as I had seen in the past. Realizing that, I now became aware that it seemed like my eyes had hardened and narrowed somewhat with such a realization emitting forth over my flat-out duck-killing shooters.

Once again after about twenty minutes of non-movement, two of my shooters arose from the blind and walked out into the decoys picking up dead and dying ducks. Then back at their blind, the other two chaps arose and stuffed the extra recently retrieved birds into two of the men's game bags in their hunting jackets. Then the men were careful to put only a legal limit onto each sport's duck strap and off they went once again. Once again, my two shooters took their time as they headed for their trailer as if looking at the same time for anyone else in the area looking on. Once at their trailer, they walked behind it and disappeared from sight once again. Sitting back behind my rock pile, I got out my notebook once again. Flipping to my page set up for my four shooters, I entered, "14 ducks killed at 9:03

with 12 shots being fired by all four gunners." I also entered below that notation, "seven divers comprised of redheads and canvasbacks, along with seven wigeon killed and retrieved."

As I now knew, I had my chaps for over-limits of ducks and for taking protected species, to wit, seven redheads and canvasback ducks. However, I was now in a quandary. I couldn't leave since I had two shooters soon to return from the trailer who would sure as all get-out see me crawling off up the hillside on the exposed and barren plowed ground. If they did and figuring I just might be the law, they would have time to return and hide the ducks just killed before I could arrive and gain control of the situation. Damn, I thought, I am stuck here until they all return and get back into their duck blind. Then I can crawl up the plowed side hill to my patrol car since their backs will be towards me and get away without being seen. With that realization and plan, I held my ground waiting for the return of my two chaps still back at the trailer.

This time my two lads did not return from their trailer until 44 minutes later. What the hell were they doing all that time? I asked myself. Oh well, it didn't matter. They were finally on their way and once back in their blind, I would make my break for it and return to my patrol car. Once there, I would head for the trailer, slink in, park behind it, and then ambush my errant shooters when they returned. A great plan that as my readers will soon see, went quickly into the "crapper"...

As my lads re-entered the duck blind, I noticed about ten ducks had lifted up from the decoys and quickly left the scene. True to form, my two shooters back at the blind had not fired a shot. It was as if they didn't want to bring any unwanted attention to themselves by a sharp-eared game warden echo locating on their carefully timed barrages of shots. Man, I thought, these guys are sure clever and killers of the ultimate degree. It would be a pleasure to cut off their water, I thought, as I quickly gathered up all my gear and made ready to leave. Taking one more look at my four chaps back in their duck blind making sure their eyes were lakeward, I began my quarter-mile crawl across the freshly plowed prairie. It was slow and dirty going but I kept at it until the sweat was just rolling off me. Then all at once just as I reached the crest of the hill on my crawl, I heard my four guys open up on the ducks once again back at their blind. Not wanting for them to kill any more birds and wanting to get to their house trailer, I just kept going and didn't look back. However, once over the hill and out of sight, I quickly rolled over then sat up. Brushing off all the accumulated dirt I was now covered with, I opened up my jacket front and shirt as well, to cool off from my long and sweaty crawl. Then I trotted the hundred yards or so back to the prairie trail on which sat my patrol car. There I got one hell of a big surprise—my damn left rear tire was flatter than a flounder! It was obvious I had run over something that my tire did not like. Quickly opening up my trunk to the sound of hammering guns once again over the hill back at my suspect duck blind, I fairly flew into my tire changing. But that took some time as well. I was on a side hill with no nearby flat place to drive and my handy-man jack kept letting the car slip off the tongue of the jack. Finally I got my tire changed. But it took a lot more sweat and a hell of a lot more loud swearing to get it done. Finishing, I threw all my gear into the car and roared across the prairie towards the highway leading to my suspects' house trailer.

Rounding a turn in the highway, I observed my four shooters emerging from the marsh walking towards their house trailer. Kicking the car into high gear, I slid the turn onto the dirt road leading to the house trailer. Then slowing down behind a nearby hill so I wouldn't 'play my hand' by my dust cloud trailing my vehicle, I slowly entered the area of the house trailer and outhouse. Seeing my four lads still coming from the marsh, I swung up my binoculars and took a look-see. All four men were sporting duck straps carrying both ducks and a few Canada geese. That must have been why I heard them shoot twice when I was crawling out, instead of their usual one time, I thought. It is hard for any duck hunter to pass up a chance on being able to take a lordly Canada goose, I thought.

Keeping my eyes peeled on the progress of my four still- unsuspecting chaps, I hurriedly re-buttoned my open shirt, zipped up my jacket and wiped off the sweat still streaming off my brow and steaming head. Then I casually leaned on the front of my car like nothing out of the ordinary had happened. Knowing I had my lads with a big over-limit of ducks, it was like blowing the smoke off the end of my pistol after making a good long shot just like they used to do in the movies. That is, I would do the blowing smoke trick off the end of the pistol barrel, once I had discovered where they had hidden their over-limits.

Finally, my four lads emerged from the Ottertail Lake marsh and rounded the end of their house trailer. It was then I caught the faint flicker of surprise running across their faces when they saw me parked behind their trailer. But I also caught a very quick recovery running across their faces just as well. These are smart 'hombres,' Terry, I thought. To only show that much recognition and surprise at seeing someone watching them, showed me I was clearly dealing with a bunch of real clever outlaws. As I was soon to discover, I wasn't far off in my theory either!

Walking over to my still-walking chaps like I belonged there all along, I said, "Good morning, Gentlemen. My name is Terry Grosz and I am a Special Agent with the Fish and Wildlife Service." They collectively didn't say a word and hardly gave me a second glance as I held up my badge and credentials. The suspicion that these are hard cases who have been here before crossed my mind once again...

"Gentlemen, I would like to check your birds, shotguns and hunting licenses," I said. The response from my four chaps was complete silence as they sat down on the trailer porch, emptied out their shotguns and then calmly removed their chest-high waders. Not a word had been spoken and it was almost as if they would take their time when it came to addressing my request. In fact, I now knew how an "untouchable" felt like in India. It was then that my antenna went clear up. *These guys are pros, Terry*, I thought. No one is this casual unless they are real pros, have a lot to hide and, have hidden it well. It was then I put on my best "investigator" face and began really watching and listening to my four lads, as yours truly — dead calm as the water behind a beaver dam — prepared to hopefully match their guile and wits.

With hardly a word except quietly among themselves, the men handed me their shotguns and hunting licenses. All checked out to be spot-on according to the laws of the land. Then they separated out their ducks and geese for me, and every man was one shy on their daily bag limits of ducks and under the numbers with their geese. As for the ducks lying before me, they were all legal as to species.

Then still holding their licenses, I dropped the bomb on my tight-assed and quiet-lipped, self-assured, gaggle of "game hogs"! "Gentlemen. Between the four of you, you have one duck apiece shy of a legal bag limit. However, I have been watching all four of you since before daylight this morning. During that time, the four of you shot into a large flock of mallards at 8:01, killing 21 of them." Then pointing to two of the men, I said, "Then the two of you brought all those ducks back to the trailer. Then 21 minutes later, you re-emerged from the trailer and went back to your duck blind where your two partners waited. When you came back, it was without any ducks in evidence. Then at 9:03, the four of you shot into a mess of feeding ducks in your decoys killing another 14, seven of which were protected species of redheads and canvasbacks. Then the same two of you brought out all those ducks, carrying them back to this house trailer in the process as well. There you remained for another 44 minutes. Then you returned to your duck blind. At 10:13, the four of you fired a barrage of shots and at 10:26, another barrage of shots killing the ducks you now have laid out in front of you. As I see it, between the four of you, Gentlemen, you have killed 31 ducks over the limit, counting these here lying before me on the ground. Now, Gentlemen, I would like the ducks you brought back to the trailer earlier if you would, please."

A man with a large, old-fashioned handlebar mustache, who had been one of the two men always bringing back to the trailer the overages of ducks, said, "Sorry, Officer, you are dead wrong. These are all the ducks and geese we have and that is that. We have no more and if you saw someone else hauling out ducks around here, it wasn't any of us. We all are good members of our local chapter of Ducks Unlimited and would never kill an over-limit. That is why we carefully came out this morning with one bird shy our legal limit just to make sure we never violated the hunting laws. After all, we work too hard all year both with physical labor and our donated monies to help the ducks. There is no way we would have killed an over-limit."

Man, that rock-hard denial sure had me in a

quandary. As that gentleman spoke, I kept not only watching his face but the faces of the others for any reflections of guilt or signs that the speaker was lying. Man, they were good. None of their faces reflected anything other than the innocence of purely driven snow and as honest a look as their mothers had made them. In fact, their faces were as expressionless as those on the face of Mt. Rushmore in South Dakota! Using all my years of interrogation and behavioral analysis I had learned at the federal criminal investigators school, various training courses taken subsequently to that federal training, and my work with all kinds of outlaws over the years, I saw nothing that gave me a single clue of their wrongdoing! That is, other than what I had seen earlier in the day.

Then I slapped on their miserable, lying carcasses Plan B. "Then none of you gentlemen will have any problem if I search your vehicles or the house trailer for those extra ducks, will you?" I asked with a "I now have all of you trapped" grin.

The man with the handlebar mustache named Tom Harkin said, "Be our guest. In the meantime, we would like to go inside and brew up some coffee and have some peppermint schnapps to nip the cold we all are feeling. Would you care for any, Officer?" he politely asked.

Now I was fit to be tied. They were so casual, I almost began to doubt what the hell I had seen earlier in the morning. ALMOST! "No thank you on the offer. I have plenty to do today and must get cracking. In the meantime, would the two owners of these vehicles open up their trunks since you gave me the right to search them so I can conduct an inspection." That they did without any hesitation and then headed for their trailer. I found nothing inside the trunks indicating any ducks had been recently stored therein. Well, I still had the trailer, I thought. Knocking on the door to the trailer, I was gracefully admitted and without any further words, began tearing that trailer apart looking for the extra ducks. And I do mean tore it apart. I had worked so many roadblocks and conducted so many searches earlier in my career, that I flat-as-a-flounder knew where to look for contraband under most conditions. Twenty minutes later, I had not found hide nor hair of the ducks! In the meantime, my four shooters just quietly visited and had a good time as they warmed up both outside and inside with the schnapps! Undeterred, I searched everything outside the trailer I could find that could have held so many illegal ducks. Nothing greeted my eyes!

Walking back inside the house trailer, I said, "Gentlemen, I can't find where you have hidden the ducks. Now, unless you want to tell me where they are so we can end this foolishness, I will have no choice but to issue each and every one of you a federal citation for a joint over-limit of ducks, to wit, 31 over the limit. Additionally, I will seize your other ducks and shotguns and hold them as evidence as well."

It was then at that point that human explosions shattered the quiet of the trailer. Every man now denied any criminal involvement, began calling me every name in the book, was going to the newspaper about my high-handed Nazi tactics, going to call their Congressman, wanted my badge number, wanted the name of my supervisor, and all the rest of the "kitchen sink" litany of "standard bomber turn" threats thrown in as well. Letting all the "I am as innocent as the driven snow" bellowing die down, I requested all the men's driver's licenses. More explosions of innocence flowed even more freely than the contents from the half-empty bottle of peppermint schnapps that had flowed earlier.

With driver's licenses in hand, I said, "One more time, Gentlemen. Show me the ducks and I will issue all of you citations just reflecting your individual overages. However, no ducks and I will cite every one of you for the collective over-limit and you are looking at a hell of a lot larger fine if you want to go that way."

Now the invectives really flew and I was called every name in the book except a "white man." However, since every man held his ground and my German "stubborn" was holding my ground as well knowing what I had witnessed earlier, I began issuing every man-jack of that group a federal citation. As I did, comments like, "We will see about this; The Governor will hear about this; My Congressman will hear about this; We will beat your ass in any court of law over this;" and the like flowed from the surly, now very heavily peppermint schnapps-covered lips of my duck and goose killing "innocents."

But I stuck to my guns. I knew what I had observed and was prepared to testify accordingly. I also knew it would be an uphill battle in federal court with me and my testimony being pitted four against one, but my German stubbornness had taken over and a good fight on the part of the critters was just part of the job to my way of thinking.

Finishing up with the paperwork, I seized all their shotguns and the remainder of their ducks amidst more howling and personal name calling. Now instead of desert-calm faces, I was met with evil sneers by the hatful. It was obvious these guys had had their ways in the waterfowl world of wildlife without officer interference for many a year previously. And Ducks Unlimited members or not, I knew what I had seen and these lads were guilty as sin! The ducks may not have had a voice that day but I sure as hell was going to have mine before it was over. Loading their shotguns and ducks into the trunk of my patrol car, I slammed it shut and walked to the front door of my car. Getting in over the still-flowing threats of having my job and suing me for everything I had, I shut the door and started up my engine.

Putting the car into reverse so I could leave the area, I started to back up and then I stopped! Putting the car back into "park," I shut off the motor. Then getting out, I stood for a long moment looking down the driveway. Once again, I looked hard at my four lads standing on the porch of the trailer. Then once again I looked past their trailer... When I did that the second time, the yelling and cursing at me stopped as if cut off by a knife. It was at that point that I kicked myself internally for not doing my job correctly the first time around. No two ways about it, my initial search for the birds had been flawed. Slamming my car door with purpose, I strode down alongside the house trailer like I knew where I was going. As for my two guardian angels, they were now humming with glee over the thoughts racing around in my thick, Germanic skull. It wasn't often I left many holes in my searches but I had on that day. And now I was going to rectify my previous poor search and close the circle. Walking eastward, I now really noticed just how quiet my accusing detractors had become back at the house trailer.

As I continued walking along on my mission, I put into practice some of the things my dad, Otis Barnes, had taught me when I was a young man. Things he had picked up as a young man when he was raised alongside Maidu Indians in California. I hadn't gone 20 feet when my eyes spotted a very welcome sign that I had been hoping for. There were fresh but dried drops of blood scattered along the driveway gravel. Drops of blood that more than likely had come from freshly gutted ducks being carried along the driveway. And those drops of blood were leading straight to the crapper or outhouse! A place I had not bothered to search when I had scoured the trailer and its surroundings initially. But why would one search a crapper? It was obviously not a place where one normally stored ducks...

Stopping at the door to the outhouse, I turned and looked back towards the four now very quiet men standing on the porch watching me intently. In fact, they all looked like a family of muskrats on top of their muskrat house, watching a deadly and hungry mink at the bottom looking up at them. At that stage, they spoke not a disparaging word but appeared to be holding their collective breaths!

Without a word, I flung open the door to the outhouse and looked in expecting the missing ducks to be there in a pile. There wasn't a duck in sight! Damn, I thought as my heart stopped! I was sure that was where they had hidden their extra ducks knowing no one in their right mind would look in a crapper for a mess of ducks. Especially a badge carrier on the hunt.

But then my sharp-eyed guardian angels directed my eyes downward and there it was! The slight trail of dried blood drops discovered earlier running across the gravel of the driveway was still there. Only now, the drops were drizzled across the plywood floor of the crapper. Then the few blood drops obviously from freshly cleaned ducks, ran up the front side of the crapper holding the seat, across the seat, and then disappeared into the bore "hole." *Surely they hadn't thrown the extra ducks down the crapper hole*, I thought as I bent over and looked therein. Nope! Nothing there lying on top of the pile of numerous unrecognizable, second-hand meals now discarded! Except... "that" and a rather large pile of fresh duck guts!

Then lifting up the toilet seat that had been hinged over the bore hole, I looked down once again. That time I was rewarded! Hanging there on five big nails that had been driven into the outhouse from the outside, hung stringers of the missing, freshly killed and gutted ducks! Grinning from ear to ear, I reached down into the crapper hole and pulled out stringer after stringer of freshly killed ducks. Every time I latched onto a fresh stringer of ducks, I tossed them out the open outhouse door for my "four muskrats" to see what the "mink" had found. Finally emerging from the outhouse with the last stringer of illegal ducks, I looked back at the men on the porch of the house trailer. There stood four of the sorriest, gloomiest-looking chaps one ever saw. Gathering up my loads of ducks, I struggled back to my patrol car under their combined weights. Opening up the trunk, I first removed the four seized shotguns and walking over to my now non-name-calling chaps, I politely returned them. Returned them since I did not now need them in light of the strings of evidence ducks I now held. Then walking back to the rear of my car, I loaded the dead ducks on top of the others previously tossed into the trunk for safekeeping.

It was then the devil got into me big time, especially in light of my lousy original search! *It was only a hunch and a long shot at that but what the hell*, I thought. Now is as good a time as any for a look-see into my hunch. After all, I had observed one clue that had flown over my head like a hummingbird earlier, when I had missed the crapper possibility. And, two wrongs do not make a right.

Removing my Buck knife from its belt carrier, I grabbed a plump mallard from their morning's kill and slit open its crop. Out tumbled cooked whole kernel corn in addition to bits of emergent vegetative matter. Then I repeated the process several more times on mallards, canvasbacks and redheads. In every crop there was freshly cooked from a can, whole kernel corn! Then shutting the trunk and without saying a word, I walked around behind the trailer house once again to the garbage pile. Thereon laid over thirty empty cans of Del Monte whole kernel corn! And surprise, surprise, there were no other cans of different types of canned food normally eaten by a bunch of hungry men in a clubhouse in that pile. Gathering up several cans as evidence, I headed back to my car. Those cans were also placed into the trunk of my patrol car without a word being spoken.

Grabbing up a metal-screened scoop used to find illegal bait in a watered area, I closed the lid to my trunk. Then without a word, I headed back to the men's duck blind out on the lake. Twenty minutes of dragging the bait scoop in and around the men's decoys revealed numerous kernels of cooked, whole kernel corn! Placing the whole kernel corn, something not normally found in a lake, into one of my empty sandwich bags I always carried for evidence, I headed back to my house trailer full of wide-eyed but now very quiet chaps. Back at my car, I opened up the trunk and replaced my bait scoop into the trunk and shut the lid.

Walking back to my vehicle's front door, I opened it and just before I entered, I turned and faced my four chaps. "If you lads want to fight this in federal court in the Twin Cities, be my guest. I am sure the federal judge and the United States Attorney will be amused at the degree you folks went to hide your illegality. In fact, when this episode of the 'ducks hidden in the outhouse crapper' hits the sports pages of the Tribune along with your names, you folks may have some tall explaining to do with the other members of your Ducks Unlimited chapter. Oh, by the way, I also discovered that you chaps were shooting over a baited area. As you well know, being good members of Ducks Unlimited and all, that is also a violation of the Migratory Bird Treaty Act, and all of you will be charged accordingly." With those words, I boarded my "iron horse" and left for other "hunting the duck hunter" adventures in western Minnesota.

I never saw my "calm as a mess of dead fish until I looked in the crapper" chaps ever again. They decided one and all, to collectively forfeit bail to the tune of $1,000 each in federal court for their massive over-limits and shooting over a baited area violations. The ducks went to needy families and my boss Bob Hodgins never tired of telling that story to anyone who would listen. Especially the part where "Big T" was all bent over, all 320 lovely, skunk-smelling pounds of him, with his head down the bore hole scooping dead ducks out from the old time crapper. That was one time when one of my farfetched adventures into the wonderful world of wildlife didn't end up turning to crap...

## BEING "SWOOPED" AND "SWOOPED"

During my years as Assistant Special Agent in Charge in Region 3, I had occasion to observe firsthand much of North America's world of wildlife. In that particular district, I had just about every kind of critter known to man on the North American continent from hellbenders (a type of very large and rare salamander) in small streams in Missouri, to muskrats in about every wetland in western Minnesota. So whenever I was able to get out and about in my district, I always made it a point to get reacquainted with "my" critters. In so doing, I always kept a sharp eye peeled for several of my very favorites, namely, members of the eagle "clan." And in those days, I kept a particularly sharp eye on the golden and bald eagles because both of those species were in trouble. The golden was trapped and poisoned at what seemed to be every turn in the road, and the bald, in addition to having problems with DDT, always seemed to be a target for those with a heavy trigger finger.

While on my week of escape from the Regional Office, I had chosen to work in the western part of Minnesota. Waterfowl season was in full swing and I wanted to get back to some old-fashioned waterfowl law enforcement. Hence, my miserable carcass being on that detail in the Fergus Falls area of western Minnesota. I had been quite successful working the run-of-the-mill duck hunter, making many cases from early and late shooting, taking over-limits of ducks and geese, as well as taking migratory waterfowl over a baited area. However, it was in the Pelican Lake area south of Fergus Falls, that I had a strange run-in with a wildlife outlaw like I had never met before. A sad day really, one which I hope I never have to repeat...

Bailing out of bed one morning before dawn, I feasted on my usual game warden breakfast, as I called it, of cold Italian salami and a Diet Coke. The evening before I had seen lots of ducks and geese working the fields and wetlands of the Pelican Lake area, and decided that is where I would put some of my time while on this particular outing. However, I had been on the road long enough and was really missing my family. So, I figured I would work the morning duck shoot and then call it quits from my week-long detail and head for home.

Driving out to the Pelican Lake area where I had heard a lot of late shooting the evening before, I carefully drove down a prairie trail towards the area in question. "Carefully" I say, because I had a bad habit of "finding" abandoned harrows in the tall prairie grasses with tire flattening results. Finally parking my patrol car near several other vehicles, I got out, stretched, and attended to a call of nature. Then gathering up all my gear, headed out for the nearby lakeshore on "shanks mare." Walking about a half-mile, I finally lodged myself in behind several sets of duck hunters as they quietly put out their decoys. Both sets of hunters were about 100 yards apart, so I just positioned myself in between the lads figuring whoever broke the law first, if they did, would get my immediate "fatherly" attention.

I didn't have long to wait. The ducks were flying heavily between the many lakes and wetlands in the area and as luck would have it, both groups of hunters were having a great day in the sport of waterfowling. Soon, I was fairly humming when it came to recording the kills from my two groups of hunters. However, I noticed that in both groups, they were careful to avoid shooting at those ducks skimming low over the water since most were of the diver variety. And as such, were hard to identify and a few were of the protected variety. That is good, I thought after a while. Here I have two groups of good waterfowl hunters who are acting responsibly when it came to their sport. And for that bit of good sportsmanlike behavior, I was pleased.

Then I got the surprise of my life. The group of hunters to my north got into a small flock of Northern pintail and managed to drop three birds. Instead of walking out and retrieving their birds, they let them lay in the water where they fell. Then I saw why. One of the birds had been shot in the head but not killed. In its dying moments, it continued swimming aimlessly around in circles in the decoys. In so doing, it created ripples in the water just like normal feeding ducks will do. And to other still-flying ducks in the immediate area, that was a pure magnet to their way of thinking. It must be safe to land there since one of our "friends" is still swimming there with those other ducks, they must have thought, and soon, here came more small groups to nature's dying "lure."

As those other ducks came into the area decoyed by their dying, swimming, head-shot brethren, I saw my duck hunters hunker down for the shoot to come. Boy, did they and yours truly get the surprise of our lives! I heard a loud swoosh, and then out of nowhere came an immature bald eagle. The eagle in a thunderous dive, pulled out at the last minute, dropped his talons, swooped up the dying swimming pintail, and then struggled for more air as it lifted the duck up and out of the water. Fascinated by what was happening, I was intently watching the eagle do what it did best, when all of a sudden, I saw it explode into a *pufff* of hundreds of feathers! It was like someone had emptied out a feathered pillow where the eagle had once been! Then I heard a loud boom-boom coming from my blind of shooters!

Dropping the field of view of my binoculars to include the duck blind, I saw one of my duck hunters now standing up with his shotgun to his shoulder. Then he fired one more time, boom! It was then I saw the struggling eagle drop heavily into the chill waters still clutching the once-wounded pintail! God, I was shocked! That dumbhead had just blown an eagle out of the sky in front of God and one of his special agent minions! I thought. Needless to say, it took a moment for me to get my breath back and then lurch into action. Jumping out from my hiding place, I let out a yell that could have been heard clear back to the Twin Cities. Then I sprinted down the hillside on which I had been hidden just moments before. Not watching where I was stepping, I stepped into a patch of soft gopher-worked dirt and took a head-over-heels high-speed tumble. Down over and over I rolled, until I came to a smashing rest against one of those damn big prairie rocks left over from the days of the glaciers. Wham, I went, and let me tell you, that stopped my rolling down the hill... In fact, I hit that boulder so hard, it knocked my .45 clear out of my holster as well as the wind from me like it hadn't been in some time. Finally getting my wind back and picking up and putting my .45 back into its holster, I got to my now-wobbly feet. The first thing I saw were my two surprised duck hunters below me staring back up at me. Moments before it had just been them, the duck hunt, the swooping eagle, and the prairie winds. Now they had a bellowing water buffalo-sized human charging down the hill at them, then fly-ing through the air head over heels, and finally smashing a prairie boulder all to pieces when he hit! Then surprisingly, the water buffalo had risen from his collision with the prairie rock and was still coming their way. And if that wasn't enough trouble, the other two duck hunters down from their blind were coming their way now and yell-ing like idiots as well... That unreal scene sure had the previously quiet prairie now on edge.

The other two duck hunters who had heard

and seen my chap shoot the swooping eagle, arrived at my suspects' blind just moments before I did. Man, they were steaming hot! They were calling my two duck hunters every name in the book and then some for killing the eagle. Then as one of my mad-but-uninvolved duck hunters went out to retrieve the now-very-dead eagle, I announced to my shooter as to whom I was. Man, you talk about a sick look on his face when confronted by two other mad duck hunters and the "badge-carrying, prairie rock-smashing, water buffalo-sized" human. It was obvious from the look on his face that his duck hunt had just turned to that smelly stuff horses leave on the highway after going through a town during a Fourth of July parade!

Suffice to say, it took a while for my other two uninvolved duck hunters to cool down. Both were Audubon "birders" and they were hot to say the least! Especially when they discovered the eagle was an immature bald eagle, one protected by the Migratory Bird Treaty Act, the Bald Eagle Protection Act, and the Endangered Species Act in those days! Untangling the mess, I took down the information from my two "birders" so I could contact them later if needed. Then as they returned back to their duck hunt, I attended to my eagle shooter. You know, the one who had his duck "swooped" by the eagle, then he "swooped" the eagle with his shotgun and several loads of number four shot. Then he in turn was finally "swooped" by two mad birders and one slightly bruised "water buffalo" type, along with his less-than-intact Special Agent prairie rock-smashing ego...

Needless to say, my lad, a local farmer's son home from college (no, he wasn't a wildlife major), realized his homecoming was one which he would now really remember! Taking down the information in my notebook as to his identifiers, I seized the eagle and after a slightly greater than life ass-chewing, took my evidence and headed back to my patrol car. Damn, it really hurt to be packing out a dead juvenile bald eagle instead of a hatful of citations and a handful of evidence ducks. But sometimes, one eats the bear and sometimes he eats you. That morning pretty well put the "hurt" on my week away from the office roaming the wilds like I used to do when I was a real agent instead of a "tweed-backed" bureaucrat supervisor...

Back at the car, I carefully bagged my eagle so it could be later used as a mount or for use by Native Americans for their cultural or religious purposes once the case was closed. Then digging out a Diet Coke from my ice chest, I had a quiet moment out on the prairie soaking up the weak sun's winter rays. As I did, I enjoyed the historical quiet the land offered to my soul. Then I loaded into my "iron horse" and headed out across the prairie towards Interstate 94 and home. Little did I realize that my week's goofy adventures in the outback had not ended...

## MOONING CAN BE A DISADVANTAGE

Back out on the Interstate, I headed for home. Soon the miles, previously spent long hours, and the prairie cold began taking its toll. Moving down the highway, I slipped into an internal quiet. Mile after soothing mile rolled by and my thoughts were of home, my wife, kids, some down time, and a great home-cooked meal.

Then I observed a car roaring by me going about 90 miles per hour. I noticed the license plate read "Montana" and contained four individuals inside all having a good time under their large black bull-rider cowboy hats, if their animation said anything as to the obvious...

Once again, I drifted off into myself and just enjoyed the quiet warmth of my patrol car as I headed home from a pleasant week in the outback. Then I noticed far down the Interstate, my "Montana" car abruptly braking and pulling off to the side of the freeway. That was when I noticed all four of its riders rapidly bailing out from their vehicle and walking towards the back of their car. As I approached in my unmarked patrol car, all four of my tough Montana lads lined up abreast blocking the slow lane of the freeway, dropped their pants, and "mooned" me as I drove by. That sure brought me out from my quiet internal relaxation.

Then the devil quickly took over in me as I slammed on my brakes and skidded down the freeway. At the same time, I hit my siren and man, did that howling son-of-a-gun wake up the surrounding countryside. In fact, it spooked a hen and rooster pheasant out of the nearby borrow ditch running alongside the freeway and sent them airborne in a flurry of feathers. Then I slammed my car into reverse and careened rapidly back towards my four tough Montana cowboys, who were all now hurriedly trying to pull up their britches. Then as I slammed to a stop in front of their car, I noticed two of my tough Montana cowboys madly hopping around like a couple of bugs on a hot rock. Stepping out from my car and quickly flashing my badge, I observed a circus of sorts. Two of the cowboys now had their pants up and presented me with the sickest, not-so-tough-guy grins a body ever saw. As for my other two tough Montana cowboys, who as I went by really didn't have much to show for their "mooning" efforts, were still hopping madly around with their pants halfway up and howling like a pack of happy to be on the trail of a Yellowstone elk, timber wolves. By now, the freeway was clogging up with other car traffic all looking on at the scene presented by four of Montana's finest.

Then I saw the problem. My first two tough guys who had managed to get their pants up and fastened, had peed all over themselves in terror when the siren must have gone off. Brother, were they a wet-looking mess! As for my other two Montana "hop rabbits," well, let's just say they had a different sort of problem. When the siren had exploded their little happy Montana cowboy moment, they had zipped up their pants like they were on fire. unfortunately, they now had other more serious problems. What they had shown me hanging down as I steamed by during their "mooning moment," had now been unceremoniously zipped up painfully in their zippers! And what they had displayed so proudly for me as I drove by, was now being seen hung up in their zippers by stopped carloads of families and the like looking on. I must note for the record, my tough Montana cowboys were sure getting a lot of laughs over their wetted fronts and zipped-up beings. Finally, two of what I considered the most unfortunate ones, gave a huge tug and with a bloody (in more ways than one) howl, got their pants zipped up...

Looking over my now bedraggled-looking cowboy lot, I smiled saying, "Guys, I just wanted to stop and say I have seen better-looking plucked turkey vultures than what the four of you displayed here this morning for all the world to see. If that is all you Montana folks have to show for prowess, I can see why all of you live in an area that is so far back in the bushes the chickens have square faces." With that, I re-boarded my "iron horse" and set her on an angle for home. *God, it was sure good to be alive and living in America,* I thought.

Come the spring of 1981, in addition to my daily office "surprises" in Region 3, I got another surprise. Only this time one that I had been hoping for years would come my way. Early on I had decided I wanted to be a Special Agent in Charge in the Service's Division of Law Enforcement. However, there were only 13 such positions (now reduced to eight) and it seemed the guys now holding such positions never left, except feet first in a box if you get my drift...

"'Big T,'" I heard my boss Bob Hodgins yell from the adjoining office, "get your tail end in here." I had screwed up once again and had been caught by Bob's eagle eye, I hustled my miserable carcass into his office.

"Yeah, 'Hodge,' what do you need," I said rounding the turn into his office.

"Sit down, 'Big T,'" he said in his usual fatherly tone of voice. "I just got off the phone with my counterpart in Denver, Harry Styles. He just informed me he has had enough and is stepping down. He figures his position will be advertised immediately and was letting me know in advance because he figured you would be a good one for the job. Are you interested?"

Man, my heart went clear up into my throat! *Hell, yes, I was interested!* I thought to myself as I saw Bob's knowing smile run across his face. A smile that damn-well knew I wanted my own district and had known so ever since he had hired me to give me Assistant Special Agent in Charge experience. Experience so I could successfully compete for just the Special Agent in Charge position of my dreams if and when one ever opened up. Then the wheels of reality spun into action. I had arrived in Minneapolis in February of 1979. It was now March of 1981. Hot damn, I thought. I will have just made the minimum two-year requirement as an Assistant Special Agent in Charge in order to qualify for the Special Agent in Charge position! Man, talk about the luck of the Irish...

For the rest of that morning, Bob and I talked about what I still needed to accomplish in the region in case I got the Denver job once it became available. Then hardly able to wait until the end of the day, I hurried home so I could share the good news with my long-suffering Bride. Suffice to say, Donna was 100% in my corner and looking forward to our next family adventure. Starting that evening after supper, I began the long process of getting together all my job experiences so I could put them to paper in a resume for the Denver Special Agent in Charge position.

Over the next couple of months, I waited for the final verdict from the selecting officials in the Denver Office of the Regional Director. Then the waiting was over. I was called by Deputy Regional Director Bob Shields from my new region (Region 6) and advised that I had been selected. I was also given a date to arrive for duty as the newly selected Special Agent in Charge in the middle of June, 1981.

One of my life's goals was to become a Special Agent in Charge in the Fish and Wildlife Service. Another of those goals was to be selected as the Special Agent in Charge for the Denver Office because of my love for the mountains and that particular eight-state region. And the last life's goal I had was to make Special Agent in Charge before I turned 40. That doesn't sound like much of a goal, doing so before the age of 40, unless one looked at all the other 13 Special Agents in Charge splashed across the country. The youngest one was in his late-40's! I did not turn 40 until

June 22 of the year. I had just made my final life's goal! I was now the youngest Special Agent in Charge in the Nation at the ripe old age of 39, having accomplished such a lofty goal in just eleven years of federal service! Little did I know of the boar's nest I was about to enter!

## ROCKY MOUNTAIN REFLECTIONS

### THE TERRITORY

After arriving in my new duty station in Region 6, meeting all the members of my staff and other division heads in the Denver Regional Office, I headed for the quiet of my new office. Sitting there quietly in my chair behind my important-looking, surplus military issue, beat all to hell when Patton took it all across Europe during the Second World War on top of his tank, steel desk (ha), I took a look at the map of the United States hanging on my wall. Then the significance of my new duty station and its responsibilities began fully manifesting themselves on my miserable carcass.

I was now in charge of the U.S. Fish and Wildlife Service's wildlife law enforcement operations for the States of North and South Dakota, Nebraska, Kansas, Montana, Wyoming, Colorado and Utah. A geographical area comprising about 750,000 square miles or an area larger than the State of Alaska! And as it turned out, the largest Service region in the United States! If the realization of that didn't 'sink one's bobber,' the fact that I had only 24 officers to face all my challenges in such a large and natural resource-rich region sure as hell did!

## OPERATIONAL PROBLEMS

As previously mentioned for an area the size of Region 6, I had only 24 Special Agents with which to do the job. Having had 11 years of prior service with the agency, I had a pretty good idea of what was really needed and the qualities of the personalities now working for me. Based on my personal working relationships with most of the lads in my district, I had what I considered less than ten hard-chargers out of the 24 agents on board! The remainder either had to pick it up in order to earn their paychecks, or to my way of thinking, find themselves another district in which to dwell. As for my six first-line supervisors or Senior Resident Agents, I had what I considered only three top-notch officers in the bunch. Two of the remaining three Senior Resident Agents were as useless as teats on a side of bacon! The remaining officer of that rank never should have been hired by the Fish and Wildlife Service as a badge carrier. Plain and simple, he was unfit for duty as a law enforcement officer. That supervisory situation alone told me I had one hell of a row to hoe in front of me...

Then because of the goofy way the Service funded our work, it only provided budgets for types of work in certain priorities in those days. Priorities for the most part, to my way of thinking, that did not take into account my law enforcement district's actual needs and protection requirements. In short, many of my new staff were basically sitting on their hind ends drawing paychecks and not doing the work needed or required for the district!

For example, one major work need not addressed by the existing budget was the region's pressing Lacey Act concerns. The Lacey Act basically provides for federal prosecution of those who take wildlife or plants in violation of other federal, state or tribal laws, and transport said items into interstate or foreign commerce. On any given day during the hunting seasons in my eight states of responsibility, two-thirds of the Nation's non-resident hunters visited therein doing what they did best. The way the Service provided funding, my region, a true Lacey Act region, received zero dollars for providing enforcement oversight concerning such activities! As a result, my officers were not allowed to spend any money or time conducting such types of interstate or foreign investigations! And the agency's stupidity didn't end there...

Looking over the property lists on the types of equipment my officers possessed, I discovered in one of the toughest geographical parts of the country to get around in, only one officer had a four-wheel-drive vehicle! All the rest had basically useless one wheel drive sedans to get around in the mud, snows, blizzards, and mountainous high terrain found commonly in the region. And most of those vehicles were traveling death traps because they had far exceeded their agency safe mileage requirements!

My equipment problems did not end there. I had a polyglot of vehicle radios, mostly from a vintage, hand-crank, steam-driven era. Hand-held portable radios were non-existent and there were few boats for water use, and of those, all were death traps because of age or inadequate built-in flotation. My outboard motors were museum specimens, most of my officers' optics were of World War II vintage surplus (one officer had a pair with only one working ocular), and I lacked any horses or horse trailers for needed backcountry work protecting endangered species such as the grizzly bear, nesting peregrine falcons, bald eagles, or wolves.

Looking over my remaining budget for the month of June, I discovered I only had $34,000 left on which to operate for the next four months. And that had to include salaries and benefits for all my staff and field officers! Hell, if we did our jobs correctly, we would have burned that much funding in gasoline alone! And that did not include any money for vehicle repairs, per diem, tire/vehicle replacement, telephones, rental costs for office space, new law enforcement equipment needs, office supplies, training, and the like...

In short, I never had enough money provided in my district's budget come the first of the year to even do the job! To emphasize that point, several years later when the budget was announced on October 1 (remember those years when Congress was able to get off their lazy hind ends and pass a budget as required by law on October 1?), I found myself for two years straight on the first day of each new fiscal year, flat broke! Hell, I couldn't even pay salaries for all the people currently on board. And that was with agents and staff who were under-graded salary-wise by the Service by at least one grade!

To my way of thinking, that funding problem for the Law Enforcement Division was easy to explain and understand. The Service leadership never did, during my 28 years of agency employment, understand how to use law enforcement as one of the prescribed five wildlife management tools. They basically hated the idea of funding law enforcement, came from colleges where the need for law enforcement was poorly understood and its values seldom taught to budding wildlife managers, were afraid of law enforcement operations' political implications, had been pinched by law enforcement officers sometime during their careers for wildlife violations, or all of the above...

Then I had various state fish and game agency 'sour as owl droppings' politics to deal with regionally. The state fish and game agencies in South Dakota, Kansas and Colorado were pretty much on-board and worked well with the Fish and Wildlife Service. As for my five other state agencies, I had a lot of work to do and numerous fences to mend. Fences to mend because my agents had not done their jobs correctly or professionally; because of our non-existent budgets we were not considered equal players in assisting the states with their problems; because of my agents or state counterparts and the hell-raising between them because of numerous personal pissing contests; lack of respect for my officers or their mission; intense states' rights issues or as in the case of North Dakota, the Chief of Law Enforcement just plain hated my guts. Suffice to say, I had a hatful of problems to overcome if the job was to be done right and we were to put the game hogs and other wildlife outlaws on the run.

As far as the Chief of North Dakota and my issues, I will now let my readers decide. When I was the Senior Resident Agent over North and South Dakota in the '70's, I decided we needed to run an Interstate Highway fish and game roadblock and compliance check during the fall hunting seasons. A decision made after I passed hundreds of vehicles on the Interstate Highway system in North Dakota displaying evidence in plain sight of numerous big game trophies heading east. I decided the roadblock could be most efficiently run in western North Dakota. There because of all the wildlife hunter traffic passing through on the Interstate to the east with few side roads to run down and escape once they discovered the presence of the game check station ahead. As a courtesy, I contacted Warden Dick Knapp, the Dickinson warden, and informed him of my plans and requested his assistance. He demurred. He did not consider wildlife violations committed in other states as any of North Dakota's business. Once again, I advised we would be working under federal authorities and I planned on swinging all issued citations through his state court so the fine monies could benefit the local rural school system accounts and the North Dakota Game Warden's Retirement fund. Once again, he demurred. Bottom line, he just didn't consider it any of his law enforcement business once a violator left the state of wildlife violation and entered the State of North Dakota.

Weeks later, a small core of my agents, North Dakota Highway Patrol Officers and local-credentialed refuge officers ran the first major fish and game inspection station on Interstate 94 near Dickinson, North Dakota (Warden Dick Knapp's home district). As history would have it, we got blown away by the volume and severity of violations flowing into North Dakota from Oregon, Washington, Montana, Idaho, Canada, and even Alaska. In fact, I finally had to ask for National Guard assistance to provide trucks to help us haul away all the tons of illegal ducks, geese, grouse, salmon (Idaho and Alaska), trout, bear, elk, deer, bighorn sheep, and moose that we seized. Suffice to say, the court in Dickinson got overrun with a ton of citations and not being forewarned or prepared by the local warden (Dick Knapp) as I had earlier requested, was surprised and a little grumpy to say the least. Embarrassed over what had just happened and his lack of interest or involvement, Mr. Knapp, who later became the Chief of the North Dakota game wardens, never forgave me. Hence his tepid support on any future state-federal law enforcement operations until his retirement many years later.

Then I had my share of problems with some of the U.S. Attorney's Offices and various federal courts in the region I had inherited. Some U.S. Attorneys didn't want to gum up their operations with wildlife cases or just flat wanted the 'Bishop's Court of the Mormon Church' (Utah) to handle federal cases! As for the courts, I had my share of problems running from getting bail schedules established in their jurisdictions like Nebraska so the officers could issue citations, to having one federal judge calling me in and telling me he did not want any wildlife cases in his court, period! Realizing that is where I had to try my federal cases to get the "biggest bang for my tax dollars being spent," I persisted with the old and cranky federal judge. In my pleas, I advised that is where I had to bring the majority of my cases because of their federal nexus. However, he persisted in advising me to go elsewhere. When asked where else I could go, he got really grumpy and advised he didn't care where I took my wildlife cases but just not into his court. Being that his jurisdiction was an important one in that a lot of future serious federal wildlife cases were going to come out of Wyoming when I finally got some better agents on station, I had to keep trying. That was when Wyoming U.S. Attorney David Frudenthal (later Governor) came to my rescue. Soon, because of his dedication and political skill, our cases were moving through the federal court systems in his state. As a sidebar, I eventually got Agents Bob Prieksat and Jim Klett into Wyoming. Between those two officers, they flat overloaded the Wyoming U.S. Attorney's Office to the point that Attorney Frudenthal had to hire extra secretaries and another Assistant U.S. Attorney just to handle the wildlife caseload! Pretty good payback for all of Dave's efforts in helping me with the cranky federal judge, I would say.

Then came the field work. I had hundreds of thousands of acres of federally controlled or administered wetlands in eastern Montana, North and South Dakota, being illegally drained by the local dirt farmers. Then I had a number of ranchers, especially those raising "range maggots" (sheep), putting out every kind of poison known to man and accessible over the counter, to kill coyotes, eagles and every other kind of wildlife heartbeat on the western ranges. Utah, Montana, western Colorado, western South Dakota and the Wyoming killing fields quickly come to mind in those problematic arenas. Then I had numerous cattlemen, guides and outfitters "shooting, shoveling and shutting up" when it came to killing almost every remaining grizzly bear in Montana and Wyoming. That was followed by Native American violators from numerous Tribes throughout the region shooting, poisoning and trapping eagles and other birds of prey for the interstate and foreign feather markets, as well as for feather replacement in their commercial dance costumes.

Then came the region's eight-state outlaw "air forces." Come winter and high fur prices, it seemed every rancher and pilot that had a set of wings and a propeller was out and about illegally aerially gunning furbearers to satisfy the demand for animal skins in the high-priced fur markets. One year alone, agents in the region seized and sold through the General Services Administration, 16 court-forfeited airplanes as a result of their pilots being apprehended and convicted for illegally gunning wildlife from the air in violation of the Airborne Hunting Act! And during one of those years, we caught a Montana helicopter pilot illegally gunning coyotes on the Charles M. Russell National Wildlife Refuge. That pilot's previous monthly gross receipts, exhumed during an executed search warrant, showed gross profits from such illegal activities exceeding $209,000! Those caught violating the Airborne Hunting Act also included outlaw federal employees in the old Animal Damage Control Division of the U.S. Fish and Wildlife Service, who subsequently quickly lost their jobs!

Hard on the heels of the aerial gunners in the region came the many thousands of big game, upland game bird, fish and waterfowl resident and non-resident sportsmen. It always amazed me, even after all the years I had experienced the illegal take of wildlife, the tons of wildlife seized by my agents and state badge-toting brethren from the above minions for illegal take and possession of sport-taken wildlife.

Hard on everyone's heels followed the oil field crowd. There are oil and gas wells in every state of Region 6. Accompanying those drilling sites by the thousands are the uncovered oil pit death traps. And into those many-times-uncovered oil pits went millions of migratory birds (especially night migrants), along with thousands of small and large mammals monthly! I later discovered through the hard work being done by some of my agents, numerous, recently drained oil pits, that were feet deep in bird and small mammal bones!

Then every time I throw an electrical switch in my home, I was reminded of the tens of thousands of miles of transmission lines in Region 6. And with those memories, I am reminded of the hard work done by my agents in patrolling and shutting down or having the companies reconfigure their transmission lines, so those dangerous fields were not doing their deadly work. In fact, I can remember one set of transmission lines in the Buffalo Basin of Wyoming that were killing dozens of migrating eagles in the fall and winter months on every mile of their lines!

Then let's not forget the hard-working miners in the region. Especially those working the open-pit, heap-leach gold operations. As my agents soon discovered, it was amazing the number of migratory song birds that paused to drink at the edge of the cyanide-laced heap-leach pits and never survived (Migratory Bird Treaty Act violations prohibiting illegal take through use and aid of poison)!

Then throw into that already overwhelming mess to be addressed by only 24 Special Agents all the illegal dove and waterfowl baiting, illegal commercial goose clubs shooting overlimits, Native Americans running illegal gill nets in Lake Oahe in South Dakota over the years, spotlighting of big game, scooping of raptor nests by unscrupulous falconers, pole traps set out to catch and kill raptors, sale of eagle feathers and their many thousands of parts and products, importations from Canada of fish and wildlife, circus animals smuggled across the borders of Montana and North Dakota and everything else in between. With just a sprinkling of the problems associated with the world of wildlife, I guess my readers can get an idea of the tasks facing yours truly when he inherited the position of Special Agent in Charge in the Denver Office.

But the misery didn't just end there. When I arrived, I discovered I had office problems as well. With one personnel exception in the Denver office, I had folks who were used to working under rather loose professional rules. It seemed arriving late and going home early was the norm. It also seemed daily ten-minute breaks had turned into half-hour breaks, and one-hour to one-and-a-half-hour lunch breaks were also the norm. Looking into the office production, I soon discovered that my office was considered the worst in the Regional Office and from their output, that was borne out in spades. For the next two tough years, I slowly documented and weeded out those less-than-desirable federal employees through imminent threat of being fired or outright release. All but one office staff employee that is. Today, that remaining little gal is a GS-12 supervisor at the National Eagle and Wildlife Property Repositories. When I arrived in Denver, she was a GS-1! Yes, federal employees can be moved on or fired, but it took just about everything I had in me and then some to get it done.

By now, some of you are beginning to think instead of that Special Agent in Charge position being my dream job, it was more like a nightmare. In some aspects you would be right. But in all the rest, it was a golden riot, especially once I managed to get a lot of good folks on board. Even more so when the bad guys began trooping by in handcuffs in streams with sour looks on their faces. And collaterally, when some of the animal populations, like the grizzly and bald eagle, began to rebound and increase after intense backcountry protection efforts. But, I digress...

## COVERT OPERATIONS

I am of the opinion that every serious law enforcement operation must have a covert operations component. Plain and simple, apprehending some of the worst violators requires the best investigative techniques to be applied, and many times to achieve those goals means the use of covert operations. Not like you are used to seeing on TV but real-life operations that require utilization of the agency's finest capabilities in order to reach out and snatch those running close to the very edge of the dark side. And in so doing that also requires agency covert "gumshoes" to run at the edge of darkness as well, in order to grab off the illegal "ridge runners." That doesn't mean just because one is going covert, agency officers can run amok like the bad guys. Hell, no! Covert operatives are bound very rigidly by agency guidelines, state and federal laws, a generous helping of common and survival sense, Department of Justice and Constitutional requirements, and everything else you haven't thought of, all rolled up into one.

Additionally, becoming a successful covert operative requires many innate properties. The individual so inclined to get involved in such dangerous and lonely assignments must possess some truly unique human facets in order to survive. First of all, they must be physically and mentally tough. That is because many times covert operations are intensely run for many months at a time, and sometimes even years. If one is to survive the rough and tumble living conditions, eating poorly and living alone under extreme stresses, he or she had better be tougher than a horseshoe nail! The operative must also be of such size, shape, nationality and commonness to blend into an illegal crowd without drawing any attention to themselves. To my way of thinking based on my own personal covert experience as a state officer and outside-looking-in observations as a federal supervisor, they must be a bit of a loner and sharper than all the other knives in the drawer. Then as added ingredients, throw into that mix one who is extremely cool under fire, has nerves of steel, is skilled in the laws of the land, and possesses the "man- or womanhood" to face down a charging bull at ten feet. And in so doing, knowing full-well that the bull more than likely wouldn't stop!

One of the first things I realized upon my arrival in the Rocky Mountain region was that I needed a strong, up-and-running covert program. There were just too many unique and valuable wildlife resources in the region not to attract those wildlife outlaws operating like a bunch of maggots versed in the business of extinction. Regionally, I had world-class antelope, mule and white-tailed deer, elk, moose, grizzly and black bear, mountain lions, desert and Rocky Mountain bighorn sheep, not to mention the already-alleged ongoing trade in eagle feathers, exotic butterflies (at a $1,000 each), antlers (Asian trade), rare birds of prey, and bear parts in the Asian communities. And unfortunately, all those critter types were readily available and numerous in the Rocky Mountain and Prairie regions of my district. That meant it was just a matter of time before I personally noticed the black hand of the business of extinction, thriving like a horde of maggots on a carcass in the summer heat in my own backyard.

Now in those days, the Service had a healthy and robust Special Operations Branch located in Washington, D.C., consisting of a number of outstanding covert operative personalities. Sitting at my desk quietly one morning troubleshooting the numerous problems in my district, I ran across in the annals of my mind the need to establish contact with that branch and a few of her "hunters of humans" personalities. I had two very close friends in that branch, namely Agents

David Kirkland and Nando Mauldin. Both men were top-notch investigators, were smoother than a schoolmarm's thigh when in action, and once they got on your trail, the only way to get them off was to shoot them. I must give each of those officers a call, I mused to myself early on during my Denver career. Mainly to see how they are doing and what their workloads looked like, just in case I would have need of their services. You know, just in case. Well, as luck would have it, I didn't have to wait long to initiate those phone calls.

A week later in early December of 1982, my secretary advised that the Chief of Law Enforcement for Colorado, Kris Mosher, was on the line. Picking up the phone, I greeted Kris and asked what I could do for him that fine morning.

Kris was a quiet, extremely intelligent and very introspective chap, who was not prone to "stampeding." As such his quiet voice said, "Can you get over to my office before ten this morning. I have a very important phone call coming in at that time and would like to have you handy to give me a hand. It looks on the surface like this call might lead to an investigation in which you "feds" would possibly be interested. I don't want to say any more over the phone but can you be here a little before ten so we can talk over this call before it arrives?"

Knowing Kris was a man of few words and had thought this issue through, I said, "I am on my way, Kris." Hanging up, I hurriedly exited the office as I explained over my shoulder to my secretary where I was going and how I could be reached. Twenty minutes later, I walked into Kris's Colorado Division of Wildlife Law Enforcement Office.

"Morning, Partner. What's up?" I asked as I shook his hand.

Getting up, Kris quietly walked across his office and shut his door. Walking back to his desk, he sat down heavily as if carrying a physical load and said, "I think I have the start of a good one, but am just not 100% sure. That is why I asked for you to be here for me. I just figured two heads are better than one in matters like this one. I received a strange call this morning from a fellow who asked if he could have a minute of my time. I advised him that I had a little time but had to go to a meeting shortly. The fellow on the other end of the line sounded real nervous and somewhat hesitant to really open up, but here is the gist of what he said. First, he asked me if any deer seasons were still open in the western part of Colorado. Something in just how he was asking that question really didn't *click* or sound right, setting off the alarm bells in my instinct department. So I played a little coy and said I thought

## 398 | Terry Grosz

there were still some seasons open but since our big game regulations were so complex on that side of the state, I would have to get out a set and check them out in order to properly answer his question. However, in light of my upcoming meeting, I would not have the time to do that right now."

It was there I stopped Kris. "Kris," I said, "you know as well as I do, all big game hunting seasons except a few depredation hunts, are now closed in Colorado and have been for some time."

"I know. But I didn't know what else to say because I still had a gut feeling something was wrong in why he would suspiciously and in such a tone of voice, ask such a question. I just couldn't put my finger on it. So I just figured I would string this person along on the phone until I could see what the hell was going on," he said.

As if Kris needed any support from me, I said, "You did good. I also am getting a gut feeling this fellow wouldn't be asking a question like that unless for some reason he was up to no good. You know, like he had inside information on a possible illegal big game hunt this late in the season for a trophy animal, and for some reason was just checking it out to see what the odds were against getting caught. Especially if all the seasons were closed and one was out and about illegally pulling the trigger. As you well know, Kris, if you want a really big trophy mule deer buck or monster bull elk from western Colorado, now is the time to go out and get one. The season has been closed for some time and the hunting pressure is off the animals. And now the big bucks and large bulls are all relaxed. Especially after the pressure of a hunting season has ended and many of the really big ones are now coming down to their winter ranges where they are readily accessible," I said. "Plus, there really aren't many folks out and about on those winter ranges this time of the year, because of the poor roads, deep snows, terrible weather and really no reason to be out there except for calling and shooting predators for the fur market," I continued.

"I know. And as you are also aware, we have a lot of poaching of those big bucks and trophy bull elk when they come down out of the high country to get out of the deep snow and onto their winter ranges," he said. "Then figuring I was getting over my head, especially if we needed to work this guy covertly, I figured I had better call you. This is more in your guys' arena and could be a federal offense if this chap is really coming out to Colorado to kill a big buck or elk on the sly and possible move it into interstate commerce. That plus, as you well know, we here in Colorado do not have a covert unit or access to one except through you fellows. And I still am a long ways away from bringing such a unit on-line just because I figure I don't have the right kind of people, the mobility of officer movement, and the proper built-in agency security for such an operation right now. So in order to keep this guy hanging in case he wanted to do some illegal big game "slumming" here in Colorado, I told him I had to go to my meeting. But I also told him to call me back in two hours and that would give me time to review our regulations and see which big game seasons we still had open. He agreed and then without another word, hung up. From the sounds of his voice, he really sounded 'hungry' if you get my drift, Terry. It was then I called you to see if you had any ideas on how to work this fellow if illegal killing and interstate transport was what he had on his mind," he continued with his typical country boy grin.

"Well, let's hope if hunting big game illegally here in Colorado is what that chap had in his craw, let's hope he didn't get 'coyote' and scare off when you told him you had to go to a meeting. I'll tell you what. Let's quickly set up for a phone trace (we didn't have caller ID in those days). Then if he calls, you stall him by telling him your meeting ran longer than you figured and you still haven't had time to review the big game regulations. But that you now have them in hand and for him to let you know where he wanted to hunt. Then in the same breath before he gets his 'suspicious legs' under him, ask if he is a resident or non-resident. Hopefully he will slip up and provide that information. Then tell him that will make it easier for you to look up what leftover licenses are currently available and what seasons, if any, are currently still open on the western side of the state. That should keep him hanging for a few minutes longer. Then when you are 'looking up that information,' that will give us even more time for a trace. When we get the trace, I will give you the high sign and you can tell him the Director just called, and needs you in his office right away. Then tell the chap to call you back this afternoon and you will set some time aside specifically to help him.

However, before you hang up, be sure and tell him it appears there are several depredation hunts open out west to keep his interest up, but you will need to check for antler restrictions and such when you get back together in the afternoon. Then hang up. Then what I will do is call that number after we get it from the trace and try to find out any more information regarding the actual identity of the caller and the like. We can also tell during that time if the number used was from a commercial business or a home phone," I said as my mind raced for ruses to use in trying to get the identification on the caller without spooking him off. Then when the number of the caller became known through the trace, I would make contact either personally or through one of my covert operatives and ascertain the reason the chap called requesting such information. And in the event he was just being cautious after being contacted, say by an outside party trying to set up an illegal big game hunt in Colorado, 'whip saw' the lad. That is, scare him to death about getting federally prosecuted for such behavior and then hopefully procure from him the information on who called him regarding an illegal hunt and ferret out all the particulars. In short, bring the lad over to our side of the fence. Then with that inside information, have 'our lad' call the person setting up the illegal hunt back in Colorado. Then have 'our lad' advise the outlaw guide (if there was one), that his wife was just injured in a serious car accident and he wouldn't be able to come. However, his 'brother-in-law' would like to come instead and if that would be alright, he would have him come in his place on the hunt. Guess who the 'brother-in-law' would be? You got that right! If I had my way, it would either be Agents Dave Kirkland or Nando Mauldin, and 'the race for the roses' would be on, big-time...

For the next half-hour, Kris and I arranged for a "trace" on the phone call through the Colorado Bureau of Investigation. Getting that set up and operational took some time but soon we had it ready to go. Then we waited for the return phone call from our mystery caller.

Finally that phone call arrived, some fifteen minutes later than expected. By then, Kris and I were on pins and needles over the issue concerning our "wild" hunches and if they had been right, or were our suspicious insides just restless and rattling around because it was lunch time...

Picking up the phone after his secretary advised this was the call he had been waiting for, Kris said, "Hello, again. This is Kris, whom am I speaking to?"

"This is Roger. Thank you for taking my call. I know you are very busy so I will get down to business right away before you are called away once again. Are there any leftover deer licenses in the Meeker or Craig areas this time of the year?" he asked in a halting tone of voice.

"Well, let me see," said Kris as he made a lot of noise with the pages of the booklet on hunting regulations near his phone so Roger could hear him rustling through the pages. Then Kris said, "Roger, are you a resident or non-resident?"

"I am a non-resident from Pennsylvania," he said as he quickly let his voice die away after what I suspected was catching himself over unwittingly giving away what state he was from...

Then Kris started talking aloud over the phone

like he was absent-mindedly reviewing various game management units in the regulations for any available leftover licenses that were still available. Finally we got the high sign that we had a trace and then on cue, Kris's now clued-in secretary came in and loudly said, "Boss, the Director wants to see you immediately. When can I tell him you are coming?" she asked loudly enough so Roger could hear her over the phone.

"Damn," said Kris over the phone in fake frustration but loudly enough for Roger to hear. "Roger, I will have to let you go. My big boss wants to see me right away and I have to go. But before I do, I will need some more time to locate the exact Game Management Units where you want to hunt to see if any late season or depredation hunts are still open. But just offhand, it looks like there are a pile of them still left open," he said.

"OK," said Roger. "When should I call back?" he now eagerly asked.

"Give me another hour and I will have all the information you need. I will also have the information handy where you need to apply for a license as well," said Kris.

"OK," said Roger, "good bye." Then there was an audible "click" as the line went dead.

"How did I do?" said Kris with a big "spider to the fly" grin.

"Not bad for an old guy," I said with a "spider to the fly" smile.

Then after waiting for about fifteen minutes, I dialed the traced phone number with a phone set up that could record the conversation. The phone rang once and was then answered by a female voice with the words, "Roger Gooding's State Farm Insurance Office. This is Pamela speaking," said the lady on the other end of the phone.

"Oh, I am sorry," I said. "I was calling a friend of mind and must have dialed the wrong number. Where is this number again?" I innocently asked.

"This is Roger Gooding's State Farm Insurance Office in Altoona, Pennsylvania," said Pamela sweetly.

"I am sorry," I said. "I really missed my dialing on this one," I said as my heart raced. Raced because as the luck of the draw would have it, Special Agent David Kirkland lived in Ridgeway, Pennsylvania! In fact, he lived not far from Altoona! How is that for the "Gods of the World of Wildlife" reaching down and lending a helping hand to "someone" who could slap the evil hands stealthily reaching into Mother Nature's cookie jar?

"That is quite alright," said Pamela. Then she hung up.

"Kris, can you believe that? This Roger chap

lives in Altoona, Pennsylvania, and runs a State Farm Insurance Office. But the really good news is this. I have a very good friend named Dave Kirkland who is an agent in our Special Operations Branch. Believe it or not, Dave lives near this guy in a small town called Ridgeway! I suggest I get hold of Dave and see if he is available to look into this matter and give us a hand. If he is available, then I can have Dave make an appointment with Roger's office. However, before Dave is to meet Roger over "some insurance business," I will give our insurance chap a call that same day. I will let Roger know who I am and also let him onto the fact that we are aware he may be setting up an illegal deer hunt here in Colorado. I will also advise him before he has a chance to get his legs under him, that he is looking at a federal offense under the Lacey Act and could possibly face a huge fine and jail time if he continues along this illegal line of thinking. Then I will really "sink his bobber" while he is still trying to figure out how we got his name and caught him in the act! I will ask Roger to get up and look out into the waiting area of his office and ask him if his ten o'clock appointment is sitting there. When he does that and responds to my question, I will advise him that the person waiting for him is also a federal agent who wants to talk to him about this very serious matter. Then I will tell him to go out and meet the agent and to make sure he does not make any phone calls back to his contact in Colorado if he had been planning an illegal hunt, because his contact's phone has now been tapped. It won't be, of course, but he doesn't need to know that. And then let him know if he attempts to make a quick call back to Colorado to warn his illegal big game hunting contact, he will be federally prosecuted for Aiding and Abetting in a federal offense. I will back up those words of caution once again with the information that his man in Colorado also has his phones bugged as well. As we both know, that won't be true but it should scare the pants off our chap and preclude him from calling. Additionally, I will also tell Dave that I will be contacting Roger right at ten o'clock so he is to be at Roger's office and ready to make his move precisely at that time as well. Then we wait to see if first of all, our goofy thinking is right-on based on our instincts and not off the wall. Off the wall because it is close to lunch time and is only our guts grumbling with hunger and not our instincts working..."

"If we are wrong in our thinking, we sure as hell are going to scare the piss out of Roger," I said with a grin. "Then I will just have to respond to the Congressional he is sure to generate for accusing him of something he was not contemplating. And in so doing, I will lay on my back like a cocker spaniel, and wet all over myself. That way, the complaining Congressman won't kick me because that means he will get the 'wet' all over his shoes. Then when he hesitates, I will get up and make good my escape. That way I will live to fight another day for those critters in the world of wildlife who have little or no voice," I said with another smile. However, I knew if our guess was wrong on this Roger chap about an illegal big game hunt, there would be political hell to pay. But if I didn't miss my guess, my guardian angels were right on the money as they fluttered around inside me over this latest event and adventure-to-be in my life. And in so doing, that was their way of letting me know I was now on to something that smelled clear to high Heaven.

Come the appointed time, Roger called back. This time Kris had his "story" ready. "Roger," he said, "I am sorry to say the regular big game hunting seasons are all closed in the Craig and Meeker areas.

And contact with my District Wildlife Managers in those two areas indicated there are no depredation hunts for deer available at this time either. Other nearby areas yes, but the areas in which you are interested, no. I am sorry you had to call back so many times but I wanted to make sure and be able to be able to give you the correct information."

"Oh, that is alright. I appreciate your time anyway. Good bye," said Roger in an off-handed and strangely happier tone of voice. Then his phone went dead.

As Kris slowly put down his phone, he said, "Damn, for such an eager-sounding voice earlier requesting information on open big game seasons, he sure wasn't too disappointed about there being no hunting opportunity available the rest of this year during that call."

"If Roger was not put off by all the big game seasons being closed as his voice tone so advertised, I would say the hunt is on! We both might be wrong in our assumptions about Roger coming out here sometime in the future on an illegal hunt, but I doubt it," I said. "Now, the real work begins. I need to get hold of David and see if he can help us. If he can, we need to set up this contact with Roger as soon as possible in order not to lose him or his outlaw guide in the shuffle. I will call Dave just as soon as I get back to my office and run him to ground wherever he is and see if he can help. If he can't, I will see if Agent Nando Mauldin can help. Either way, both officers are top-notch hunters of humans and will get to the bottom of this in a heartbeat," I continued.

Heading back to my office, my head literally swam with "catch-dog" ideas. This could possibly be my first covert operation in the region as the new Chief of Law Enforcement if my hunch was right on this Roger thing. Having so little to go on other than a strong inner gut instinct that I was on the right track had to count for something, I thought. But, I had been wrong before, so I was still cautious. After all, the last perfect person we had on earth was 2,000 years ago, and we had hung him on a wooden cross. Since that time, there had been a lot of us lesser mortals wrong on many things. Hopefully this wasn't one of them...

After a few tries on the phone, I finally managed to run down Agent Kirkland. Dave listened quietly to my tale and thoughts about Roger. When I had finished, Dave said, "You know, Terry, I have made many large cases on less than what you have chosen to run with on this one. But I am also inclined to suspect you might be onto something. As you know, there are a lot of big game trophy animals in Colorado as well as a slew of other western states. And we are finding out here in the Special Operations Branch that the illegal commercial market hunting of large animal trophies is on the rise and getting stronger every year that we have been in the business. I will set aside some time to give you and Kris a hand and let's see where this one will take us. Additionally and just for your information, this Roger Gooding is a Safari Club high-up member and I have some intelligence that he is a real big game trophy hunter. Also, some intelligence leads from our branch in Washington indicate he is truly a big game killer to be reckoned with. Not that any of this is directly connected to your suspicions but that information at this time is certainly fortuitous. I say we go with what we have and see what develops," he continued. With that, I agreed.

Hanging up with Dave, I immediately called Kris with the news that we had a super covert operative in our camp ready to go. I could almost hear the bear trap clunk shut in Kris's pleased response over the news...almost.

The following day found Kris in my office. We were moving fast, almost too fast but to my way of thinking, time was of the essence. If Roger had decided to participate in an illegal big game hunt in Colorado, and based on Dave's intelligence as to our suspect being a real killer, I figured he was not too far away from a "trigger pull." With those thoughts breathing down our necks, like a good gun fighter of old on the dusty streets of Laredo, I 'pulled my trigger' first.

After our earlier discussion and setting up a plan of attack, Dave had made a ten o'clock appointment with Roger Gooding at his State Farm Office in Altoona for the fourth of January, 1982. Figuring time zones and allowing for Dave's arrival, Kris and I pulled our "rabbit out of our hat" on the appointed day.

Dialing Roger's office phone number, I was once again greeted with Miss Pamela's pleasant voice. Past the greetings and salutations, I advised Pamela I needed to speak to Roger right away about his upcoming big game hunt in Colorado. She put me through to Roger immediately. That in and of itself said something to me about my way of thinking about the possibility of an illegal hunt. It was obvious Pamela knew of her boss's upcoming hunt and the value he placed on such ventures. Why else dropping everything else and patching me through to him in such speed?

"Good morning, Dale. Any weather changes in store to screw up our hunting trip?" came Roger's strident voice over the line.

*Damn,* I thought, *Kris's and my suspicious thoughts were smack dab on!* I now had confirmation on a pending hunt and the first name of his outlaw guide named Dale!

"Roger," I said. "This is Terry Grosz. I am a Special Agent working for the U.S. Fish and Wildlife Service in Denver, Colorado. I am on to your pending illegal big game hunt in western Colorado with an outlaw guide named Dale. I wish to make you aware of a few, very important things so don't hang up! First, as you are well aware, such a hunt this time of the year is illegal under Colorado law and soon to be a federal law violation if you go through with it. Second, if you look at your schedule, you have a ten o'clock appointment waiting for you in your outside office. He is there as we speak. He is also a Special Agent for the Fish and Wildlife Service. As such, he has been sent to interview you today regarding this matter involving this planned illegal deer hunt in Colorado. Third, your phone has been tapped and if you attempt to make a quick call to Dale and tip him off that the feds are onto him and your little venture, you will be charged with a crime for interfering with a federal investigation. Dale's phone has also been tapped as well, just in case you try something funny after I finish speaking with you. Also, this conversation is being recorded. Do you clearly understand all of what I just said?" I asked.

For the longest time, all I heard was surprised, heavy breathing on the other end of the phone line. Then I heard what I would characterize as a very surprised and emotional voice saying, "I understand."

"Now," I continued, "have Pamela call Agent Dave Kirkland into your office and say nothing else to anyone. Do you understand?"

"Yes, I do," said the voice now breaking with even more emotion.

"Good," I said, "and when the agent comes

into your office, immediately hand him your phone so I may talk to him."

Then I heard Roger's shaking voice instructing Pamela to call in his ten o'clock appointment right away.

The next voice I heard on the phone was Dave's. "Dave," I said. "When I called Roger and had his secretary advise him I needed to talk to him about his pending hunting trip to Colorado, his first and exact words to me were, 'Good morning, Dale. Any weather changes in store to screw up our hunting trip?' So Dale is the first name of our outlaw Colorado contact. I also advised Roger as to who you were and that you were going to interview him about this hunting trip to Colorado. I also told him that his phone was tapped so he didn't attempt to call his outlaw guide before you got together with him. Also, be sure and tell him Dale's phone is tapped as well. That should preclude him from calling and tipping him off that we are onto them. You can take it from there. Call me when you are done." Typical of Dave's undercover demeanor, he quietly hung up without saying one more word.

For the next two hours, Kris and I sat on pins and needles hoping Dave would be successful in his interview. Finally, Dave's call came and Kris and I each got onto a phone so we could hear what Dave had uncovered during his interview with Roger.

Dave began by saying he was now speaking from a secure phone line and was not in Roger's office. Then typically all business, Dave's information flowed like the yolk from a broken egg. "Roger is scared to death, especially after I advised him over all the potential state and federal trouble he was in. I also made sure I graced my information with the possibility of the loss of his business through customer losses, if anyone discovered the kind of illegal commercial market hunting he might be involved in. Especially when a state like Pennsylvania with its over one million hunters, discovered one of their own was crooked! By so doing, that sufficiently drew him up short when the potential loss of his customers and business was linked to the conversation. Then acting like I knew a whole lot more than I did, I told him to tell me the whole story about what illegal business he was in with the Colorado connection and not to leave anything out, so I could compare his information against what I already knew."

"Roger began by telling me he had met a fellow from Colorado during a Safari Club meeting held in Denver. It was during that meeting that one Dale Leonard, had advised he could supply guaranteed hunts involving 30-plus-inch mule deer bucks, full-curl bighorn sheep rams, and seven-point bull elk if he was interested. Roger told him that he was more than interested but was fearful of being caught if anything was illegal about the hunts. Leonard had advised Roger he had nothing to fear if the big game hunts were illegal. He went on to advise Roger that he, Dale, was a Denver police officer. And as such, had general access to intelligence going on in the ways of other agencies investigating such illegal hunting operations. He had also advised Roger that his brother, Carl, was a Colorado game warden and he had access through that source on any covert operations they might have working illegal hunting schemes. (That proved to be totally untrue. Carl Leonard, even though Dale's brother, was a loyal and totally above-board law enforcement officer.) Lastly, Dale told Roger that the really big bucks, rams and bull elk could easily be taken during the closed seasons because of the lack of hunting pressure, hence that is why he hunted during that time of the year versus during the regular hunting seasons."

"Roger then advised he had shown interest in participating in such a hunt to Dale, and the two of them had begun working out hunting arrangements for the late winter-early spring of 1982. A hunt that called for either a record class mule deer buck or a bighorn sheep ram. Then he, Roger, had returned to Pennsylvania with the assurances that Dale would contact him when he had located either a trophy mule deer or monster bighorn sheep ram."

Continuing, Dave advised both Kris and me that he had set up the next series of events with Roger over an upcoming hunt with Dale Leonard. Dave advised he had instructed Roger to call Dale Leonard on the phone at his Denver Police Department phone number that very moment. His instructions to Roger were that he was to say that his wife had been in a serious vehicle accident and was in pretty rough shape. As such, he did not think he could go on any big game hunt in the near future until his wife had healed up. However, he was to tell Dale that his brother-in-law was also an avid big game trophy hunter and would like to go in Roger's place if that was alright. Roger then advised if Dale thought that was alright, he would contact his brother-in-law and make him aware of the situation so he would be ready for Dale's phone call. Then he was to give Dale, Agent Kirkland's undercover phone number and name as that of his brother-in-law. With that, Dale and the "brother-in-law" could make their own arrangements.

That phone call was made as Dave had so instructed. Once Roger had contacted Dale and advised him of the situation, Dale had expressed sorrow over Roger's wife's car wreck on the slick Pennsylvania winter highways but understood.

Dale further advised that he would contact Roger's "brother-in-law" just as soon as he had located either a trophy buck or bighorn sheep ram. Then after a few more pleasantries, the phone call was terminated.

Dave then instructed Roger that if he said anything to Dale about the switch, he would be indicted along with Dale when the time came. Dave also advised that when he met Dale on a subsequent big game hunt, that if anything was wrong signifying that the word had gotten out on his true identity, he, Dave, would know immediately. And once again, Roger would have to stand trial for interfering with a federal investigation.

Dave concluded the discussion with the evaluation that everything had gone off as planned during that phone call, and he was heading home to get his gear ready in case Dale called him sooner than later. Dave also felt he had sufficiently scared Roger enough that he would do only as he was told.

Kris and I agreed on Dave's plan and I advised Dave that I would immediately start the investigative case reporting process for procuring the needed clearances and buy-monies necessary for this investigation from the Washington, D.C., Office. I also advised I would have the buy-monies sent to my office in Denver unless he advised otherwise. And with an investigative case report sent to Washington the next day, my very first commercial market hunting investigation in my new region was off and running. And what a run that would soon turn out to be! As I expected, things moved pretty fast, showing just how deadly was the Colorado circle the commercial market hunters moved within.

On January 14, Dale Leonard telephoned Dave requesting a letter inquiring as to what big game animals he, Dave, might be interested in taking in upcoming hunts. Dave complied with information relative to his interest in trophy mule deer and bighorn sheep.

On January 28, Leonard advised his accomplice, one Jerry Marjerrison in Montana, was currently checking out the hunting conditions in the Plains, Montana, area for bighorn sheep. Now we had information showing there were at least two subjects in this illegal commercial market hunting ring and that it had grown from one to two western states in my region! Leonard had also requested in that latest contact, that Dave send him $2,500 as a down payment on a possible $7,500 hunt if the bighorn was a trophy animal. In keeping with Service policy in not flashing around lots of money and "creating a market," Dave only sent $1,000 as the first down payment.

On February 26, Leonard telephoned Dave

with the information that Marjerrison had located nine trophy bighorn sheep rams for the upcoming hunt.

On March 4, Leonard and Dave made plans for the upcoming bighorn sheep hunt. Keep in mind, every time a letter was sent or a telephone call was made, it was documented and entered into the latest case files.

On March 17, Leonard telephoned Dave and advised he was to fly into Missoula and there they would meet. Dave was also instructed that all the bighorn rams were trophy class and to plan on spending $7,500 for the hunt. Leonard also advised that he would be supplying a .270-caliber rifle that Dave was to use. For my big game readers' benefit, I had a district policy that all covert operatives were to shoot to miss, or shoot not the trophy animals but lesser-horned or -antlered animals if possible. That was so when the investigation broke, the local sportsmen would not be so mad over the fact that the biggest and best of their game animals were being taken during covert operations. Bet they didn't cover such restrictions on like-in-kind TV programs, did they?

On March 28, the day for arrival in Missoula, Dave advised Leonard that he was bringing a "trusted" friend who was also interested in killing trophy big game animals. Leonard, a high-ranking Denver police officer who should have smelled a rat if he was any kind of investigator over such a late entry into an illegal operation, never wavered over the unknown hunter. How is that for being so greedy to make an illegal kill for such a large sum of money, that your survival defenses are non-existent or ignored. Goes to show all you legal fair chase hunters what your wildlife officers are up against, trying to control this illegal killing and trade in animal species. As you would expect, Dave's friend was none other than undercover agent Bob Standish. Another "deadly federal catch-dog" if there ever was one...

On the afternoon of March 28, undercover agents Kirkland and Standish met Jerry Marjerrison at the Missoula airport. From there they were transported to Plains, Montana, where the agents were put up in a motel. Then they were met by Leonard and the four of them went out drinking and to dinner. During the drinking session, Leonard excused himself advising he had to go to the bathroom to get rid of last night's rich dinner. In the meantime, the two undercover agents and Marjerrison continued to drink. After a long pause, Leonard returned to the group and the foursome drank some more, then had supper. After that, the four split up. The agents went to their room where Dave checked his still-locked luggage. As a precautionary covert operations trick, Dave had wedged one of his hairs from his head between the top and bottom of his hard-sided luggage. It was an old trick used by covert officers to see if anyone had opened up their luggage in one's absence. Sure as God made cats as dogs' best targets, the carefully placed hair was missing! Someone had entered their locked motel room and had opened up the agent's luggage. After checking out the contents of the luggage to see if the "hunters" were "undercover agents" and finding nothing out of the ordinary, the thief in the night had carefully put back all the clothing, locked the luggage and quietly exited the motel room! Dave told me later he had just grinned over his "missing hair" discovery.

Knowing Leonard was a cop, Dave had expected to have him use a pick set from the Denver Police Department and unlock their flimsy motel door lock. Then Leonard had proceeded to pick the locks on the undercover officer's luggage. However, the concealed hair trick had exposed the thief. When the luggage was opened up, the carefully placed and unseen hair had fallen out and the thief was a "had dad"...

But the cagey Denver Police Officer was not done. When the lads had parted after dinner and drinking, Leonard had carefully picked up several drinking glasses used by the two undercover officers and had their fingerprints run through the FBI files to see if they were listed as federal officers. Since our agency had early on realized the dangers in listing our officers in the FBI's data base as federal officers, it had opted out from such a program. Therefore, our somewhat careful Denver Police Officer was once again thwarted in his efforts and none the wiser as to the actual identity of the two out-of-state "hunters." Just goes to illustrate once again to my readers, just how serious and deadly this business of putting those in the business of extinction out of business through covert operations, really is!

On March 30, when the Montana game wardens were holding a district meeting and absent from the field, the two outlaw guides and the two undercover agents went bighorn sheep "shooting." Note I didn't use the term "hunting" which denotes fair chase. Shortly after entering the field, Dave was presented with a shot at a bighorn ram. He shot and purposely missed. Not wanting to advertise their presence on a bighorn sheep wintering ground with the noise of lots of shooting, Leonard grabbed the rifle from Dave's hands and with one shot, killed a large bighorn sheep ram. Then the ram's massive head was removed and the rest of the animal left to rot! The sheep's head was carried out by Leonard and hidden in Marjerrison's vehicle and the men left the area. Later that day, Dave wrote out a personal check for $6,500 to Leonard. Leonard in turn had Dave write on the check that the check was for "lumber purchased."

On May 29, Leonard drove to Montana, picked up the bighorn sheep head and returned to the Denver area. There he turned the bighorn sheep head over to outlaw taxidermist Kevin Fruit, whose taxidermy studio was also located in Denver. Once again, another member in this deadly market hunting ring "raised his head" and had been identified.

Notified by Leonard that Fruit now had Dave's sheep head, he was also apprised he needed another $1,000 for a black market cape for the mount. Once again, Dave sent Leonard the required amount.

On July 12, Dave called Fruit on the progress of his bighorn sheep mount. Kevin advised it was almost ready and told Dave he needed $286 for his taxidermy services which Dave paid.

On September 5, Dave received his mounted bighorn sheep head via freight at his home in Pennsylvania. Now we had two outlaw guides and an outlaw taxidermist involved in a commercial market hunting operation who all were now facing a blizzard of Montana state and federal charges! And our little original, slimmest-of-slim "gut hunches," based on an unknown caller's eagerness for a possible closed season hunt in

Colorado, was just getting started. Plus, the long arm of the law was now closing in on a crooked cop, an outlaw guide in Montana and an outlaw taxidermist. All who were willing to sell their souls for a few pieces of silver... How many more outstretched hands for the tainted silver pieces would there be before the sun finally set? That information wasn't long in coming.

Before all was said and done in this my first regional covert investigation, "more flies came to the spiders." Soon illegal closed season deer hunts in western Colorado were taking place as outlaw guide Dale Leonard stayed busy leading more and more folks on such hunts. However, it seemed as he did, more and more covert operatives were introduced into such illegal poaching and market hunting practices.

After Dave had participated in several more illegal hunts, he was able to introduce Agent Nando Mauldin into the mix as well. And when that was done, the agents discovered just how brazen Dale Leonard had become. When the agents would fly into the old Stapleton International Airport in Denver, Dale would be there to meet them. Meet them, that is, in a marked Denver Police cruiser. Then they would be transported to their motel room in the police car. Such was the degree of brass and arrogance Dale showed in his disregard for the laws of the land. Then Agent Mauldin was introduced to one Steve McClung from Colorado Springs, who was also involved in the Leonard-Marjerrison commercial market hunting ring. It seemed McClung's specialty was the illegal killing of the Pikes Peak wintering bighorn sheep population. In that, McClung would provide illegal bighorn "shoots" on the wintering grounds in the Pikes Peak area for willing buyers. After the selected animal was shot, it was caped out and minus the horns, the rest of the critter left to rot.

In fact, Steve's method of operation was unique. He would photograph the wintering rams and then list the price of each ram in the picture on the back of the photo. Then he would send a pile of photos of the wintering rams with the price tag on the back to a willing outlaw shooter. The outlaw shooter-to-be would look at the ram and the price for killing that particular animal on the back. Then the outlaw shooter would make his pick as to which animal he wanted to purchase and kill. That specific photo would then be sent back and later, when the illegal shooter arrived, he was led to that ram in the picture and it was taken. Such were the type of pictures sent to Agent Mauldin.

Soon Agent Mauldin had quietly ensnared Mr. McClung into our web of game hog poachers after an illegal bighorn sheep shoot. Once again, the bighorn sheep trophy head would be smuggled under the cover of darkness into Kevin Fruit's taxidermy shop for mounting. And once again, when mounted, the illegal sheep would be shipped to the shooter.

As Kris and I moved the investigation to a close before this trio of poachers had shot off all the biggest and best of Colorado and Montana's big game species, we faced a knotty problem. That problem being that of bringing District Wildlife Manager Carl Leonard into the arena of those who knew about the investigation of his brother. That meant informing Carl of the impending arrest of his Denver Police Officer brother and the eventual destruction of Dale Leonard's law enforcement career.

In my office that fateful day was Kris Mosher, Carl Leonard and myself. When we closed the door, you could see the many questions flying across Carl's face as to why he was there in front of his Chief of Law Enforcement and the Service's Special Agent in Charge.

"Carl," I said, "you are probably wondering why you are here. As such, I will get right to the point. Your brother is involved in an illegal commercial market hunting ring and is soon to be arrested and prosecuted for numerous Colorado, Montana and federal wildlife charges."

The look on Carl's face said it all. You could

tell he had suspected that his brother was not totally above-board when it came to the legality of things in the world of wildlife. "What has he done?" he slowly asked.

For the next twenty minutes or so I explained the length and breadth of the investigation. When I finished, Carl said, "I figured someday he would run afoul of the law. Well, that is his problem. What do you need me to do to further this investigation?" he asked in a matter-of-fact tone of voice.

Both Kris and I knew that Carl was one damn fine and loyal officer and his words and reactions had just shown that to the nth degree. "We will be using your expertise during the upcoming take-down but you will not be arresting your brother. I will be doing that, as well as being involved in the searching of his home for evidence," I said.

Carl just nodded and then said, "I will go wherever I am needed in this investigation and you need not worry about this being leaked to my brother. He has made his own nest and now he can lay in it," he quietly said.

A week later, we dropped the hammer on the outlaw lads in Montana and Colorado. In Dale's case, since he had a temper, we hit his house right at daylight just as he was leaving for work. We had in the arresting party his supervisor, who approached Dale in the pre-dawn as he was walking over to his vehicle to go to work, and confronting Dale, asked if he was armed. Dale replied in the affirmative and then his supervisor told him to hand over his pistol which Dale promptly did. Then he was arrested and whisked off to the lockup to be interviewed. A subsequent search of his home produced the .270 rifle that had been used in so many of the killings, as well as many evidence photos of not only his but his group's illegal kills.

Later during the subsequent trials, all the main operatives in that commercial market hunting ring were found guilty, fined and served time in prison. Additionally, Dale was dismissed from the Denver Police Department prior to serving his time in incarceration. All four main characters in that investigation have since disappeared into the dustbins of time...

During the next 16 years of my service as the Denver Special Agent in Charge, there were dozens of other covert operations conducted in my region by both the Special Operations Branch and my own officers as well. Those investigations involved the smuggling of wildlife; illegal sale of resident species of wildlife; illegal sale of big game meat through outlaw meat suppliers; illegal gill netting of fish by Native Americans on Lake Oahe in South Dakota; investigating rings of Native Americans who were killing and selling eagle parts and products; outlaw guides hunting in the national parks; investigating illegal falconers scooping eggs from nests; as well as the illegal sale of falcons and everything else under the sun that was illegal and would turn a buck.

My last covert investigation run in the region before my retirement was one of the largest investigations of its kind ever run by the U.S. Fish and Wildlife Service. It took me over seven years to set up and conduct such an investigation from start to finish. It involved over 100 commercial market hunters primarily in the States of Colorado and New Mexico, snaring everyone from the routine killer to active-duty police officers! In order to safely culminate such a large takedown, it eventually involved over 275 arresting and investigating officers! It also involved much in the way of questionable politics from sitting Colorado Governor Roy Romer and the always-present, liberal-biased news media. In the end, 105 out of the 108 defendants apprehended were convicted, regardless of the disgraceful politics and news media bias heaped on the heads of the state and federal investigating agencies. The three of that number not prosecuted were because they had smoked their minds on crack cocaine and didn't even remember where they were. So the prosecuting attorneys dismissed those three cases.

But if you want to read the real-life particulars about that operation, my readers will have to read Chapter 5, "Operation SLV," in my book titled, The Thin Green Line. That story alone will show the unwashed just how important and yet how dangerous running covert operations can be. It will also illustrate the tremendous killing pressure our wildlife resources are facing and just how tough and relentless our officers protecting such resources can be, pursuing those pulling the triggers. The American people owe an awful lot to such a dedicated core of operatives who sacrifice not only their family lives but many times their own lives as well, so that many of you readers today can enjoy the wildlife wonders the good Lord set forth on this planet for your enjoyment!

## ROADBLOCKS

Another little-known wildlife law enforcement tool is the roadblock, or as those of us in the profession like to call them, "wildlife inspection stations." In my old region, comprised of the States of North and South Dakota, Nebraska, Kansas, Montana, Wyoming, Colorado, and Utah, over two-thirds of the Nation's non-resident hunters will visit and hunt during those many states' fall and winter hunting seasons. That represents one hell of a pile of people rattling around at one time.

And not all of them are honest or legal when it came to following the wildlife laws of the land.

Consider this, that in all of North America there are only about 10,000 county, state, tribal, military, provincial and federal wildlife officers. And once one gets rid of all the sick, lame, off-duty and lazy officers, including some supervisors, there are only about 3,000 officers afield on any given day in all of North America! Bottom line, that means a lot of your wildlife natural resources are going to get the chop, since there is no one close at hand to protect all of them... That means if one is going to provide the best wildlife protection they can, they must resort to any means available, just as long as it is legal, in order to get the job done. Hence the sometimes utilization of the roadblock law enforcement tool on secondary highways and major Interstates!

In my particular case, I would take a good hard look at all the state-regulated and sanctioned hunting and fishing seasons authorized in my region. Then after taking a hard look at all the season opening dates and the states involved, I would look at a road map. There I would look at all the major highways draining the nearest states of their hunting traffic when such sportsmen headed home. That is, I would look at those major arteries draining out from my resource-rich hunting states leading back to the major metropolitan areas in the country. With that information in mind, I would look for a major choke point where primarily most of my exiting hunting traffic would cross or traverse on their way home. Then I would look for a large rest area in which a major check station could be set up and operated safely. With that in mind, I would figure out the starting big game hunting season dates and add five days after that opening date. That "five days after the opening date" would be my starting date for a major highway roadblock. The reason I used that formula was because by the fifth day after the season opener, my successful hunters would be heading home. By then they would be tired of their own cooking, be needing a bath, be out of camp whiskey, and if they didn't get their game shortly into a cooler, the hard-sought-after meat would start to spoil.

With that information in hand, I would place a call to my state chiefs in the region requesting supporting manpower from their cadres of officers. I would call the state fish and game chiefs (except North Dakota — remember that chief hated my guts) because their officers were cross-credentialed with federal credentials so they could legally search and seize any illegal game traveling in interstate commerce (i.e., along the freeways). Additionally, such operations provided a very concentrated crash course for officers on search-

and-seizure techniques plus a chance to practice their interviewing/interrogation techniques, and provide an opportunity for those officers of the brotherhood to get reacquainted. Finally such operations would allow officers the chance to work many good wildlife cases in a short period of time, allow for the exchange of techniques, provide excellent, concentrated rookie training, as well as give those critters without a voice who had been illegally taken some form of settlement.

Then the real work started. Copies of all the adjoining states' laws had to be acquired en masse so any season/bag limit questions could be quickly answered once the inspections began. Then my resident officer nearest the planned check station had to get DOT and state highway patrol clearances. Additionally, fire, emergency services and storage facilities had to be arranged for on-site utilization. Then calls were made to all the state and federal "legal beagles" in the immediate area of the roadblock to ascertain what legal limitations they would individually require. Then if possible, arrange for attorneys and judges to be on-site to respond to legal issues arising over the legality of the inspection, search techniques being utilized, and on-site case settlement if the transporting offenders so decided. Arrangements were also made for house trailers to be on-site for administrators who would be processing all official documents so those apprehended could be quickly sent on their way. Trailers would also be placed on-site for use by local judges holding court. Then contact would be made with local area churches, so when eatable game was seized, a simple phone call to a church, stake house or parish would quickly produce a needy family. Then the seized, previously documented game would be distributed to those needy families approved by their church groups so the meat could be properly utilized. As in the case of every check station operation, information would be gathered to ascertain that the nearest town had enough eating and sleeping facilities available for the sudden influx of sometimes over 250 officers who would be running the check station! Lastly, arrangements would be made for nearby cold storage lockers to store some of the seized carcasses not yet distributed to the needy and for acquisition of dry ice. Dry ice would be kept on the check station site so after examination of the truck- and trailer-loads of meat being unloaded, examined and cleared, it could then be added for safe shipment of the meat to its ultimate destination.

Arrangements were also made for prisoner containment and confinement. On any roadblock, there are always arrests made of illegal aliens, drug runners, drivers of stolen vehicles,

# TERRY GROSZ

In most cases involving a major Interstate roadblock, there would be state highway patrol cars available to act as chase cars on those "boneheads" running the roadblock, arresting and transporting prisoners, searching for potential penal code violations, and many times a "bear in the air" was also utilized. With the use of an airplane, many times it would fly the secondary roads checking for runners who had heard of the roadblock ahead of their direction of travel. Numerous vehicles having something to hide would move off the major Interstate and travel secondary roads to avoid being caught. When such runners were identified, highway patrol cars usually roaming the area would pursue the runner and bring them back to the check station for inspection and clearance.

All that hard work and preparation brings up another issue. After about 2 ½ days, the truckers on their CB's and the local restaurants would blab to everyone who would listen about the nearby roadblock. In so doing, many wildlife outlaws would drive clear around the impending roadblock in order to get their ill-gotten booty home. Sometimes even going so far as driving clear through another state to avoid such wildlife inspections. Once again, the airplane finding such runners (many times seeing the game in the back of pickups or being carried in trailers) would have local highway patrol or adjacent wildlife officers stop and check them.

By now my readers should be learning that a roadblock is never a simple thing. There is so much that can go wrong that one has to cover all his bases when setting up such operations. Then there were the local politics to deal with. Many times, terrified, afraid-of-their-shadows local politicians would stop by and have to have their hands held, fearing something would go wrong and their personal skirts would be dirtied if they didn't.

Then there was the cost. When I set up a major roadblock, I figured it would cost me at least $50,000 for three days of around-the-clock work! However, you can't let that figure scare you. If a roadblock was properly run, especially during the right time of the open hunting seasons, we would average about $30,000 in fines daily! Not a bad return, especially when the fine monies many times went into the adjacent counties' local school accounts for educational purposes. Plus, imagine the future deterrent such an operation would have on the outlaw hunting community once the word of such an extensive operation got out and about.

But in a way, that was also sad. Sad because it

spoke to the lack of hunter ethics in some of the so-called sportsmen running around in the back-country looking to kill and bring home a critter. Many roadblocks netted about 12-19% violators of all the folks coming through the check station! Additionally, about 15-17% of those violators breaking the law were law enforcement officers of some sort. And as for illegal fish and game seized, it many times ran into the tons! And those figures didn't take into account the minor infractions, which usually resulted in a damn good ass-chewing and then the offenders were sent packing home with a stern warning.

Being from such a huge resource-rich region, I would try to run at least one to two major Interstate roadblock operations a year. Additionally, there would also be about four or five secondary highway roadblocks run per annum throughout the region as well. This was made possible because almost all of my state fish and game agencies realized their value and the many pluses it represented for their officers in the learning departments. That was not to mention the "learning departments" of the sportsmen who ran over the legal line drawn in the sand when it came to future behavior. Additionally, most of my state counterparts contributed supporting monies for such operations, hence the large number of such operations run each year. Last but not least, think of the overall public educational value gleaned. Imagine what the thousands of sportsmen who passed through such operations had to tell their buddies when they got back to work the following week. Especially in light of exiting one's vehicle at such an inspection station and seeing 200-plus wildlife officers checking dozens of other vehicles amid piles of every sort of game animal or fish under the sun all laid out in plain view. Also, imagine the deterrent it had on the outlaws who were caught and shared such tales of woe with their other outlaw buddies once they got back home. I remain convinced the threat and regularity with which we ran such large-scale inspections in our resource-rich region had some cautionary effects on the outlaws during the numerous big game seasons. If nothing else, I am sure our roadblock histories made many outlaws take their illegal activities elsewhere, or at least, make them pause or look over their shoulders before they pulled "that" trigger. Then sweat the possibilities of being apprehended, fined or jailed somewhere later down the road, plus lose their ill-gotten gains.

As I said earlier, running roadblocks had many officer advantages. If for nothing else, it gave many hard-working officers the chance to relax, learn new techniques and reconnect around their many buddies. That also included the playing of high jinks on one another if and when the chance prevailed. That chance was always soon in coming because the very nature of wildlife officers is that of a coyote. In short, they, as a group, are opportunists. And such high jinks made for easier days and their 12-hour shifts working on the hard asphalt of a rest stop under all types of adverse weather conditions Mother Nature managed to throw at them, be it during one's day or night-time assignment.

## THE POWDER RIVER ROADBLOCK AND AN AMBUSH OF SORTS

For example, I can still remember with a grin on my handsome puss, a secondary roadblock that was held in the Powder River region of the State of Wyoming. Even as these words are written, I still hold a favorite place for memories in my heart for the times I worked with the Wyoming Fish and Game officers on such details.

They are a hard-working, driven lot and most of those officers were as tough as horseshoe nails. Out West, ethics involving hard work, honesty, bravery, common sense, and friendship are highly cherished by those of us in the wildlife law enforcement brotherhood. To one exhibiting such high ideals and performance, being called a "good hand" means everything. Those guys I worked with over the years in Wyoming were "good hands!"

Back to the Powder River roadblock. It was not a large roadblock by today's standards, utilizing about 25 state and federal officers in total. However almost to a man, those officers on that roadblock were close friends and gunners in and of their own right. Finding a small rest area between Shoshoni and Casper on state highway 20, all the required arrangements were made and with a few signs advising "wildlife check station ahead," we were open and ready for business.

I remember initially being assigned to work with Wyoming officer Jim Oudin. Now Jim was older than glacial dirt, hard-working, frank as the point of a spear, and had a love for the world of wildlife beyond compare. Since this was a Wyoming-sponsored roadblock, I played second fiddle being a federal officer, when Jim and I swung into action as a team. Jim would do the meeting and greeting of those who were swung into our checking station, conducted most of the inspections and license checks, as I just basically looked on to see if anything missed his old but eagle-sharp eyes. Then if something "heavy" needed to be lifted, the two of us would get involved until the inspection was complete. If any violations were detected, Jim would handle them through the issuance of a state citation or warning. Me, looking beautiful and wonderful, I basically stood there looking like a giant-sized sandhill crane in all my self-importance and majesty. Well, I was there for another reason as well. If the criminal violations required some "legal heavy lifting," I would be there as a federal officer to swing said violations into federal court. But for the most part, Jim just tolerated my federal presence, did as he pleased, and if the crap hit the fan, would accept my pitiful assistance as a last resort. I think my sharper-eyed readers can see this train wreck coming...

Once we got up and going along with all the other officers, we were soon lost in our individual work efforts conducting wildlife inspections at our assigned check station site. For the most part, the American Sportsmen behaved themselves and stayed above the letter of the law. If not, being that you were in Wyoming, the violators were wrapped in a green cowhide and left out to shrink wrap in the sun. Well, not really... However, there were always a few who did not get the lawful message and Jim and I managed to give all those lads a "legal haircut" that matched their degree of stepping into the slop!

Like our counterparts, we made several cases involving folks killing the wrong sex of critter, snapped up those taking the wrong species, lifted the hair on those who didn't bother to tag their kills as is required by law, or ran to ground those who were using another person's hunting license or tags. But as luck would have it, our part of the check station was not being blessed with the really big or gross violations that some of the other officers were experiencing. Then during lulls in the action, as is the case with Alpha male wildlife officers, notes were compared. As it turned out, Jim and I were not in the running with the rest of the pack of Alpha males and that began to concern Jim. Especially when Jim was soon compared to a Beta female due to his lack of meaningful production. Then some of the other more-blessed officers began taking friendly pokes at Jim for being too old, unable to see good violations in front of his aging eyes, or being saddled with a useless, super-sized, slow as the second coming of Christ, federal wildlife officer (me)...

Then I could see the friendly ribbing was beginning to get to my partner. Me, no biggie. I was used to all the abuse my fellow officers hurled my way because I was so slow and low to the ground. However, in the case of my fellow federal officers giving me holy 'Ned,' I was still the boss... And to mess with the BIG boss would get them a crummy assignment somewhere down the road. So overall, I was pretty safe. But in the case of my partner Jim, I could see he was rising to the competitive bait being tossed his way by his state counterparts. Rising to the bait like a large carp would to a big gob of worms on a warm day in some quiet backwater in a mill pond.

Then it was our turn to conduct an inspection, as a large, expensive four-door sedan slowly pulled into our part of the check station. I could see the sedan held two grizzled old chaps all dressed up, and in the backseat sat two nicely dressed older women who were obviously their wives. Right off the bat, my law enforcement senses told me there was nothing wrong with those chaps. As such, I would have given them a quick cursory check and sent them on their way. Yup, I figured. No violations here. However, Jim now, after all the friendly ribbing from his mess mates, seemed to be on a tear. He professionally checked the men's licenses, asked where they had been hunting and if they had any game. The driver replied they had killed two deer and the animals were in the trunk of their car. Jim asked to see the critters so he could check their licenses against the deer tags and the driver obliged. Out of the vehicle stepped the driver in his suit and tie, and walked back to the vehicle's trunk. Here he unlocked the trunk and then stepped back so Jim could check the two deer. Well, both deer had been totally commercially processed, packaged, wrapped and neatly placed into four large cardboard boxes. The boxes had been filled with dry ice and taped solidly shut.

For some reason, Jim suspected something was not right. He whipped out his trusty Buck knife and began opening the neatly wrapped boxes. Then he started taking out all the commercially wrapped packages of deer meat and began counting them to ascertain if the number of packages represented only two deer. By now, I was getting a little uncomfortable. It was plain in my mind that the boxes didn't represent any more than two cut and wrapped deer. But my partner kept digging and placing all the meat from the boxes out onto the ground.

Now, my two well-dressed old hunters were getting a bit testy over Jim's more than thorough inspection. Soon they began to grumble about the meat packages getting dirty and all the dry ice being carelessly unpacked. Then the two women in the back seat of the car got into the act. They bailed out from the car and wanted to know why the hold-up. Well, since Jim was busy with counting out the packages of meat for the tenth time, guess who the hell they turned their wrath to? You got that right! The big silly-looking sandhill crane standing there with his thumb in his nose. After all, it only took one officer to count out the few packages of deer meat. The two ladies, now getting cold, began jawboning at me about the unreasonable search and the time it was taking. Well, it was to get worse! The two men had descended verbally into Jim and now in his old age, lost count of the deer meat packages. So he had to start all over once again in his counting. By now, a mess of other wardens and federal officers not having anything else to do but gawk at me getting my ass chewed and my frustrated partner recounting all the packages, began gathering all around like vultures on a roadkill. Then the two old men descended on me having gotten nowhere complaining to Jim who could barely hear anyway. Now being the abject center of attention, I did all I could to explain, somewhat lamely I might add, why my partner was now recounting for the 97th time the few little packages of meat. Now I was really catching hell over the inspection from all the old folks who were getting meaner by the second. Trying to deflect the angry grumbling with soothing words, I got it all shoved up my ass by two now very cold women and a pair of grumpy old men!

By now, all the Wyoming wardens and my federal officers were having a good laugh at their 320-pound, very high chief federal agent, most beautiful of the lot, most highly decorated, getting his ass handed to him by four old farts who were cold, mad and demanding my name and badge number so they could write their Congressman and complain.

Looking past the four outraged old farts for

some help from Jim, I was thunderstruck! He was nowhere to be seen! Realizing he had stirred up the wrong bucket of "dog-doo," he had quietly disappeared into the crowd of friendly Wyoming wardens, leaving me to handle the unnecessary mess he had solely created. And he had left all the packages of meat scattered all over the ground in his haste to disappear into the crowd of smiling, friendly and on his side Wyoming game wardens before the grumpy old men caught him escaping!

Suffice to say, I somewhat cowardly set about replacing all the now-dirty packages of deer meat back into the boxes. And in so doing since I hadn't seen which box each package of meat had come from, got them all mixed up. By now, the smelly brown stuff Jim had created had spilled all over my shoes, on top of my used-to-be-beautiful head, and had heavily coated the fan, if you get my drift... and, still no Jim in sight to provide aid to his beleaguered partner. Finally getting all the meat back into the original boxes and placing them into the trunk of the old farts' car, I closed the lid. But not before they had acquired my name and badge number so they could write their Congressman about me. And only me, because Jim had disappeared into the crowd of officers and being of normal size, had happily blended into the on looking group of men. But there was one damn silly sandhill crane-like looking fool who was more than remembered. Then in a furious huff, my four old people left the Powder River check station less than cordially.

Looking around for my partner so I could take a leg of his in each hand and make a wish, I saw his laughing face finally emerging from the protective crowd of his warden buddies. Boy, did that group of Alpha males have a roaring good laugh at the sheepish antics of the 6' 4", 320-pound Special Agent in Charge sandhill crane type with no hind end feathers! You know, the one with a flaming red face from embarrassment and a ripped clear off last part over the fence by four grumpy old people...

By the way, I did get a Congressional letter demanding my explanation of unprofessional behavior and conduct on the Powder River roadblock. The old people could not remember what Jim had looked like, but they sure as hell had my 6' 4", 320-pound number memorized! Try writing a response that satisfied the Congressional type, when one already had his ass handed to him numerous times at the roadblock by the four old farts. By the way, two days later while working with another partner, I was able to hand Jim a piece of his "last part over the fence." And as it turned out, it was a rather large and satisfying piece at that...

## STINKING DEER AND MORE

The next day as was our procedure, I was assigned another Wyoming officer as a partner. This time my partner was a supervisor named Terry Cleveland. Terry was one of the finest state officers I ever worked with. He was a gentleman, a great friend, a very experienced officer, very ethical and a John Wayne bigger than life figure with a Clint Eastwood "quiet." Now there was what one would call a "good hand." He wasn't a crapper like I had the day before, who had abandoned me in the face of "battle" with two grumpy old men.

Again, once the roadblock was open and ready for action as the hunters streamed into the check station area, my partner and I went to work. Soon Terry and I were up to our elbows in dead critters, smelly sportsmen, many times poorly cared for game animals, and the usual number of violations. By midday, between the two of us, we had caught a handful of violators with the usual types of violations. Violations such as untagged animals, over-limits of game birds and trout, game killed in the wrong game management unit, using the wrong license or tags using someone else's license and tags, and the like.

Not being one to duck a "high jinks" opportunity myself, I had a grand one drop right into my lap. As Terry was finishing up the paperwork on a previous violation, I had a truck turn into our check station space. *Oh brother!* I thought. Here came a Ford 3/4-ton truck with two occupants. On each front fender were tied two massive mule deer bucks with at least 30"-spreads on their antlers! As the truck pulled into our space, I noticed both deer had not been gutted and were wrapped completely in heavy clear plastic with only their heads sticking out from the wrapping! *What the hell?* I thought, as the truck braked to a stop in front of me. Then the smell hit me. The two deer were spoiling! The heavy sickly sweet smell of rotting meat pierced my senses and turned my stomach to the point that I was now suppressing dry heaves! These two dummies had taken two very fine mule deer trophies and in their total ignorance, failed to gut either animal! Then they had tightly wrapped their deer in heavy plastic, which basically allowed the two deer to cook in their own juices and body heat! Then the two men jumped out from their truck and walked over to me as I tried to contain my dry heaving from the heavy, sickly sweet and oppressive smell of spoiled deer meat.

Holding up my badge and credentials, I advised the men that I was a federal officer working with the Wyoming Fish and Game Department at the check station. Then I asked to examine the men's hunting licenses and the tags on the deer.

By now, my dry heaving was becoming very hard to suppress, as the warm sweet smell completely enveloped my being. Stepping back from that ring of smell, I waited as the two men quietly dug out their licenses. Then the driver handed me his license as did the passenger. Even the licenses recently handed to me smelled to high heaven! By now the sickly sweet smell was so strong and enveloping that my eyes were beginning to water as I reviewed the licenses. Then I asked the silent men where were their deer tags?

Boy, did I ever get the surprise of my life! The driver spoke to me about the deer tags and I didn't understand a single word he said. "What did you say?" I asked. Once again the man spoke to me and I couldn't understand him. Then it dawned on me after the second man spoke and I didn't understand a word he uttered either. Both men were deaf mutes and were speaking in their normal muted tone of voice!

"Gentlemen, your deer are rotting," I said. "You need to get that heavy plastic wrap off the carcasses and get them off the front of your truck and away from the engine's heat. Then you lads need to take those deer over there under those trees and dress them out. They already smell like they have spoiled, and in Wyoming, wasting a game animal will get you a ticket. So I suggest the two of you get cracking," I said.

Both men began complying as I took their hunting and driver's licenses along with their deer tags over to where Terry was just finishing up the last of his paperwork. Walking up to Terry as he closed up his workbook, I saw him wrinkle up his nose and look hard at me. Then he said, "Damn, Terry, didn't you bathe last night? You smell like death warmed over and spoiled in the hot sun," he said with his characteristic, poking fun at me smile.

That was when the devil took over in me and I thought, *OK, if you want to take a whack at the big one just because I smelled badly from being so close to those rotting deer, I will just play a little trick on your last part over the fence.*

"Terry, those two lads over there gutting their deer under the trees, have a wanton waste violation. Since you Wyoming guys are writing all the tickets, I suggest you go over there and interview those chaps regarding that violation," said "the spider to the fly"...

Terry gave me a funny look since I was obviously opting out in helping him with my two chaps but being the good officer he was, off he went. Then I grabbed off a few officers who were waiting for their turns checking sportsmen and quickly told them what Terry was about to get himself into. Man, did that ever gather a crowd to see the very popular supervisor walk into the "deaf mute-stinking deer mess" he was heading for...

As Terry walked downwind towards the two men cleaning their deer, I saw him stagger, then quickly step back from the two men and begin dry heaving. That brought a roar of laughter from the rest of the officers and yours truly who were looking gleefully on. Then when Terry walked around the smell to talk to the two men about their rotting deer, he got the same surprise I had gotten earlier. After trying to understand what was being said and not being successful, I saw Terry take a quick glance towards me and mouth the words, "you son-of-a-gun" (or something worse along those lines...)! Again there was more laughter from those of us standing around watching Terry in his agony as he got out a pad and pencil so he could communicate with his two deaf, mute chaps. Of course, having to write out all his questions so he could communicate since he didn't know sign, only made him stand closer in the sickly sweet smell of the rotting deer and the two smelly men. Me, on the other hand, since I felt Terry had the situation well in hand, did my sandhill crane trick and just stood off at a distance, upwind of course...

Sometime later, Terry finished his violation paperwork with the two men and their stinking deer. Then he got even with me as I had to help him seize and move the rotting deer off to one side. But all of us got even later when we had lunch. Terry had lingered too long around the two men and their rotting deer and now smelled just like them. So as we all had lunch, we made Terry eat off to one side since his clothing had been impregnated with the awful rotting deer smell. Typically if something was wrong with any one of us badge carriers, then the rest of the Alpha males will strike. You know, like making one eat sitting off by himself. That got even funnier when all the other patrons of the commercial eating establishment got a whiff of Terry as well.

## "RAISE YOUR RIGHT HANDS"

Later that same day in which Terry Cleveland and I had the misfortune to run into the two men with the rotting deer, another unique issue arose. Several sportsmen arrived at the check station complaining about five men in a camper who had seen the check station sign and stopped along the roadside. Then the five men from the cab-over-camper vehicle bailed out and began throwing out numerous sets of deer antlers and antelope horns onto the side of the road! Then, the complaining sportsmen advised, the vehicle had attempted to cross over the median and return from the direction in which it had come. However, their long cab-over-camper had bottomed out in the median and was now stuck. With that word, a nearby Wyoming Highway Patrol officer was hastily dispatched. Twenty minutes later he returned with a large cab-over-camper rig from out of state in tow.

As it turned out, the team of Dave Bragonier and Jim Oudin got to inspect that vehicle and question its occupants. Soon, they had a pile of deer antlers and pronghorn antelope horns scattered alongside their check station area. From the looks of the freshly dried blood and tissue still hanging on the antlers and horns, the animals had been recently killed.

About then Terry and I received another vehicle to be inspected and we had to quit looking on at the drama being carried out at Dave and Jim's portion of the check station. As was the instant case, we were soon lost in our own vehicle inspection endeavor. Our work continued unabated for about another forty minutes, when out from the corner of my eye, I observed Jim Oudin walking my way. You know, the officer who had abandoned me the day before, in the presence of the two, mad as hornets, old farts.

When he arrived, he signaled that he wanted to talk to me. Leaving Terry as he inspected another hunter's game, I walked over to Jim. "What's up, Partner?" I asked.

"Dave and I have a tough one," he said

somewhat sheepishly. "We have been working on those five guys who dumped all those deer antlers and pronghorn antelope horns alongside the highway and have gotten nowhere. As they were being escorted here by the highway patrol, they had time to get their stories together. And their stories are now lock solid. They claim they hunted in Wyoming but got nothing. Then they discovered the pile of great-looking deer antlers and antelope horns alongside the highway as they were heading home and picked them up as souvenirs. No matter how hard we have tried, they won't budge off that story and now have Dave and me stumped. We both suspect they took those animals and just left the meat back at the ranch where they hunted, or to rot where they killed the animals. But I will be damned if we can break their "just found them alongside the highway" stories. Seems we could use your help if you have the time," he continued sheepishly.

Now I just had to chuckle. Both Dave and Jim were officers from the old school. They were as tough as horseshoe nails, strongly states' rights oriented, proud as an eight-point bull elk, and tough as a bull moose in rut. Both of those officers could have been Mountain Men. And for all I knew, in their past lives had been. But now here was the funny part. These two officers, proud as hell and never "stump broke," were now asking for help from a lowly federal. *Man, I thought, they must have had to swallow one hell of a lot of pride to admit defeat and come crawling to me, a lowly federal sandhill crane (but a beautiful-looking one) for assistance.* Suffice to say, I had to chuckle inside at their new-found discomfort.

"Damn, Jim. If the two of you couldn't break that group of outlaws, what the hell makes you think I can?" I said with a small smile slowly creeping across my young Robert Redford-like, beautiful-looking mug. Or maybe that fine day, I looked a lot like "The Duke"...

"Well, I don't know. But we are convinced they have broken the law and don't want to let them go without trying everything in the book in order to break them," he responded somewhat down in the mouth once again at having to ask for assistance from a lowly federal officer.

"Well, I can try. But don't get your hopes up. If the two of you have been questioning them for the last half-hour and have gotten nowhere, I am not so sure I can either," I said. "OK, let me try," I said, as I got an off-the-wall idea and headed for my parked patrol vehicle for some paperwork to support the goofy idea I now had swimming around in my head. I, too, was a hunter of men and didn't like any outlaw escaping from my clutches if I could help it. So, I was more than willing to give it a go towards breaking down the men's suspicious stories. Arriving at my truck, I opened up my briefcase and took out some paperwork I always carried. Then Jim and I headed for his group of suspect lads. As I approached, I closely examined the faces of my five suspects for any signs revealing guilt or nervousness.

Picture this. Bad guys with something to hide, seeing a game inspection sign ahead as they traveled down the highway after figuring they were home free. Then realizing they might be in a legal crack with all their antlers and horns if closely examined. Consternation quickly ruled, and they had stopped and dumped out all their illegal goods alongside the road to avoid apprehension. Then while trying to escape even further, found themselves being run to ground by the surprising appearance of a highway patrol officer. After that run of bad luck, they were quickly returned to the game inspection station they had tried to avoid. Shortly thereafter, they were surrounded by two hard-bitten game wardens and grilled for thirty minutes on something they did wrong, but tried in the process to maintain their fresh as a newborn baby innocence... Then after being grilled by the two game wardens, have one of them head off for reinforcements and seeking that of a federal officer at that! Then seeing this smallish Wyoming game warden returning with a monster-sized man walking alongside, presumably the terrible federal agent! And then thinking what had started out as a get-together with a nothing state game warden (in their minds) and now the. entire federal government was on their asses! No two ways about it, that had to curdle a normal wildlife outlaw's milk!

As I approached, I could see all of the above flashing across their now somewhat, eagerly looking at me, yet concerned faces. That was good, I thought, because that sign of worry would give me an edge. That, and I was sure all of those five men now had a heartbeat running at the same rate of a hummingbird, namely 180 beats per minute. Especially, now that reinforcements were on the way.

Walking up to the group of men, I noticed every one of them was looking at me like a Rhode Island Red chicken looking at a fat grub on the ground in the chicken pen. With that observation under my belt, I went into my act. I stood there for at least a minute, individually closely examining each and every man standing in front of me for the psychological effect that would have on their rapidly beating hearts and tightened anal sphincters. And they tried to do the same with me. Then I boomed out, "Good afternoon, Gentlemen. My name is Terry Grosz and I am a Special Agent for the U.S. Fish and Wildlife

Service. I, like other federal officers on-site, am here assisting the State of Wyoming at this wildlife inspection station.

We are here because once any illegal wildlife moves in interstate or foreign commerce, it is also a violation of federal law, namely the Lacey Act. Violation of that law provides for fines and incarceration in the federal penitentiary if one is found guilty. It also provides for the seizure of any equipment used in violation of that law, namely knives, rifles and vehicles utilized in taking or transporting said illegal game. Now as I understand it, you fellas have been talking to these two Wyoming officers and have denied any wrongdoing relative to the pile of antlers and horns you were seen dumping alongside the highway. However, the circumstances you men are operating under when you tossed the antlers and horns of freshly-killed animals and had to be brought back here under highway patrol escort, does not bode or look well. As I also understand it, you fellows have been offered the opportunity to confess and settle up with the State of Wyoming through the simple issuance of citations. That you have chosen not to do. So under the circumstances, these state officers have now deferred prosecution and requested my involvement as a federal officer because of the interstate transport nexus of illegal game. And have in so doing elevated this simple state violation into that of one to be prosecuted in federal court in Cheyenne, Wyoming."

After all that "gunsmoke" blown into the air, I now pulled out my "Winchester."

"All you men raise your right hands," I barked out. Man, all five of their right hands shot up into the cool Wyoming fall air as if shot out of a gun. "Repeat after me. I do solemnly swear to tell the truth, the whole truth and nothing but the whole truth, so help me God." To a man, my lads squeaked back to me the words I had bellowed out. Then I said, "Now, Gentlemen, I am handing out to each and every one of you a federal document. In it, I want every one of you in detail to write out exactly what you did, which laws you broke and which antlers or horns personally belong to you. Then I want each and every one of you to sign the document after you have read the warning at the bottom. Notice that it spells out that if you are lying in what you affix to this document, it constitutes perjury. An offense that is punishable for up to five years in prison and $5,000 in fines. Do all of you have a clear understanding of what I just said?" I asked in a rather stern tone of voice as I grinned mightily inside at my "semi-tough guy" behavior...

"Yes, Sir," all my lads uttered in concert. By now, from the looks on each of their faces, I could

tell my bombast and bellowing like a water buffalo stuck in a rice paddy words had struck an
inner chord in the bottom of their very black-
hearted souls.

Then I handed each man a government form to
be filled out and signed under the threat of perjury. Following that solemn action, I saw to it that
each man was separated and sent to a different
picnic table in the rest area so they could write
out their confessions in peace without "help" or
collusion from the other members of their party.

Without a word, Jim, Dave and I stood sternly
looking on at our charges, as they wrote out their
confessions. Me, I had to try really hard to keep
from laughing out loud. Laughing out loud over
my five unfortunates' hastily writing out their
confessions and even more importantly, over
the wool I had pulled over the eyes of my two
Wyoming game warden friends. Especially that
damn devil Jim, for what he had done to me the
day before by allowing the four old farts to chew
off my hind end without him helping me out...

About twenty minutes later, all five of my men
had finished writing out their confessions. As
they came up to me, they individually handed
me their documents. I read their statements and
made sure they had signed the documents. Then
satisfied with their work, I handed the documents over to my two still-thunderstruck game

wardens. As it turned out, every man of the group of five confessed in writing to numerous Wyoming wildlife violations. Even to the point of where they had hunted and where they had illegally left their untagged meat hanging in a rancher's barn. Two Wyoming wardens were quickly dispatched and I later learned they had caught the landowner red-handed. And in so doing, had seized all the illegal meat plus some not accounted for by proper hunting licenses and tags that the crooked rancher had himself shot!

Then I advised the five lads that if they settled up with the State of Wyoming, I would not seek federal prosecution. This they quickly agreed to do. Then without further ado, just like this had been a walk in the park, "I holstered my still-smoking pistol in my holster after making a great shot," like in the John Wayne movies. Then I casually walked back to my portion of the check station like it was an everyday event. As I walked away from Dave and Jim, I chanced a quick look back. Both wardens were still standing there with a "how the hell did he do that after all the work we had done had failed" look and I smiled. Now we are even, Jim, after that hosing you gave me yesterday with the four old farts, I thought inside with a huge grin...

An hour later as I worked at my portion of the check station, Terry Cleveland gave me the high sign. Then he said, "Look over your shoulder."

Turning around, I saw my five outlaws coming my way like a runaway horse with the bit in its teeth. Together, they had happily just signed off on over $5,000 worth of state tickets for their illegal deeds. But here they came with a dedicated walk in their steps. Approaching me, the leader of the group said, "Officer Grosz. We just wanted to say thank you. You were more than fair with us and we wanted to show our appreciation even though we caused a lot of you folks some trouble. So, thank you!" And with that, all five men trooped by in front of all those officers still working the check station and shook my hand. Then with a wave of their hands, they entered their truck and left the scene. Man, I thought, there you go, Jim Oudin. Just leave me there undefended again with four old and mad as wet hens folks and see what else can happen to your miserable carcass.

Then Terry Cleveland walked over to me saying, "One doesn't see that too often. Especially when you just took over $5,000 from their hides. I guess you could say you and Jim are now even after what he did to you yesterday," he chuckled with a knowing grin on his face.

I had to grin inside as well. After all, it is not often a lowly federal could put one over on those hardened old bastards Dave Bragonier and Jim Oudin. Like I said earlier, I put my smoking pistol back into my holster after making one hell of a good shot, just like John Wayne had done so many times in his movies.

In my seventeen years as the Special Agent in Charge of the eight-state Rocky Mountain-Prairie region, I initiated over 25 major roadblocks. Additionally, I would imagine way over $500,000 in fine monies were generated from those roadblocks catching those walking on the dark side of the world of wildlife. But in addition to nicking the hides of those violators, we had many thousands of fine sportsmen pass through our portals during those roadblocks. Good sportsmen who had obeyed the laws of the land and stood by, condemning those we were in the process of apprehending.

In today's Fish and Wildlife Service as these words are being written in 2011, and as I understand it, hardly any major federally sponsored roadblocks are being utilized as a wildlife law enforcement management tool. That is too bad. So much was gained by such operations and so much is being lost by their absence. I guess in today's world of wildlife, it is a matter of one's supervisors or leaders that if they haven't been there, they don't know how to lead.

## FRIDAY NIGHT LIGHTS

My oldest son Richard was 6' 7" in high school, quick as a weasel and strong as a bull. So being such a stork, it was only natural that he played basketball at his high school in Evergreen, Colorado. And it was only natural that my Bride Donna and I went every Friday night to his school team's basketball games in the fall. Especially since they had a winning record and were headed to the state basketball tournament for the first time in the school's history!

So come every Friday night during basketball season, Donna and I would quickly feed our family, then head out to the game so we could get a good seat in the bleachers. Once there, we soon would be surrounded by the parents of the other basketball players on the Evergreen team. Everyone knew everyone else and soon we formed a strong rooting section for our team.

One November basketball game I remember very well. All of our usual families had gathered in our rooting section in an already filled to capacity gym. As we waited for the exciting game to start against a bigger high school rival, I noticed three men heading for our rooting section. I casually knew all three men who didn't have any kids on the team but were just there for the game's excitement. Oh, did I mention that all three were defense attorneys? Now all my quick readers out there who know me and my writing style, don't go jumping to conclusions. I just wanted all of you to know the men were defense attorneys, that is all... No, I didn't say they were the weakest link in the chain of democracy...

As it turned out, the three attorneys moved into our section of the bleachers and sat directly behind Donna and me. No big deal. They had paid their entrance fees, purchased their popcorn and were entitled to sit wherever they wanted. However, God had chosen where they were to sit that evening as my readers will soon see. Now there, you readers who know my writing style and my conservative bent, go once again. All of you can see a wreck coming. Just stop it!

Then the game started and soon all of us were into the action out on the floor. Well, that was except for one rather large, Robert Redford-the-younger, like-looking fellow sitting in the bleachers. You know, sitting in the bleachers minding my own business... Well as it turned out, I just happened to be sitting in front of the three attorneys. And, it wasn't my fault that I just so happened to be listening to their semi-private conversations. Well, how the hell could I not overhear what they were saying? After all, they had chosen to sit right behind me some two feet away and were talking like they owned the place! And by now, most of you readers know, all federal agents are always listening to what is going on around them.

"Damn, those are sure nice-looking cowboy

boots. Where the hell did you get them?" asked one of the three attorneys sitting behind me.

"I just purchased them in Mexico. They are Justin Boot Company knock-offs and man, they sure are good-looking and comfortable. And they only cost me $100 American," said the attorney wearing the nice cowboy boots.

"What kind of leather is that on your boots?" asked the first attorney.

"They are damn good-looking, aren't they? They are made from sea turtle and sure are, like I said, comfortable as all get-out," replied the second attorney and wearer of the boots.

Hearing the words "sea turtle," which just happened to be an endangered species, and one that could not be legally imported into the United States as cowboy boots, had me perking up my bat-like ears. It was then that I saw my loving and knowing Bride looking hard at me with those beautiful blue eyes of hers. She of course knew I was a federal wildlife officer, understood that one could not import endangered species contrary to the Endangered Species Act of 1973, and knew I was carrying my badge and credentials on my person (as I always did) that evening... She also knew I was always on the "hunt," like a dog who sticks his head out the window in a moving truck... That was even though we were more or less on neutral ground at our son's basketball game. And her beautiful blue eyes just telegraphed: "Just for once can we not create a ruckus and for a change enjoy the game?" Loving the hell out of my wife and not wanting to create a scene, I tried to relax and not listen to the conversation ongoing behind me. Note the operative word, "tried."

Then attorney 'number one' said, "How did you bring those boots into the United States? Aren't sea turtles supposed to be endangered?"

The attorney wearing the beautiful sea turtle boots said, "Yes, they are endangered. But all one has to do is purchase them on the Mexican side of the border, where by the way they are illegal as well. Then just wear them across through Customs (before the days Immigration and Customs were joined together into the Immigration and Customs Service, or ICE as they are known today). The Mexican Government doesn't enforce the law and if you wear them across the U.S. border, all you have to do is tell the dumb Customs bastards that you have had the boots for years and they are pre-Act. Those damn dummies don't know any better and will just let you walk across. Hell, I did that twice this last trip and now have two brand-new pairs of sea turtle boots. In fact, the other pair is still in my Suburban out in the parking lot. I didn't have time to go home and unpack from this last trip or

I would have missed this game. So, all my clothes and new boots are still in my car."

It was at that point my beautiful wife and Bride of many years just looked at me like a small rabbit would look at a very large long tailed weasel from two feet distant. She damn well knew "those just-spoken words" were like someone had just dragged a gob of worms (the sea turtle boots) across in front of a rather large carp (me) and the worms were soon to be history!

Turning, I looked at the attorney's boots. Yep, they were green sea turtle boots, I thought.

Then looking up at the sea turtle boot wearer, I said, "I couldn't help but overhearing your conversation about your new cowboy boots (remember, the attorney didn't know me from "Ned" or a pair of also-endangered hawksbill sea turtle boots). What did you say they were made from?"

"They are from a sea turtle. Good-looking boots, aren't they?" the attorney "target" responded with a "proud as a new dad" look on his face.

"Damn, they sure are. Didn't I hear one of you say they were endangered and couldn't be brought into the United States legally?" I asked like an admiring dummy of the nth degree.

"Sure did," he replied.

"If that was the case, how did you get them across the border without being caught? I am asking, because I would like to have a pair like that. And I need to know how you did it so I can do it the same way and not get caught when I go into Mexico the next time," I said.

That was when I realized my wife wasn't the only one listening to my "sea turtle boot" conversation. Everyone else sitting around both me and the three attorneys knew who I was and what I did for a living. Every one of them realized I was now on the hunt and a strike was near at hand. So all were listening to the game going on on the hardwood between Evergreen and Mullen High Schools with one ear and eye, and the other eye and ear on the ongoing 'game' being played out and unfolding in the bleachers.

"It is simple," continued the target attorney in response to my question on how he got his restricted boots across the border. "As I was just telling Dave here, all you have to do is purchase a new pair of boots, put them on and wear them across the border. That way you scuff up the soles of the boots as you are walking, making them look like they are old and used, and when Customs recognizes the boots as being sea turtle, they will stop you. Then you tell those dummies that you have had them for some time and then show them the scuff marks on the soles. Since all they are looking for are drugs, they will believe you, and let you cross the border into the United

States slicker than all get-out. And then, voila, you have a beautiful pair of sea turtle boots that can't be purchased here in the United States," he said with a helpful neighborly grin.

I had deliberately asked the dumb-assed questions because I wanted to establish culpability on the part of my sea turtle boot wearer. In short, I wanted to make damn sure he knew he was knowingly in violation of U.S. law. By asking such a question, my attorney "friend" had just crossed the "t" and dotted the "i" to my seeking his level of knowledge and guilt regarding the endangered species import and smuggling violations.

Reaching into my shirt pocket, I withdrew my federal badge and credentials. Opening them up, I flashed them so my sea turtle boot wearer could see the "gold." Then I said, "My name is Terry Grosz and I am a federal agent with the U.S. Fish and Wildlife Service. As such, I overheard your conversation relative to how you came by the sea turtle boots. Then I also asked you the question as to how you came by them. In your answers, you informed me as to how you smuggled them into the country. Which by the way, smuggling is also another federal offense in addition to your seeming violation of the Endangered Species Act. As such, you cannot possess anything illegally smuggled into the country like the sea turtle boots you are currently wearing. I also overheard you bragging about another pair of sea turtle boots you recently smuggled into this country which are still out in your car. Now, I am going to seize the boots you are wearing and the other pair in your car because they are contraband and evidence of a federal crime. Since I am going to be doing that, I suggest we quietly leave this area so as not to embarrass you any further or create a circus among your friends. And I suggest we do that right now," I continued in a tone of voice that would be clearly understood even by a defense attorney.

With that, I stood and indicated with my hand he was to take the lead in leaving the bleachers. As for the still-in-shock looks on his and his buddies' faces, they were priceless! Then, understanding my earlier voiced meaning, my sea turtle boot wearer arose and left the bleachers. Me, I casually followed so as not to create a show. But before I left, I could now see that all our basketball friends in the bleachers were looking at me tailing my 'catch' and none of them were watching the game. In fact, as I walked by our team all under the basket getting ready for my son to make a free throw, their eyes to a man were on my little caravan as well. Most didn't have the foggiest as to why I, their number one rooting fan (and noisiest), was leaving such an important game. However, from the look on my free throw shooting son's face, he knew what his old man was up to.

Once outside in the parking lot, I took down the attorney's personal information in my ever-present notebook. Then I seized his new pair of sea turtle boots from his car and had him remove the ones he was wearing! Keep in mind, this was wintertime in Colorado at about 9,000-feet elevation. Finishing up with the paperwork, I advised him that the clerk of the federal court would soon be in contact with him regarding this matter in the form of a mail-in citation. Since he had no questions, I gathered up my two pairs of seized sea turtle leather boots and headed back into the gym so I could finish watching the basketball game. Which from the sounds the crowd was making, the game was a good one.

Walking down the sidelines of the basketball court carrying my two pair of recently acquired sea turtle boots, I noticed my section of the bleachers were all eyes. Walking back to my seat, I placed my two pairs of seized boots by my feet for safekeeping. Then I began watching the game once again, like nothing out of the ordinary had happened. However, I was well aware there were about twenty sets of eyes from my section of the bleachers now fastened onto the two pair of boots sitting by my feet. And the original owner and wearer was not "standing" in them!

Then the "funny" for the evening really happened. Shortly after I had seated myself, here came my bragging sea turtle boot smuggler walking back to his seat. The funny thing was that he was walking without any shoes on his feet! He was now walking awkwardly in a pair of oversize rubber boot galoshes! As he "slopped" his way down towards where he had been seated before he had been scooped up from his place in the bleachers and marched out from the basketball court, all eyes from that section were now on him. Or should I say, his awkward set of boots and the fumbling way he was walking, trying hard not to step out of them with his stocking feet.

Finally standing at the bottom of the bleachers on the gym floor and not looking at me, he beckoned for his two other attorney friends to come down. They did and without a word being spoken, all three of my legal beagles left the basketball game and presumably went home. Then and only then, my section of the bleachers came alive with questions as to what had happened and for a moment, the important basketball game was forgotten. In the meantime, my poor mortified Bride just looked at me, realizing her husband was a hopeless mess. Just like "that slobbering dog who was hanging his head out the window of a moving pickup."

Several weeks later, my sea turtle boot smuggler attorney forfeited bail to the tune of $1,000. As for my poor sea turtles who were killed and sold under the crooked auspices of the Mexican Government so my attorney could have two pair of Justin Boot Company knock-offs, "they" ended up with some form of revenge. As for me, that little episode did nothing more than strengthen my opinion that defense attorneys are still the weakest link in the chain of democracy!

Oh, by the way, Evergreen beat Mullen for the first time in their life during that game and my son scored 16 points and had 10 rebounds. Later the little outclassed Evergreen team went to state and finished third. Truth be known, they damn near beat Montbello, the team that went on to win the state basketball tournament. An ankle injury to one of Evergreen's first string centers cost them that closely fought game. When the winning team was later interviewed, they to a player said the toughest team they played all year was Evergreen.

Regardless of that fact, my attorney "friend" had to long remember that game between Evergreen and Mullen. Or should I say, what little of it he saw before he got cold feet in a not-so-beautiful pair of oversize rubber galoshes because he "somehow" had lost his cowboy boots...

## BEANS, BULLETS AND BLACK OIL

Ever since European man arrived in North America, he has exhibited a unique behavior that has transcended even unto this day. Fleeing land controls, persecution of many kinds, stifling folkways, limited freedoms and restricted resources on the European continent, historical humankind headed west across the Atlantic to what promised to be a better life. Once in the Americas, they were overwhelmed by the "as far as they could see" open spaces, vast resource riches and almost unlimited freedoms. Faced with such bounty, the new inhabitants quickly threw off their motherland's restrictive mental shackles and since that day, have been rabid consumers of just about everything within reach. Their wasteful consumption of the land's wildlife resources was so ferocious and destructive, that infant conservation laws out of concern for the resource began appearing on society's books in America as early as 1690! And that was when we had few people throughout the land and the technology of destruction was in its infancy.

Even with fledgling concerns and limited conservation practices, many examples of man's destructive resource excesses continued like a whirlwind-driven fire across the prairie grasses. Destruction of vast flocks of passenger pigeons numbering in the billions for protein-hungry cities quickly passed the point of common sense harvest and into the black hole of extinction in just a few years! Paralleling that insane inhumanity was the rapid elimination of East Coast virgin oak forests for sailing ships' masts. Collaterally, runs of Atlantic salmon were quickly eliminated through the construction of dams, overfishing and pulp mill pollution. Then as the masses of humankind spilled westward, destruction of white pine forests through wasteful logging practices and numerous deadly forest fires followed man's heavy footprint. Expanding ever westward soon resulted in the almost total destruction of millions of buffalo for the leather trade, which, in turn destroyed many Native American peoples' way of life. Pelagically, man's excesses led to the shooting and clubbing almost to extinction of millions of Northern sea otters and fur seals for the value of their furs. That was soon followed with the destruction of millions of acres of native prairies by newly arriving European immigrants through wasteful land-use practices. Finally conquering the oceans, man soon overhunted the great pods of whales for their oils, baleen, ambergris and ivory. Those and many other destructive resource-use practices marched numerous species of flora and fauna almost to the brink of extinction by the Nation's seemingly uncaring and destructive peoples.

By the late 1800's, many resource species were in severe distress or alarming decline. It was then many enlightened people began clamoring for control over such rampant destruction and commercial market hunting of the Nation's resources. But the taste for species destruction and its collateral profits was so great that hardly a dent was made in America's vast runaway destructive resource practices. However, by 1900, the killing and clamoring for conservation controls had both reached an extreme fever pitch in our still-fledgling, East Coast society.

As part of this new awareness within the conservation movement, American presidents began setting aside great reserves of forest lands, wildlife sanctuaries and wetland preserves. Collateral with land set-asides, Congress initiated and supported the creation of new conservation laws to curtail the almost out of control destruction of our now quickly dwindling natural resources. Simultaneously, state and national organizations were created to act as watchdogs over the resources of the land and to implement the new conservation laws. Created at the same time within these organizations were numbers of professional land managers who began overseeing the management and enforcement of these new land creations and those remaining species inhabiting them.

One such land management agency, created in 1871, was ultimately called the United States Fish and Wildlife Service. This organization was created to manage those lands set aside as national wildlife preserves along with their many fisheries resources. Within that agency's scope of duties came care and management of those species on said lands, provision of the research into the management and life history complexities of plant and animal communities, and the development of conservation regulations relative to their protection, care and management.

Lastly, the agency was tasked with the regulation and control of those human communities utilizing such landed resources. It is to that federal 'tool' of wildlife law enforcement and some of its problems, that I wish to examine further.

In the early 1900's, the first of such statutes and regulations handed down to the Fish and Wildlife Service were designed to protect, preserve and enhance many species of fish and wildlife. Done so in part because of the still-great injustices being committed against the now specifically identified and federally protected species of fish and wildlife. Lastly, said protections were handed down because of the still-great hue and cry from the American people demanding protection for what was left for future generations to experience and enjoy.

To facilitate protections granted under those new statutes and regulations, a branch was created within the Service called the Division of

Law Enforcement. It is currently staffed with well-trained professional criminal investigative officers at numerous ports of entry and duty stations across the United States and its territories. However, wildlife inspectors at the ports of entry and limited numbers of Special Agents staff those positions in an almost vain attempt to control the many illegalities being perpetuated against the Nation's valuable plant, fish and wildlife resources.

In 1970, when I came on board with the Service as a United States Game Management Agent, there were 178 officers nationwide enforcing approximately 20 federal wildlife statutes. Today, there are a little over 200 officers nationwide enforcing over 40 such wildlife statutes. Custer had better odds in managing his historical situation then do today's Special Agents, and look what it got him...

As if the numbers disadvantage wasn't a large enough albatross to carry, there existed a major philosophical issue that is an even darker element to this lack of officers and their protective nationwide net. The American people, all 300-plus million of them, do not like being regulated in any way, shape or form. This country was founded on a fierce spirit of individualism, independence, lack of controls, freedom, vast resources at every turn in the road, and what is generally perceived among the masses, as a God-given right to do as one pleases...

If that philosophical difference was not enough of a burden for the 200-plus federal officer corps trying to hold the line against overuse (bear in mind these federal officers are working in conjunction with state wildlife officers, who are the first line of defense), how about this: The Service's senior and mid-level management folks (like many of those in state management organizations), are largely comprised of somewhat liberal personalities trained in today's academic institutions by those of like philosophical ilk. Most such managers are trained and skilled in the general principles of fisheries and wildlife management.

However, there are five basic wildlife management 'tools' as laid out by visionary Aldo Leopold, long considered the father of modern-day wildlife management. One of those "tools," not the biggest or best, mind you, but just one of those "tools," is law enforcement. Truth be known, you can pass all the regulations you want. But if you don't enforce those regulations, that is, provide a "law enforcement bite" for those straying across the line of legality who have their hands in Mother Nature's cookie jar, you will have nothing left in your resource container. In the scheme of wildlife management, one needs a regulatory function along with all the other tools or you will find your wildlife management programs slowly being destroyed, or at best, left wanting.

Today, unfortunately, many wildlife, fisheries and land management resource managers use only four of the five wildlife management tools set down by Leopold. Part of the problem is that most liberal universities have never "been there" or "done that." Therefore, they don't fully understand the need for and use of law enforcement as a wildlife management tool. As a result, law enforcement is not really taught at most levels of higher education because of its "use" misunderstanding or the distinct dislike for teaching such a discipline.

To my readers, to be a good law enforcement type, one has to be multi-faceted. That profession requires a high level of intelligence to address the legal complexities of the profession, lots of brass, a backbone, curiosity, and being a hunter of humans. It also requires one carry a force of dedication in one's heart and soul, that transcends the fear of injury, death, lawsuit or public criticism. To my way of thinking, based on my 32 years of experience in the law enforcement arena and sporting three degrees in wildlife management, I feel many of today's wildlife managers lack many of those traits necessary to make exacting use of all five wildlife management tools.

Additionally, there is an even more insidious

nature that surfaces among the national corps of wildlife and fisheries managers. There are many senior managers within the ranks of the Service who are down-right afraid of this law enforcement tool. That is, they don't know how to use the law enforcement tool. Truth be known, some have been "pinched" by law enforcement, are afraid of the negative politics law enforcement generates (problems created when a politician or important person is apprehended), were never academically trained in its use, just don't give a damn, don't appreciate its value as a management tool, or all of the above...

To any of you senior managers within the Service or from another fish and game agency who identify with these lines, or in reading them find yourself grimacing over being "targeted," examine carefully what I have said. If you are one of those souls only using four of the five wildlife management tools, you may see the hands of your management style negatively reflected in the life histories of those species you are managing. To those of you managers using the law enforcement tool, the critters and I thank you for your courage and insight. So do those yet to come regarding that matter as well...

The above rambling finally brings me to the current gist and issue of this story. Since there are so many senior managers lacking or practicing this fifth wildlife management tool in very high places controlling the purse strings, guess where the money isn't going?

When I became an agent in the Service in 1970, the Division of Law Enforcement was flat broke! When I ended my career in 1998 due to mandatory retirement, the Division of Law Enforcement was still strapped for operating funds. During my entire 28-year Service career, I can remember only three years when I had enough operating funds in which to just meet the federally mandated requirements of the job! And those three good years were because I broke agency policy and personally went begging to a Colorado Congressional type for enough operating funds so my professionally trained officers could do their jobs.

I even bitterly remember three years during my entire career when Congress finally got off their dead backsides and had a mandated budget passed for the start of the fiscal year by October 1! And in each of those three years in the middle '80's when a budget arrived on the anointed day of October 1, I was $30,000 in the red before my officers did a lick of work! I didn't even have enough funding to pay for the salaries and benefits for the staff I currently had on board! The rest of the branches of the Service had money in which to operate but there was none for the Division of Law Enforcement! And that was when the Division of Law Enforcement was governed by an Assistant Director for Refuges and Law Enforcement. You know, being led by one of those senior managers who was only familiar with or respected four wildlife management tools instead of five...

Part of that blame flatly rests with the Division of Law Enforcement itself. Field law enforcement was one of the greatest professions within the Service. Most field-grade officers were on a 'vision quest' as law enforcement officers and dearly loved their profession, allowing for such peregrinations. As such, few officers ever tried to infiltrate the senior Service officer ranks or showed an interest. And one can't lead unless he has been there and we weren't. We, for the most part, happily remained in the field with our critters and our freedoms. As a result, our division suffered because of our lack of subsequent involvement and lack of leadership at the highest of levels.

Secondly, personality issues were rife within the higher ranks in the Service. The then-Chief of Law Enforcement of that era, Clark Bavin, could be a "hind end" at the drop of a hat and was not respected by many other Service higher-ups. Instead of sharing law enforcement information that could be safely shared, he chose not to do so.

In short, he lost a valuable opportunity to train other senior Service managers in the how's and why's of law enforcement operations. In fact, many times when others outside the division expressed an interest in what law enforcement was doing, they were rebuffed. Rebuffed with such words as "that if they weren't in the loop and had a need to know, then they weren't told anything." That behavior in part led to a dearth of funds when the buckets of "golden guilders" were passed around to the various divisions at the start of the fiscal year, if you get my drift...

Lastly, most of the Service senior management leadership had no idea as to how law enforcement fit into the wildlife management scheme of things in the world of wildlife. And those in senior leadership positions in law enforcement never bothered to educate them! Since senior management, on both sides of the aisle, had little or no information as how to utilize that fifth or last management tool, it was left unused in the tool box.

I personally had one mean-spirited Assistant Director who was in charge of Refuges and Law Enforcement, tell me in anguish towards the end of the fiscal year that he had thousands of dollars he couldn't find a way in which to spend. He knew if he didn't spend those monies, they would have to be returned to the General Fund at the end of the year. Not being the dullest knife in the drawer, I piped up saying, "Dave, Law Enforcement sure could use those monies for new vehicle purchases. Most of the division's vehicles are running on their last legs and we sure could put that money to excellent use."

Assistant Director Dave Olson turned and looked hard at me for a moment. Then he said, "I will have my refuge managers paint their equipment barns one more time this year just to use up that money before I give one red cent to you guys in Law Enforcement!" Suffice to say, Dave did not have the foggiest idea as to how law enforcement fit into the management scheme of things in the world of wildlife management, nor did he care. So with that background history and flurry of words, I am now finally ready (whew...) to get on with the story at hand.

One fine day towards the end of my career, Chief of Law Enforcement John Doggett (a good guy) asked if I would represent the division at a Congressional hearing into the lack of funding for the Division of Law Enforcement. It seemed that a hearing had been called by Members of Congress after a host of non-governmental organizations, American sportsmen and Congressional staffers favoring the use of more law enforcement, had raised the lack of funding issue to high Service levels. The ongoing thinking among those groups and organizations was that the Service had a ton of money in their operating budgets and the Division of Law Enforcement, as usual, had none. Basically, the hearing was to establish why the Division of Law Enforcement had little or no money, when the rest of the Service was more or less funded.

However, one needs to understand that a hearing could be a double-edged sword. If one said the wrong thing to the Congressional group, there could be hell to pay. Especially if one ended up embarrassing some of the fat cat politicals within the agency or any of the 'lesser Gods' between the political Titans and the rest of the agency heads.

I advised the Chief that I would be glad to do what I could and after a few other pleasantries, the call was terminated. Sitting back into my chair, I got to thinking. Then it hit me. Out of the rest of my seven senior counterparts, I was the closest to retirement... That was why the Chief had called me asking if I would stick my head into a possible political grinder. Then I got to smiling. I was a senior law enforcement representative, had numerous years of experience under my belt, and had basically been there and done that. I had received numerous awards for my excellence of service and represented a very active district. So, I thought, what the hell. I figured the division had nothing to lose because we had nothing as it now stood, in the way of respectable funding, and everything to gain. So, I had agreed.

I also knew another reason as to why I had been 'personally' called. I, too, had friends in Congress who were worried that the Division of Law Enforcement was being kept down by the Service leadership through inadequate funding. As a result, I had been approached by several key funding Congressional staffers on the 'Hill' and asked if given the opportunity, would I speak frankly to the issue of inadequate funding. To that I said, "Ask me the right questions and I will give you the correct answers." Shortly thereafter, Chief Doggett called me regarding my availability for a Congressional hearing and I was on the "team."

Several weeks later, I was notified through official channels that Acting Director Dick Smith, Law Enforcement Chief John Doggett and Special Agent in Charge Terry Grosz would be testifying on the Hill before a Congressional committee regarding charges of underfunding the Division of Law Enforcement by the Service. I just grinned after reading that missive. It was amazing the power those unnamed staffers held over their elected bosses. Well, not really. It seems to me that those folks quietly serving under their 'commanders' are always the ones doing the work and getting few if any of the rewards or recognition...

Come the day to testify found me in Washington, D.C., all dolled up in a suit and tie if you readers can imagine that! Catching a cab with Chief John Doggett, we went to the Main Interior Building where we met Acting Director Dick Smith. From there, the three of us caught another cab over to the Hill to testify.

Dick was in his usual grumpy mood at having to testify in front of what he considered a mess of dingbats in Congress on a fool's mission. Chief Doggett was holding his breath so he wouldn't get swept off his feet into this political mess and killed in the process. Me, well, I was retiring in a few years, had a good work record and was having one hell of a good time knowing what was to come...

Arriving in the Congressional hearing room set aside for such endeavors, the three of us were ushered inside. At one end of the room sat a high up row of chairs and long table staffed with serious-looking Congressional types surrounded by their staffers. Out on the floor of the hearing room at a lower level, sat three chairs behind a table littered with microphones and bottles of water for those of us folks on the hot seat. Dick took his seat on the far right, Chief Doggett sat in the middle, and I occupied the chair to the left (can any of you readers who know me, see me sitting on the "left" philosophically, ideologically or otherwise?). Then the meeting was called into session and the three of us were duly sworn in. Having gone through such foolishness before, I figured I would just watch Chief Doggett and Acting Director "Grumpy" Smith and then follow suit.

Dick being the Acting Director of the Service, was called upon first by the chairman of the committee.

"Mr. Director. The purpose of this hearing today is to discuss the troubling word we have received from many of our constituents that the Division of Law Enforcement, under your direction, is so largely underfunded that it cannot do the job Congress has mandated. I have before me numerous complaints from not only our people back home but national organizations as well. All of them are complaining that a serious funding shortfall exists within the Division of Law Enforcement. As a result, the Special Agents of the Service are no longer present in the field in such numbers as is required to do their jobs professionally. Would you care to comment?"

Grumbling, Dick said in a voice loud enough to really be heard and possibly intimidate the Congressional folks, said, "There is no problem. The Service supports the President's budget and there is sufficient funding in which to operate. In fact, if the Division of Law Enforcement would manage their funding better, we wouldn't be here today!"

God, I thought, what a tough old bird. Totally wrong in his response and assessment as to law enforcement's problems, but as Acting Director of the Service, within his rights to say what he wanted.

With that, it was obvious from Dick's body language that he had nothing more to say regarding this matter. Now to those of you who have never met Dick Smith, you are in for a RIDE and a real managerial treat...

First of all, he is a big fellow and prone to intimidate. His booming voice is always loud, abrupt and intimidating to those folks possessing the heartbeat of a hummingbird. Especially so to those who are subordinate staff to his "Eminence." I had the "pleasure" of sitting through a meeting with him in Marana, Arizona, at a training session one year. There he confronted all the Nation's Special Agents in Charge for the Service. In that session he asked why we, the Special Agents in Charge, all GS-14's, hadn't put in for our GS-15's (grades of rank within government service). (The higher the grade — the higher the rank, the higher the pay, and the higher complexity of the job.) Not one of my counterparts said a word out of fear of being eaten alive by the Acting Director and head of the Service. That is except for yours truly. I am also a big guy, and always loud and noisy according to those normally associated with me. I had also been shot at six times in my career and only hit once. I had been in more fights then I need to account for with hard cases, was raised under hardscrabble living conditions, raised by a single mom, and logged in the woods for three years with a ton of logger hard cases before I graduated from college. Suffice to say, I was not easily intimidated by Dick, the "blow hard."

Looking him hard in his glaring and challenging eyes at that Arizona meeting, I said, "We have not put in for our '15's' even though we have earned them, because you would just tear up the paperwork or refuse to sign off on them." (He was finally forced to sign off on our GS-15's many months later, when I asked for a desk audit—an audit that came back as a solid GS-15 or possibly a GS-16, if we still had that rank.) That, and I had the strong backing from one of the best Regional Directors in the Service, Ralph Morgenweck, to execute and support such a move. However, prompt action was not taken, costing me about $6,000 per year in lost retirement because of Dick's late signing off on my promotion and me having to mandatorily retire before I had my three high years in that higher grade...

"You got that right!" bellowed Dick. "As long as you guys get extra pay for the odd and unusual hours you work and have a government vehicle at your beck and call, I will never sign off on your raises. Get rid of your cars and your extra pay for the long hours you work and I will sign off on the '15's,'" he growled.

How is that for being a good, supportive and understanding senior manager? Especially in light of the fact that Congress had established, not the Service, the extra pay schedules and uses of patrol cars for every federally credentialed law enforcement officer in government! Oh well, Dick never did like anyone in the Division of Law Enforcement or support its mission within the world of wildlife management, so I guess I could understand what I considered his out-of-line and juvenile behavior.

But just for the record, I personally liked the old, grouchy, overbearing son-of-a-bitch. Loud and the like, he would at least make a decision, unlike many other Service higher-ups I had known. And then he would stare hard at his challengers over his decision making process and later laugh at their weaknesses when not challenged. But I digress... Now, back to the Congressional hearing and all its warts...

Shaking their collective heads at "Soapy" Smith's defiant response, the Congressman turned hopefully to Chief Doggett. "Mr. Doggett, would you care to comment as to our previously stated concerns?"

Without missing a beat, John sat up straight in his chair and said, "Mr. Chairman, the Division of Law Enforcement supports the President's budget. However, we do have a small problem with our level of funding but with better management of our resources, I am sure we can overcome our problems." With that, he leaned back into his chair and it was also apparent from his body language that he had no further testimony to offer.

It was also obvious from the many panicked faces of the supporting staffers up on the podium with their bosses, that they were getting concerned. They had told their bosses of the division's frightful funding plight and had garnered their political support for a time-consuming hearing. But now in the face of the non-supporting testimony from two of the Service's top leaders, they were now looking like "Chicken Little" in the eyes of their bosses! That feeling was more than reinforced to my way of thinking after looking at the harsh looks the Congressmen were now giving their staffers out of the corners of their eyes. Kind of like, "Why are we here today?" Turning from looking sternly at his staffers, the chairman looked hard at me. "Well, Mr. Grosz, that leaves you. Since your two superiors said nothing contrary to the President's budget, I suppose you are in the same boat?"

By now four staffers were really looking hard at me, as if to say, "Please bail us out of this mess, Terry." Looking over at Dick and John for any final guidance, I got none. They were just staring straight ahead and hoping all this Congressional scrutiny would soon go away. *OK*, I thought, *damn the torpedoes!*

"Mr. Chairman," I began in a voice that denoted I understood the seriousness of his issue and question, "we in the division of law enforcement need beans, bullets and black oil!" With those words, I just sat there looking back hard at the chairman. I could hear a buzzing of surprise from the gallery behind me of non-governmental organization representatives, who were there to emphasize the pathetic funding state the Division of Law Enforcement found itself in. The Congressmen and their staffers, upon hearing those words, just sat there with dazed looks on their faces as they tried to interpret what the hell I had just said.

"What did you just say, Mr. Grosz?" asked the surprised chairman as he leaned forward over his large table, as if not wanting to miss a word in case I uttered them again.

"Mr. Chairman, we in the Division of Law

Enforcement desperately need beans, bullets and black oil," I again responded, only this time in a tenor of voice everyone understood. Then I just sat there like a great big toad waiting for an opportunity at a big juicy fly if he did not show the proper common sense at flying by my miserable carcass too closely...

Again, there was more hurried talk between the Congressmen and their staffers, as well as some coming from the full gallery of supporters behind where I sat. Looking over at "Soapy" Smith and John out of the corner of my eye, I could see neither of them were moving. They, too, were sitting there like two big toads only they were looking like they were hoping I would just go away. You would have thought they were poured in cement.

"Mr. Grosz. Would you care to explain what you just said," asked the chairman in a voice indicating he was now totally confused over what the hell was happening.

"Certainly, Mr. Chairman. We in the Division of Law Enforcement have had hard fiscal times ever since I joined the Service in 1970. I can't remember hardly a year up to this date in time, in which we have had enough money to professionally do our jobs." Man, with those words, you should have seen the looks on the faces of the Congressional staffers! They were not only priceless but much relieved. As for "Soapy" Smith and John, their faces were still poured in cement...

Continuing, I said, "Mr. Chairman. When I said we needed beans, bullets and black oil, that is just what I meant. The division does not have enough current funding to provide us the ability to conduct routine field investigations. In short, our current level of funding does not allow us to maintain our Congressionally mandated presence on-line or in the field. There is never enough funding for normal uses of per diem, nor money for fuel or replacement of our worn-out vehicles! As an example, I am driving a vehicle today that has 206,000 miles on it! By agency policy, I am supposed to surplus that vehicle off the line because of safety reasons, at 50,000 miles! I cannot afford to do so. Additionally, I have other officers whose vehicles are so worn-out, they are trading them to our refuge division counterparts for their worn-out vehicles. We are doing so because refuges' worn out vehicles have less mileage on them than do our vehicles! and contrary to the testimony you have received here today, I do know how to manage my money! I challenge anyone to look at my books and show me where I have mismanaged or carelessly misspent one single taxpayer's darn nickel." With those stinging words, I looked over to where John and Dick sat looking for a challenge over what I had just uttered. Now both of those chaps looked like their faces were like those carved on Mt. Rushmore... Seeing nothing coming from those two chaps in the form of guidance, I continued with my obviously out of step testimony when compared with that from Acting Director "Soapy" Smith and "I hope they don't see me" Chief Doggett...

"Furthermore, Mr. Chairman, over the last few years, the Division of Law Enforcement has lost six of its top 13 Special Agents in Charge and six of its Assistant Special Agents in Charge positions because of budget shortfalls. Leadership positions one does not do away with when enforcing complex laws and managing difficult covert and overt operations. During that same time period, we have gone from 45 Senior Resident Agent positions or first-line supervisors to just 22. Once again, victims of the Service-directed budget axes. Again, first-line leadership positions sorely needed when enforcing Congressional mandates. And that was with the number of major complex laws we are Congressionally mandated to enforce going from 20 to 42. Lastly during the last 20 years, the overall Service numbers have increased numerically by about 73%, while the Division of Law Enforcement's numbers have been reduced by about 7%! Yes, Mr. Chairman, the division of law enforcement is in serious financial trouble. As I stated under oath earlier, we need beans, bullets and black oil!"

With that, I sat back in my chair and waited for what was to happen next. Dick and John, somewhat discredited by my testimony, continued looking straight ahead like nothing out of the ordinary had happened, as they remained "cemented in position"... Meanwhile, the Congressional folks huddled among themselves for a few minutes talking quietly. Then I noticed that one of my staffer friends, holding his hand so Dick and John could not see what he was doing, gave me a thumbs-up...

"Well, Gentlemen," continued the chairman as he turned around and faced the three of us, "this certainly has been an interesting day. I have never heard such a request for relief presented in such a fashion, but it is one that will stick in my mind for the rest of time. Thank you, Mr. Grosz, for not wasting our time today and being ready with such facts and figures that my colleagues and I can carefully ponder over at a later time. At this time, I cannot advise as to what our actions will be following your testimony, but Acting Director Smith will hear from this committee shortly."

With that, the chairman closed the hearing. Then the Congressional group arose and left the chambers. Seeing our job was done, Dick and John got up to leave and I followed. As I passed the gallery of folks who were there that day in support of the Division of Law Enforcement, I received several "pats" on my shoulder for my testimony. That was followed by several others saying, "Good job!"

Dick, John and I quietly walked to the outside of the Congressional Building to hail a cab so we could return to work. It was then that Dick abruptly turned and said gruffly to me, "You folded!"

"What!" I said.

"You folded!" he repeated sternly.

"What the hell do you mean, I folded?" I shot right back.

"You are supposed to say you support the President's budget. You didn't do that. You folded!" he continued in a grumpy tone of voice.

"Horse Pucky," I replied. "No one told me I had to say that, much less lie in front of a Congressional committee. I was raised to tell the truth and if I didn't know the answer, go find it out and then let the one asking the question know the correct answer. So don't tell me I folded. I told the truth and you of all people know it! We are struggling for our very lives in the division because you and others within the Service won't adequately fund us. And you of all people, not only know that to be true but are in part directly to blame for our situation. You have little or no use for law enforcement and you know it. No one knows that better than you and those around you, so don't bark at me," I continued now getting a little hot under the seams.

Our cab arrived at that moment and Dick without another word got in. As John and I started to get into the same cab, Dick held up his hand stopping us in the process. "You folded. You aren't riding in this cab. Get yourself another one," he grumbled. With that, he told the cab driver to move on, and then he was gone.

At first I was pissed at Dick's childish behavior but then thought better and started to grin like the meathead I am. I had gotten the old buzzard's goat and in the process, had alerted Congress as to how the "cow had really taken a dump in the clover." Hey, that looked like a victory to me. Plus, what the hell could he do to punish me? Make me a Special Agent in Charge in the Service with little or no budget...?

"Let's go," said John as he hailed another cab trying hard to conceal a really large smile. Then he said, "You can't talk to the Acting Director of the Service like that," he said with a grin.

"I don't know why not. When he acts like a horse's hind end, then he can be treated like one," I responded with an equally large grin. After those words, Chief Doggett didn't have much more to say to his out-of-control Special Agent in Charge "in need of beans, bullets and black oil" as we rode back to his office. I am sure Chief Doggett caught hell later from Acting Director Dick Smith for my testimony. But knowing John, who was also a practicing attorney, all he did was just grin because that wasn't his first rodeo either. John, being a staff officer to Acting Director Smith, couldn't always say what was really on his mind or the truth for that matter. But his rebels in the field who were under the protective cover of their Regional Directors, who were line officers, could and would have their say if they had any sand. And having the great Regional Director like Ralph Morgenweck that I had, I never heard another word about my big mouth at the Congressional hearing. I am sure that if any insults came my way over that "hoorah" from the Acting Director, my boss just deflected them by asking Dick if he was still coming out for their annual sharp-tailed grouse hunt later that fall in the Sandhills of Nebraska...

If I remember correctly, Congress did manage (read the staffers) to get the Division of Law Enforcement another half-million dollars in which to operate the rest of that fiscal year through an add-on. But that small add-on figure didn't surprise me, knowing Congress's penchant for not wanting to micro-manage the operations of a governmental agency to any great degree. The half-million dollars was not really big money in light of the division's "starving" condition but it was happily received. "Happily received" because it paid for a few more tanks of gas or a motel room and a hot shower, instead of having to sleep in a cold, surplus FEMA house trailer or on the ground while out on patrol by the Service's hard-working agents.

I only wish you readers could have seen the faces on the entire Congressional committee, Dick Smith and John Doggett included, when I responded to the chairman's question with the statement, "We need beans, bullets and black oil!" I wonder if they can still hear me laughing over my unusual response, as I write this story many years later?

As an aside, several months later I was presented with the Service's Meritorious Service Award, one of its highest, for almost 30 years of Service excellence. Oh, by the way, Acting Director Dick Smith and the Secretary of the Interior, Bruce Babbitt, signed off on that prestigious award. I wonder if the Service has an award for "beans, bullets and black oil...?"

## TO KILL AN ELK AND THEN SOME

Throughout my days as the Special Agent in Charge of the Rocky Mountain-Prairie States Region, I always had an ace in the hole up my sleeve, other than that of my Bride. There were times the crappy administrative work in the office would get to me and I would long for the great outdoors. A "longing," that when experienced, always provided for a wandering soul like me. You know, great sunrises, sunsets, long hours, crappy food, awful weather, muddy roads, the bugle of a bull elk, and best of all, the chance to throw my rope over the neck of a hard-running and killing wildlife outlaw so he could be brought before the "mast"...

When I did get a chance to just plain run like I did as a game warden or field agent in the days of old, I would head for North Park near the Town of Walden, Colorado. Therein resided two damn fine Colorado game wardens in Steve Porter and Kirk Snyder. They were two pretty hard-charging, tough as horseshoe nails-type officers, who either afoot or on horseback were pretty damn good catch-dogs when it came to throwing a loop over the heads of fast-moving wildlife outlaws.

One of this pair of backcountry, badge-carrying hard cases was Kirk Snyder. Now, there was a guy who was as sharp as a fox and crazy as a loon, all in the same breath. Not much bothered him and when it did, he just plain plowed right through the danger like it didn't exist. And as a case in point, his patrol truck always showed the effects of battle. Be it a broken windshield, smashed-in front fenders or dented doors. Dings and such that came from chasing poachers at a high rate of speed into the trees or rocks. No two ways about it, the critters had a totally dedicated man for all seasons in that one.

As an aside, in August of 2011 during cow elk season on Mexican Ridge, I had the occasion one afternoon to see Kirk and his son, a chip off the old block, Keenan. During that visit, Kirk, Keenan, Rich (my son) and I had a great time together. As it turned out, Kirk was making ready to meet with a friend and then the two of them were to scout out some trophy mule deer territory in another game unit in Colorado. The next evening as Rich and I sat tiredly in our motel room after the day's elk hunt, we received news that Kirk had just been killed! As we later discovered, he and his friend had been hit by a bolt of lightning while out scouting mule deer habitat for a big trophy buck. Kirk had been killed instantly and his friend knocked out.

Prior to that, Kirk had retired and was working on the E.B. Shaver Ranch in the North Park of Colorado. In that, he had been bringing back a large tract of heavily overgrazed land through his management skills. As a result of Kirk's hard work, the land was coming back with a rush from its overutilization. The mule deer were everywhere, elk were plentiful and even the hard-pressed sage grouse were making a great comeback. Had Kirk lived, being the great land manager he was, the land would have soon been returned to just as God had originally left it. A real legacy if you will.

Kirk was cremated by family wishes and his remains scattered over his beloved Mexican Ridge by his great little wife, Kathy, family and friends. I lost a dear friend, and the critters and I lost one hell of an advocate and catch-dog son-of-a-gun. I guess it just goes to show that God doesn't want just all weeds in His garden. So long, Partner. Hopefully I will see you on the other side and we can share a cigar and cast our eyes over the lands and the critters we tried to care for...

The other North Park officer was a sawed-off, muscular stump of a man named Steve Porter. He may have always looked like he was standing in a hole because of his shortened stature but believe you me, now there was one rawhide leather-tough individual. Be it doing 150 push-ups in a row or single-handedly wrestling an illegally taken moose into the back of his patrol truck. He was another one who was as tough as they came. Every man-jack who ever met him instantly knew, here was "the man." And if they didn't, they had a life's lesson coming in short order...

I am not sure how I ran across and was be-friended by these two "charmers" but it was a marriage made in Heaven. When I was ground down to a nubbin with my damn paperwork and needed a few days afield, I would head for those two officers' district in northern Colorado. When I did, the two of them realized I was trying to escape from the drudgery of running an eight-state region. And when on their home turf just wanted some downtime chasing louts in the field taking wildlife in such a manner they shouldn't have. I can still remember once I topped Willow Creek Pass during big game hunting season, I would get on my radio and make contact with either one of those officers. In that conversation, I would ascertain where they were thin in the ranks and needed a gap filled by another badge carrier. Then I would head that way and revert back to my old ways as a game warden or field Special Agent, in my unmarked federal vehicle, with a chew between my cheek and gum and a cigar firmly clenched in my teeth. And in so doing, making life interesting for those humans I was hunting. And the neat thing was my two friends didn't care if I made a case or not. They just liked the presence of a federal in their valley among the unwashed. And if they ever needed a strong back hefting illegal game into the back of a truck, they knew where they could find a will-ing partner.

Usually such peregrinations on my part evolved around Colorado's big game seasons. I loved the cold weather (being thicker in girth than most Yogi Bears), beautiful color-turning aspens, bugling of the bull elk, and the sounds of rifle shots ringing throughout the crisp fall and winter air. It was also the season my two friends needed the most help because they were usually overrun with all sorts of elk, deer, moose and pronghorn hunters. And as I have often said, in every societal group, you will find the good, the bad and the ugly. Since I was such a good-looking son of a buck for a bull elephant type, I always set my sights on "the bad and the ugly" elements oozing around in the leaf litter. And when working in the vast reaches of the North Park area, one could always find that kind of off-color clientele scurrying around in the toadstools. Now, don't get me wrong. Most of the hunting community I ran across were pretty good chaps. Oh, there were always those who would screw up now and then, but it was those with an evil glint in their shooting eye and a heavy-handed trigger pull that I was interested in. And that being the case, the hunt in North Park was on like a crap shoot in a back alley...

## THE LUCK OF THE DRAW AT LAKE JOHN

One fine fall day, I picked up my son Richard

and the two of us headed for the North Park area. That was to be a special day. It was early in my career in the State of Colorado and I was on one of my rare days off. My son and I were on an elk hunt in the Mexican Ridge area, a short distance from Rabbit Ears Pass. And the weather, well, it was just like the kind that John Denver always got one of his "Rocky Mountain Highs" from...

Arriving in our hunting area, I let my two hard-working game warden buddies know where I was and to call me (I always wore a handi-talki on my belt on their radio frequency) if they needed any assistance. Rich and I were the size of a couple of well-built cavalry horses, and I made that call in case either of my two friends needed any help in dragging out an illegally killed moose. Then it was off to the hunt.

The morning produced nary a critter and Rich and I came back to my truck for lunch. After lunch, Rich headed one way in the timber and I another. We both were good rifle shots and had an understanding that if one of us shot, the other would head that way because more than likely, an animal would be down. I had walked no more than about a 100 yards down what we called "the cut," when I walked headlong into a five-point bull elk coming my way "carrying the mail." Dropping my sandwich and Diet Pepsi, I shouldered my .340 Weatherby rifle as my now-surprised elk ran just a "hellin'" through a heavy stand of aspen some 40 yards distant. Seeing an opening in the aspen in the direction the elk was now moving like a scalded cat (my apologies to all you cat lovers out there), I made ready. In an instant, the elk filled my scope and I fired. The elk never broke stride but just thundered downhill at amazing speed for such a large animal. However, I was shooting a 250-grain, spire-point bullet that was moving at 2,900 feet per second and shortly thereafter, I could hear the elk giving forth his death rattle.

I found my elk and as I filled out my kill tag, my long-legged 6' 7" son arrived. Then the work began. I don't know how many of you readers out there have killed a rolling fat, early fall elk but it is like killing and moving your neighbor's horse... Two hours later, we had the elk in the back of my truck and were heading for a place where we had permission to hang and skin the animal. Another hour later and we had the animal skinned out, covered with an elk bag, and were letting it cool out and glaze over. Once again to all my readers out there, especially you potential elk hunters, elk are rolling fat and thick down through the shoulders. It is imperative to always get the skin off and let the animal cool out. If you don't, it will spoil, especially in the early fall temperatures. And to lose any elk meat, well, to those of you lucky enough to kill and eat an elk, you know of what I speak. There is no finer eating, especially a nice, well-seasoned steak, fried spuds, good greens of any kind, and a shot of Jack Daniels around a campfire. But I digress...

In the meantime, Rich and I, now covered with elk snot, hair, blood, sweat and dirt, gathered up our knives and saws as we headed for nearby Lake John. On the way, we lit up two strong Italian cigars and treated ourselves to another ice cold Diet Pepsi. Arriving at Lake John in the early afternoon, we got out our knives, saws, axes, ropes and the like. Our plans were to take such gear to the lakeside and wash all the blood and coagulated fat from our equipment so it would be ready for the next day's hunt (Rich still had an elk and deer tag to fill and I still had an unused deer tag). Then once cleaned, I would put an edge on our knives so they would be ready for the next critter as well.

It was then our beautiful day off turned to that proverbial brown stuff splattering through the fan... A vehicle drove up to where Rich and I were casually cleaning up our gear and stopped. In it were two adult men and a young woman. They bailed out, unlimbered several lawn chairs and fishing rods, and commenced fishing not twenty yards distant from where we were parked. As it turned out, the fishing was red hot and soon our folks were reeling in those typical Lake John "hogs." Soon they were approaching their limits and now our fishing folks were nervously eye-balling Richard and yours truly as we continued cleaning all our gear. As it turned out, I had been closely watching our fishermen and had been keeping track of the numbers of fish and times each person had caught them (day off or not, the game warden in one never sleeps...). Realizing we might have a violation in progress as a topping to our successful elk hunt, we slowed down in the cleaning of our gear so we could keep watching without alerting our fishing like a bunch of pelicans, neighbors.

Finally we had cleaned all our gear that needed cleaning and I had set the edges on our knives. As luck would have it, our fishermen were just getting into that "catching thing" with gusto. Realizing we now stuck out like sore thumbs since we were all done, I told Rich to grab our rifles and bring them to the back of the truck which he did. Then we began cleaning and oiling our rifles, slowly.

Finally one of the men who had already limited out on the number of trout legally allowed, came wandering over to where we sat. Striking up a false conversation, I noticed him looking over all our cleaned gear, blood up to our elbows, and hunting rifles. Excellent cover I might add, especially if one is checking you out and you are a badge carrier. Then we made small conversation regarding our successful elk hunt and finally satisfied, he wandered back to his fellow fishermen. Satisfied that we were OK and of no threat, he commenced fishing once again! Soon our "friend" had taken five very large trout over the limit and was hiding them throughout his vehicle in case the local game warden happened along...

Finishing the cleaning of our rifles, we placed them back into their scabbards. Then I walked over to make conversation with our fishermen and look at their fish. Sure as God made grasshoppers good fish bait, all of our chaps either had limits or in the case of our "friend," had exceeded them. It was then I had enough of this "game hog" thing and rolled the gold on our chaps. Man, you talk about catching all of them with their pants down! They had looked us over carefully and had decided we were nothing more than a couple of nitwit hunters cleaning up our gear after a successful hunt. Well, that is where the charade ended over the raid on Mother Nature's cookie jar, and the starkness of reality set in.

I ended up chewing off the last parts over the fence on two fishermen who had limits and were still fishing. Since they had not taken over-limits,

I let them off with a damn good ass-chewing instead of citing them for attempting to take. However, for the lad taking and hiding a large over-limit, that is where the rubber met the road. I took down all his identifiers in my notebook and told him that I would be contacting Steve Porter. Shortly after making that contact, I informed him that he would be soon receiving a citation for his five trout over the limit.

When I said those words, my fisherman went into a tailspin. When he finally wound down, I discovered he knew Steve very well, was a friend and had never been caught by Officer Porter. Our chap was just sick, knowing he would soon be facing his friend in a small town where information traveled fast. And in that small town, fast-traveling information would be the word of his capture for stupidly taking and possessing an over-limit of trout at Lake John.

He was right. When Rich and I met Steve later that evening and told him who we had caught with an over-limit of trout during our gear-cleaning operations, he broke out laughing. It seemed Steve had suspected his "friend" all those years he had known him and yet, had never caught him. However, that evening was a fine kettle of fish as far as Steve was concerned. He was pleased over my catch of the day, and just couldn't wait until he saw his friend the next day to issue him a citation for taking and possessing an over-limit of trout. This may not have been a great elk over-limit case but as I discovered that evening, to Steve's way of thinking, it was just as "tasty." Especially, since Steve's friend had been just that. A friend to his face and an outlaw when Steve was not looking. To Steve's way of thinking, this catch was outstanding. Outstanding because his "friend" would now always have to be extra careful around anyone, when he pulled in a trout illegally or pulled a trigger when he should not have...

## A PICKUP LOAD OF CAJUNS AND THREE DEAD ELK

Driving through the main gate to the Rocky Mountain Arsenal National Wildlife Refuge in Commerce City, Colorado, I stopped at the headquarters office. Standing there waiting for me was my oldest son Richard, with his traveling gear and a big smile. Rich was a wildlife biologist and credentialed law enforcement officer for the refuge. On that day, he had been assigned to me for more law enforcement training by Refuge Manager Ray Rausch. And training for the next several days would consist of assisting the Colorado wildlife officers during their big game hunting seasons in the northern part of the state. Throwing his gear into the bed of my truck, Rich unlimbered his 6' 7" frame and stuffed it into the front seat. There waiting for him was a fresh pouch of Levi Garrett chewing tobacco, a fresh Italian Toscani cigar, and a cold Diet Pepsi. With such traveling fare stuck between our cheeks and gums, cigars firmly clenched between our lips, and cold Diet Pepsi's trickling down our gullet, off we two wildlife warriors went.

Heading west on I-70 and up over Berthoud Pass, we soon found ourselves moving through several high mountain valleys, then up and over the top of Willow Creek Pass. While there on the high ground of the pass for better radio transmission, I called Colorado Wildlife Officer Steve Porter. He soon responded and I alerted him to the fact that Rich and I would be in country for the next three days and asked where he could use some extra help. Elk season was in full-swing, the country was full of resident and nonresident hunters, and Steve and his compatriots were hard-pressed to hold the "Thin Green Line."

"We could use your help in the Rand-Owl Mountain area and all the backcountry in between," came Steve's voice back over the radio. "We have a few other visiting wildlife officers in the area helping out so if you need assistance, you can call on them as well for backup. However, I doubt you will need any backup though, since you have that stork Rich riding with you," he chuckled.

"If that is the case, then Rich and I will be working the area to the west of the Rocky Mountain National Park boundary between the Rand and Gould areas if that is alright," I responded.

"That would be perfect," Steve replied.

"We will give you folks a whistle if we get in over our heads or need some assistance if we run into a big one," I replied.

"Good hunting," said Steve and then he was gone.

Looking over at Rich, I said, "Well, we just got our marching orders, so let's get cracking." With that, we turned off at the Rand Cutoff and headed for the prime elk hunting hinterlands between the Rand "outpost" and the Gould "hole in the wall." On the front seat sat a 60-power spotting scope, two pairs of binoculars and two charged up portable radios ready for use. With cigars firmly clenched between our lips, a fresh chew of Levi Garrett between our cheeks and gums, two cold Diet Pepsi's riding on the dash carrying rack, and two sets of "peeled" eyes, off the two of us happy warriors went into what surprises colorful Colorado would bring us. We didn't have long to wait for the devil to rear his ugly head.

Heading northeast on the Rand Cutoff towards Gould, we crossed the Illinois River and headed up onto the adjacent plateau. About four miles down that road, we ran across a fenced high mountain meadow bordered on the northeast by the Rand Cutoff Road and another unmarked graveled road heading towards the northwest. All along the roads that we traveled that afternoon were vehicle after vehicle of obvious hunter-orange elk hunters scurrying to and fro. Passing the long high mountain meadow, I chanced a look out to the northwest into the middle of that huge grassy pasturage.

"Damn, Rich. Would you look out there in the middle of that meadow by that swale. Damned if that isn't a small group of elk lying down right in front of God and everybody!"

"Sure as hell, Dad. Turn down that gravel road running alongside the meadow. Maybe we can stake out an area where we can watch over that bunch of elk and see if anyone poaches them," he continued as he looked through his binoculars over the area containing the elk.

Turning down the paralleling gravel road, I drove about a half-mile and then stopped behind a large stand of covering willows. Out to the west of where we sat, lay six cow and calf elk about a quarter-mile distant. They were perfectly camouflaged in the meadow's tall grasses and as we sat there watching the area like a couple of rather large buzzards, several groups of anxious hunters drove by. Not one time did any of the dozen or so vehicles carrying elk hunters spy that small group of elk lying down in the middle of that meadow. Not one! Boy, you talk about a bunch of elk hunters being blind as a bunch of bats in the bright sunlight...

Rich and I sat there like a couple of rather large-sized spiders waiting for some elk hunter to fall into our trap but we had nary a taker for over an hour. Tiring of our little non-fruit bearing trap, we loaded up and moved on.

Back to the Rand Cutoff I drove and then I remembered that there was a big game hunting outfit located on the south side of that big meadow. Figuring no one had checked them for some time, Rich and I headed that way. Arriving at a small dirt road leading into the clubhouse, we turned north and headed into the timber. We hadn't gone a quarter-mile when I noticed a large dust cloud coming our way. Moments later, a Jeep station wagon came steaming down our narrow dirt road in such a reckless manner that I had to quickly drive off the road and into the sagebrush to avoid being hit! As the Jeep careened by, I could see two individuals in the front seat and one man in the back. However, the man in the backseat was the one who really caught my eye in all the dust and flying Jeep. He was talking on a hand-held portable radio and its antenna was stuck out the side window. Then the Jeep was gone in a flying cloud of dust like the devil was right behind him.

"Rich," I said, "the man in the backseat of that Jeep had a portable hand-held radio and he was talking on it! If I don't miss my guess, someone has spotted that small bunch of elk out in the middle of that huge private meadow and is onto them."

With that, I power spun my patrol truck in a tight circle in the now-flying dirt and sagebrush, then headed back towards from whence the Jeep had fled. Roaring down that narrow dirt road, I saw Rich undo his seat belt so he could make a fast exit from the cab if he had to in order to grab off some soon to be apprehended poachers. Hitting the Rand Cutoff, I power slid the truck across the road towards the small hole in the wall Town of Gould and in the direction of the fast-fading road dust from what I assumed had been made by our earlier rapidly moving Jeep. Hitting our step, we soon were at the edge of the huge fenced-in meadow alongside the Rand Cutoff road. Gathered at its gate were three vehicles and a mess of elk hunters milling around who did not have permission to go out into that private fenced-in meadow after the elk. It was apparent they had also spotted the resting group of elk out in the field as well.

Sitting there in that "parking lot" was our Jeep with the chap holding the portable hand-held radio. He was looking out into the meadow at a fast-moving yellow-colored pickup with a camper shell. The door to the camper shell was up and the tailgate was down on the rear of the pickup, as it hurriedly hurdled out across the meadow towards the resting elk. Sitting on the tailgate hanging on for dear life as the pickup bounced out across the meadow were three elk hunters, soon to become "shooters!" It was apparent to me that the Jeep and the pickup in the field were together. And from the looks of it, the Jeep with the man holding the portable radio had called in his buddies to take a whack at the elk in the field. Then our buddy in the Jeep had held back as his other hunting party members had sped out into the field after the resting elk.

It was then Rich and I observed the pickup in the field slamming on its brakes. In so doing, the three shooters riding on the tailgate were thrown into a jumbled pile inside the bed from the rapid stop. Then all three men bailed out from the bed of the truck and began blazing away at the elk, which were now standing up in the field in alarm.

Hearing the shooting coming from the pickup, the elk now fled for the timber located on the west side of the meadow. Then the three men bailed back onto the tailgate and the yellow pickup once again bolted out across the meadow trying to cut off the six running elk. As it did, we could see all three shooters in the back of the pickup hurriedly trying to reload their rifles. Finished reloading, we could now see through our binoculars that our three shooters riding on the bouncing pickup's tailgate, were once again hanging on for dear life!

By then Rich had opened up the gate on the fenced pasture and I streamed through with my patrol truck. But before I did, the landowner and the Jeep operator tried to stop me from going into the meadow realizing I was now after his shooters.

"Hey!" he yelled.

Slapping on my brakes as the landowner ran to the side of my truck in an attempt to keep me from chasing his three shooters, I flashed my badge and barked out, "Federal officer. Back off! We are after those chaps in the vehicle chasing those elk. And it is now too late to use your portable radio to warn them. They have already violated Colorado law by using their motor vehicle to aid them in taking a big game animal. If I get out there and I discover that you used your portable radio to warn them that we are here, you will also be arrested for Aiding and Abetting in the illegal taking of a big game animal."

Thunderstruck at the turn of events, the landowner quickly stepped back from my truck. Then Rich bailed into the cab and before he hit the seat, we were off after our pickup full of outlaws.

Outlaws, because it is illegal in Colorado to chase wildlife with a motor vehicle and attempt to kill any big game with the use and aid of that motor vehicle.

Racing out across that field in pursuit of our shooters, we saw them stop one more time. Again, all three shooters were hurled into the bed of the pickup in a jumbled pile because of the violent stop made by the operator of the pickup. After getting untangled, our three shooters once again bailed out from the back of the pickup and began unloading their rifles on the now-terrified and running elk. Then in that melee, we saw three elk stagger and go down under the onslaught of flying lead from our three shooters!

Then the crap hit the fan! The loud howl of my siren split the cool fall Colorado afternoon air. It was only then that our three shooters realized the jig was up and the law was at hand. All three men scrambled for the back of the truck in order to flee, only to have a 6′ 7″, 270-pound hard-as-nails federal officer interrupt their little escape attempt. By the time I slid to a stop and had put my patrol truck into "park," Rich had all three men unarmed and was cradling their rifles in his arms. Stepping over to my lad driving the truck, I identified myself and quietly acquired his pickup keys and loaded rifle. Then I had him step out and join his partners at the rear of the truck. There both Rich and I identified ourselves and apprised the four men of the illegality of their ways. That was, the three shooters for shooting elk with the use and aid of a motor vehicle, and the driver for aiding and abetting the three men in the back of the truck with their big game violations.

Then as I acquired all the men's hunting licenses, tags and driver's licenses, Rich hotfooted it across the meadow to make sure all the elk were dead, not wounded and still suffering or attempting to flee in a crippled state. Examining the men's licenses, I observed that all four men were from Louisiana. Hell, I thought, instead of enjoying a "mess of mud bugs" Louisiana style, I now had a catch of coon-asses or Cajuns, Colorado style.

Laying my armload of fancy Weatherby rifles carefully across the tailgate after I had unloaded the same, I just shook my head. Every one of those fine rifles had been scratched and gouged in the mad race across the meadow in their attempt to kill a few elk. Man, I thought, it must be nice to have so much money that one can abuse a fine $2,500 rifle like a Weatherby Mark V.

Then I advised the men of their rights even though I didn't have to because I had observed them personally breaking Colorado big game hunting laws. However, I did so because I wanted to hear from the men their reasons why they had chased the elk instead of going after them by fair chase. All four men admitted they had been hunting for four days and had not killed anything. They were going back home to their business in a couple of days and basically stated that they were getting desperate to kill an elk. So when the chance surfaced to run elk down with a truck and kill them, they had jumped at the opportunity. Then I asked the men if they realized having come from another state, that they were breaking Colorado law. There was a lot of shifting their weight from one foot to another and finally one of them said, "Well, it is illegal where we live, so I gather it would be illegal here as well."

Having the information I wanted, I said, "Let's go, Lads. We have three evidence elk to gut before they spoil and load into my truck."

With that, we all loaded up and drove over to where Rich was in the process of gutting out the elk. As we did, I put out a call for Steve. When he responded, I said, "Steve, we have four out-of-state chaps we just caught chasing elk with a pickup and shooting the same. They have three elk down. Since they are non-residents, they must by Colorado law be booked into jail or pay the fine for such endeavors on the spot. I need you to dispatch one of your chaps to personally handle the paperwork if you would, so Rich and

I can once again hit the road in our undercover truck looking for other sports breaking the law."

"Dang, Terry. I turned you and Rich loose to have a good time and not get into big trouble. Appears from what you just said, you have a good one. I will dispatch one of our lads who is closer than me to give you an assist. Give me your exact location so I can guide him into your area," he replied.

I gave Steve my location, then whirled around to the sound of shooting coming from the gravel road that paralleled the fenced meadow in which we had sat on earlier from the north! Grabbing my binoculars, I observed two other elk shooters, blazing away at the still-running elk from far up the meadow. Then the elk burst into the timber and disappeared. Realizing they had broken the law, my new shooters got back into their vehicle and fled the scene. Suffice to say, Rich and I were too far away to chase, catch, and bring those chaps before the mast as well...

By that time, Rich had all three calf elk gutted and ready for loading. As Rich and I loaded all the elk into my truck, one of my shooters walked over to the dead animals and remarked, "I always thought elk were bigger than that."

*Boy, what a hunch of rubes,* I thought. Young of the year animals are usually smaller than their parents...

Then is when the rubber met the road. I sat all my chaps down and advised a Colorado officer was on his way to give me a hand. It was then that I also advised the men that Colorado law required for them to pay for their violations on the spot or they would have to be incarcerated. With those words, I expected the men to come unglued at that disturbing news. However, they never even batted an eye or took in a deep breath of surprise!

Finally one of the men asked, "What will the damages amount to, Officer?"

Doing some quick figuring in my head, I threw out the amount of bail they would have to pony up to avoid going to jail. There was hardly a discernible movement among them, as none of the men blinked at such a high figure! Damn, I thought. Either these are some of the dumbest Cajuns I have ever run across, or the bouncing truck ride shook all their brains loose in their jaunt across the field. Hell, the figure I spelled out for them would have provided for an all-expenses-paid moose hunt for me in British Columbia!

About then the cavalry arrived in the form of an out-of-district Colorado officer, who was helping Steve and Kirk with their elk hunter coverage. Once there, with all the particulars laid out including what Rich and I had observed and were prepared to testify to, as well as the four men's confessions, the officer began his paperwork. When finished, he spelled out the amount of bail the men would have to pony up on the spot for their violations. That figure, he advised, would have to be paid on the spot or he was obligated by law to take the entire party to jail.

The men heretofore sitting on the grassy ground, arose in unison. Then they dug into their wallets and handed over to the Colorado officer $6,600 in crisp $100 bills! With that, Colorado was happy and Rich and I headed off to Walden and the fish and game warehouse. Once there, we hung and skinned the three evidence elk. While there, Rich and I got another revelation. The shooters had shot the hell out of the elk. There were bullet holes in just about every part of the calf elks' carcasses. Some sportsmen, I thought... If my kids had shot up such a great eating animal like that, they would have gotten a stick "whanged" across their heads!

With that, a pinch of chew between our cheeks and gums, a strong Italian cigar firmly clenched in our lips, and holding a cold Diet Pepsi, Rich and I sat about drumming up some more business at hand that fine colorful Colorado fall day.

## A COLD WINTER DAY AND A SAGEBRUSH SHOOTOUT

One cold November morning, Chris my younger son (may God rest his soul) and I loaded up into my truck. Then we headed from our home in Evergreen up over Berthoud Pass and eventually into North Park in northern Colorado. Once in North Park, we cut across the southern end of the valley on a gravel road en route to Mexican Ridge. Today was one of my rare days off and my younger son and I were going to hunt elk. There was about eight inches of snow on the ground, the sky was heavy with dark storm clouds and it was about twenty below zero. Just right for a winter elk hunt, I thought. At least if we killed one, the meat wouldn't spoil...

Arriving at our hunting spot on Mexican Ridge, we grabbed up our hunting gear and attaching my federal portable radio to my belt, Chris and I started our hunt. To all you big game hunters out there, it was a wondrous morning. Cold, crisp, elk tracks and their distinct, close-at-hand odor everywhere, and no one else in sight to ruin a great father-and-son outing. It couldn't have been a better start to a day even if we had tried. My younger son Chris, was a bull of a man. Just six feet tall, built like a brick outhouse and strong as a bull. He was a perfect elk hunting partner. Especially if one was to dump a horse-sized animal like an elk, drag it uphill, and then strong-arm load it into the bed of a waiting pickup...

However, we hadn't gone a hundred yards into the snow-covered, quiet as a deer mouse leafless aspen and coniferous forest, when my portable government radio came to life. It was the Jackson County Sheriff's Office trying to raise either of their two Walden resident game wardens. Their efforts were in vain, however, because as later discovered, both of their resident game wardens were at that very moment engaged in investigating an illegal moose killing and were not available. Realizing the Jackson County Sheriff's Office wouldn't have called their local officers without pretty good reason, I responded. Taking my radio, I pressed the transmit key as my son gave me "that" look, that our father-son elk hunt was probably "in the crapper"... It was.

In response, the sheriff's office advised they had several citizens calling about an ongoing elk killing along the Rand-Gould Cutoff Road. It seems a large herd of elk had been run off nearby Owl Mountain by a bevy of elk hunters and had gotten confused as elk will do. In their terror and confusion, they had run from the cover of a dense coniferous forest onto the nearby open sagebrush plains. In so doing, they were now in front of God and everybody else with a Winchester, a heavy trigger finger, and anyone possessing an empty elk tag with a hungry look in their eyes.

I advised Jackson County that I was a federal officer and probably about thirty minutes away. But would respond if they felt I could get there in time. From the sound of the tone and tenor of their reply, it sounded like there was a real old-fashioned elk killing going on by numerous individuals in two pickups. One that needed a badge carrier "look-see." And no two ways about it, they would be obliged if I could somehow intercede. Looking over at Chris, I could already see that he was interested in rolling that way and suffice to say, so was I. The father-son elk hunting would just have to wait for another day, I thought, as the two of us ran for our distant parked vehicle.

Roaring off Mexican Ridge, we were soon retracing our steps across the cutoff road at the southern end of North Park. Thundering off the gravel roadway and onto state Highway 125, I headed south towards the small mountain town of Rand. Once there, we stormed onto the Rand Cutoff Road that ran towards the wide spot in the road called Gould. It was on that road near Owl Mountain, where the supposed elk slaughter was currently taking place. By now, the ambient temperature had dropped to about 25 degrees below zero and the low-hanging dark gray clouds threatened snow. Flying down that road in four-wheel drive, high range, Chris and I stormed across several low lying hills over the snow-covered roadway.

Cresting a hill in the road, I was confronted by a dead elk that had not been gutted, lying in the

534 | TERRY GROSZ

middle of the road where it had fallen after being shot! Slapping on my brakes to avoid hitting the animal, I finally skidded to a stop about ten feet from the fresh carcass. Quickly looking northward across the sagebrush flats, I could see numerous tracks in the snow, where a large herd of elk had recently run alongside the road. From all the dirt that had been kicked up onto the freshly fallen snow in the "track field," one could detect that the elk herd was running in a state of panic. Bailing out from my truck, Chris and I moved the freshly killed elk from the center of the road so we could proceed. It was then that I noticed someone had placed an elk tag on the side of the dead animal letting any finder know that the elk was someone's personal property. A quick examination revealed that the tag had not been validated. Quickly placing the tag into my shirt pocket as evidence, Chris and I then involuntarily ducked as a number of bullets came screaming our way! That was quickly followed by the sounds of multiple rifles, firing out of sight from beyond the next slight rise in the road. Quickly loading up back into my truck as more bullets screamed our way, we were then greeted by a herd of elk in the sagebrush streaming our way at a dead run. It was obvious they had just fled from the shooting beyond the next rise in the road. Their tongues were hanging out and several were running with flopping front legs because of broken shoulders from poorly placed shots. That "picture" was framed under a haze of steaming, panicked elk breath in the frigid winter air! Gunning my truck down the road, I stormed over the next rise in the road. What greeted my eyes was something right out of the history books! Simply, it looked like a buffalo hunt right out of the pages of history, except the dead animals now scattered across and all over the ground were elk.

Racing towards me were two pickups! In the back of each pickup was a man firing a rifle into the tail end of the racing elk herd running along-side the road in the sagebrush! From each passenger window of the pickups racing towards me poked a firing rifle as well. In an instant it became obvious as to what was happening. As the elk frantically ran in the sagebrush alongside the road trying to cross back into the timber on the other side, the pickups were parallel racing, pacing the animals' flight, cutting off their escape back into the timber. Then the elk would drift out into the sagebrush just out of rifle range. Then when their fear subsided, they would once again make a run for the timber. Then the shooters in the two pickups would start up once again. As the elk herd approached, they would once again chase the elk away from their escape route with deadly rifle fire. Then the terrified elk herd once again drifted out into the sagebrush out of range, only to stand and rest before they made another attempt to get back into the heavy timber across the road. In so doing, that moment in time gave the shooters time to reload their rifles once again. And from the looks of things, this killing scenario had been going on for some time. Plain and simple, it was a modern-day slaughter!

Slamming on my brakes in the middle of the road to prevent any escape of my modern-day "buffalo hunters," I bailed out. About then the drivers from the two racing pickups saw me and slammed on their brakes. By stopping in such an abrupt manner, both men shooting from the back of the pickups were slammed off their feet and fell down in the back of their trucks. Then the truck in the rear being so close to the first truck, skidded into the rear end of the lead vehicle! When it did, it smashed in its front end, destroying its radiator which then "complained" by emitting great clouds of steam into the frigid winter air. By the time they both had skidded to a stop, I was alongside the first truck holding up my federal badge and ordering the driver to shut off his engine. Since the truck in the back now had a smashed front end, there was no urgency in getting that driver to shut off his damaged engine because he couldn't go far.

Then to make sure all my "gunners" and God

understood what I wanted, I ordered, in a polite tone of voice of course, for all the men to drop their rifles, exit their trucks, and stand out onto the roadbed. It took a few seconds for my bellowed disembark order to sink in, but finally all six of the men obeyed and dismounted. I once again identified to all who I was and in such a manner that even God understood. Then my shooters were commanded to "sit" on the roadbed! By then, I had every man's attention and they did as I ordered. Then I removed all six rifles from the vehicles and found all of them fully loaded on a public way, a violation of Colorado law. Quickly unloading the rifles, I handed them to Chris who just as quickly placed all of them into the cab of my truck for safekeeping.

Then for some reason, my eyes wandered down the road. Lying on the road or directly alongside were three freshly killed elk spread out over the next hundred yards or so! Then looking out over the snow-covered sagebrush, I observed five other brownish-tan objects lying scattered along the elk trail in the snow! *Damn,* I thought. *These six guys have left a trail of dead elk for as far as I could see!* Then from the direction I had just traveled, I heard the thundering herd of about 100 elk or so, cross the road safely and run into the timber. In that crossing, I observed at least three crippled elk with broken front shoulders struggling to keep up with the herd! Turning to my six shooters who were now as quiet as a nest of field mice pissin' on a ball of cotton, all I could see was shame running across their faces.

"Gentlemen, I have identified myself to all of you as a federal wildlife officer. I now need for all of you to dig out your hunting and driver's licenses along with your elk tags," I ordered. Then I got the Jackson County Sheriff's Office on my portable radio as the men were in the act of complying, and advised I had their elk shooting subjects in custody. The Jackson County Sheriff's Office advised they would pass on that information to the first Colorado wildlife officer that came back on the air. I also advised the dispatcher as to our collective location. Then I walked over to each pickup and provided Jackson County their license plate numbers. It was then, that I noticed that my elk shooting chaps' vehicles were licensed in Louisiana.

Then Chris and I gathered up all the licenses and tags from my shooters for the paperwork that was to follow. Looking back down the road at the elk carnage, I had Chris grab his hunting rifle and head that way. I did so, so he could locate all the dead elk and kill any still out in the sagebrush that appeared to be wounded beyond repair. As Chris disappeared down the road, I looked over all the driver's licenses that I now held from my errant shooters. The licenses indicated all my subjects were non-residents from Louisiana.

Stuffing the licenses and elk tags into my shirt pocket, I took a look at my watch to get the time of arrest. Then I turned and advised all the men they were under arrest for possession of loaded firearms in motor vehicles on a way open to the public, for taking big game animals with the use and aid of a motor vehicle, shooting from a public highway, aiding and abetting (the drivers), untagged elk, and for taking over-limits of elk. Over-limits being deduced from the nine dead elk scattered along the roadway and sagebrush I had initially observed (the limit being one lawfully taken elk per person).

Then my son returned. "Dad," he said, "I count eleven dead elk and at least six blood trails heading north out across the sagebrush. I didn't see any crippled elk that needed killing though because those that were wounded had run off into the sagebrush."

I just shook my head. I had seen killing frenzies in my lifetime before when it came to dove, trout, salmon, ducks, and geese but never anything like I was seeing that day. Especially when it came to such noble big game animals! "Chris," I said, "take your knife and start gutting those animals before they start to bloat (harder to gut once they start to bloat). Since there are so many, start with those closest in and just open up the skin so their paunches can expand. That way it will make gutting easier when we finally get to them. Be careful you don't open up their paunches though. Then when you get to the last one, fully gut him and then work back this way gutting as you go. Hopefully by then, we will have some more help from the wardens as they finish their moose investigations." Chris just nodded and off he went. "Chris," I said. "Take your rifle just in case you run across any crippled ones you missed that need putting down." Returning, he grabbed his rifle and then off he went into the frigid winter air at a trot.

Lord, how his mother and I miss that young man to this day. In fact, how is this for irony — exactly six years ago to this very day as this story is being edited, my son passed away after suffering a massive heart attack at the age of 35!

About then my portable radio crackled to life and it was Game Warden Steve Porter. He had finished his moose investigation and was heading my way as fast as he could. Twenty minutes later he arrived and I briefed him on what Chris and I had observed and were prepared to testify to. Then I handed Steve all my shooters' licenses and elk tags. Leaving him in command of my six now-long-faced chaps from Louisiana (this was a state case), I grabbed my cutting and gutting gear and headed out to give Chris a hand in the elk-gutting duties. About thirty minutes later, Deputy Chief of Law Enforcement Dave Croonquist arrived. He, like me, had been elk hunting in the area and had also been carrying a radio. Upon hearing the earlier radio traffic, he had rolled in the direction of the action just like me. When he arrived, Chris, Dave, and I made fast work of gutting all the dead elk we could find. Then Chris teamed up with Dave and they began running out all the blood trails running across the snow in the sagebrush for any cripples left alive. As it turned out, about two hours of cold tracking produced nary a crippled elk. Elk being such big, fast-moving and tough animals, they had apparently shrugged off their light wounds, fleeing to the north, never to be seen again. And if they didn't make it, eventually succumbing to their wounds, the coyotes and any stray wolves in the area would eat well, I thought...

Finished with my "cutting and gutting" detail, I headed back to give Steve a hand as needed. Now, as I have stated in an earlier story, anyone from out of state who breaks the hunting laws in Colorado, must pay on the spot or be incarcerated. Standing there, I heard Steve advise the lads from Louisiana that the total of their fines came to just a tad over $34,000! Those lads never blinked an eye!

Then one of the men said, "I guess we had better get Grandpa."

Surprised, I said, "Who is Grandpa?"

"He is the leader of this group of hunters from back home," said one of the captured six.

"Where is this 'Grandpa?'" asked Steve.

"He is back at our camp with our other buddies on Owl Mountain," responded the man now doing all the talking. With that, Steve got directions and dispatched one of the now just-arriving, supporting fish and game officers to the Owl Mountain campsite. Forty minutes later, our fish and game officer returned with an older-looking gentleman. The older-looking gentleman got out and walked over to where Steve and I stood. Clutched under one of his arms was a three-pound "MJB" coffee can...

Walking up to us, he introduced himself and asked how much was the fine. Steve told him it was a tad over $34,000. The older man with the coffee can never said a word. He just took that coffee can, popped off the lid and shook it over the hood of my truck. Out dropped a tightly wound wad of American money wrapped in a rubber band! In fact as Steve and I were soon to see, that damn can was clear full of tightly rolled up $50 and $100 bills! And there was another like roll still in the bottom of that coffee can!

I just stood there in shock! I had never seen that

much money at any one time. Then the cop in me took over. "Where did you get all that money?" I asked, thinking maybe there was a drug money connection somewhere in all that mess.

"Well, our group of 17 hunters come here every fall to hunt elk. Before we come, everyone puts $500 into the can. We do that because sometimes our boys get a bit rambunctious and go over the edge of what the law allows. If they get caught, we take the fine money out of this can and pay what we owe. If we don't get caught, the money stays in the can for next year. Then when we get together for next year's out-of-state big game hunt, we all pitch in another $500 and into the can it goes, and so on," the old gentleman quietly said like it was nothing.

"How much do you think is in that can?" I asked like the dummy I am.

"Oh, I don't know. I s'pose there is about $200-300,000," he said without any hitch in his "giddy-up" or change in the tone or tenor of his voice...

I just looked at Steve and he at me in amazement. No wonder wildlife law enforcement is almost a lost cause, I thought... There is no way the critters will ever win if we are facing such attitudes from the outlaws in the hunting public...

Then the old fellow walked back to the damaged pickup, took a close look, shook his head and then walked back to me. "Say, Sonny, is there a place in Walden that sells trucks? I would say from the looks of it, our damaged truck is pretty well spent and we will need another one to use until this one is fixed. Is there a place in Walden where we can purchase another new one?"

Still in amazement, I told the gentleman where he could go to purchase another truck in Walden. Then with all their fines paid in full, off they went back to their camp on Owl Mountain, all piled into the back of their remaining good truck. Of course with nary an elk or elk tag to their names I might add. I saw a bunch of them later in Walden and they did indeed have a brand-new truck. Like I have been known to utter, Custer had better odds than we in the wildlife law enforcement field...

Then our work began. Besides the two moose the wardens had seized earlier, we now had eleven elk to skin and clean up. The entire group of wardens, Chris and I, returned to the fish and game warehouse in Walden. There six of us labored until three the following morning skinning out and cleaning up the evidence moose and elk. And just as fast as we cleaned up the critters, Steve called in locals who were living on the poverty line. Then when they arrived, he gave each of them their winter's supply of damn good eating moose and elk meat.

By the way, the next day Chris killed a monster

eight-point bull elk that took six of us just to load the carcass into the back of my truck! And then we, too, had our supply of wonderfully good eating Colorado elk for the winter. Only ours was legal...

## SPECIAL AGENT BILL SKAR

Bill Skar came to the Division of Law Enforcement from the Refuge Division. He was the only Special Agent I ever lost to eternity while on duty...

I first met Bill when I was the Senior Resident Agent over North and South Dakota in the late '70's. Working with agents Ed Nichols, Joel Scrafford, and John Cooper, we comprised a traveling squad whose duties that summer (in addition to all our other duties) were to train badge-carrying refuge officers in law enforcement procedures. It was at Squaw Creek National Wildlife Refuge during one of those summer training exercises that I first met Bill. He was a mountain of a man. I figured he was at least 6' 9" tall and weighed at least 300 rock-solid pounds. But his size was not the first thing I saw about him that caught my eye. He had a wonderfully genuine and infectious smile and manner that I still can picture in my mind's eye even today, some 25 years later after his passing. He was just a big old, grinning bear of a man, with a sense of humor and a love of life to match.

The four of us agents, when I first met Bill, were

trying hard to train a gaggle of refuge officers in the summer heat of northwestern Missouri in 1975. The first three days of the weeklong training session had gone badly and strangely, to say the least. It almost seemed that the entire class was very resistant to learning anything about law enforcement. In short, they just sat back in class and almost appeared to be sullen at their best. Then during one of the practical exercises in the humid heat of Missouri, I finally saw the light.

In our class were two different refuge managers whose entire refuge badge-carrying staffs made up the rest of the class. As was our teaching style, we lectured law enforcement principles, standards, and application in the mornings. Then we had the lads apply those techniques learned earlier that day in the field on car stops, field inspections and the like. It was during one of those car stops that I finally saw the light of our problem. The two "puke" refuge managers as I was soon to discover, did not want to be there. Nor did either of them support any kind of law enforcement activity on their refuges. As I saw it, they in turn had poisoned the entire class by holding it hostage. Especially since all those chaps in class worked for those two refuge managers. In short, the two sour on law enforcement refuge managers had put out "the word" and every one of the rest of the class fell into line since they worked for those two knotheads. And of course, to disobey their bosses meant a lot of misery later on when we left...

Sensing my two senior refuge managers were at the root of the problem, I assigned the two of them to make a "dangerous" car stop during a practical exercise to test my theory. Going through the exercise, the first thing they did was make a poor car stop by not off-setting their vehicle for the protection its metal body offered, as they had been trained. Then one of the managers sauntered up to the car full of "bad guys" like they were his buddies and did not observe the proper and safe approach techniques. His back-up, the other refuge manager, instead of quickly coming up on the blind side of the vehicle being stopped, took time to light up a cigar, then he made his grand approach!

It was then that I hit the roof. I immediately stopped the practical exercise with a very loud, "hold it right there!" My bellowing like a freshly gored bull sure as hell got everyone's attention in short order. Since we law enforcement chaps had been assigned to put on these training exercises and had the full support of the class's top area offices' supervisors, I lowered the boom!

"You two refuge managers are excused from any more participation in these training exercises! In essence, both of you are fired from any associated duties during the rest of the training and have flunked your annual law enforcement training. Therefore, in violation of the duties described in your position descriptions, both of you are disallowed from carrying a badge until you recertify! That means until you successfully pass one of these law enforcement refresher classes!" I bellowed. By now, I sure as hell had everyone's rapt attention over my loss of cool!

"You can't do that," yelled back one of the errant refuge managers. "We work for refuges not law enforcement, and you have no authority over us," he snapped right back.

"You just watch me," I growled, still pissed over their lack of manners and intelligence. "All I have to do is pick up a phone and call the Area Manager who assigned us these tasks. Then we will see just how long you last," I shot right back. "Now, the two of you can leave and I don't want to see your mugs back here anytime during the rest of the training!" By now, you could have heard a mouse passing gas from 100 yards away in the prairie winds over the shock of my "explosion"...

"Explosion" because my officers had given up precious time from our districts to provide training for our refuge counterparts. Additionally, we had given up our whole summer training across the region and in so doing, that meant that our families paid the "absence" price as well. Plus it was humid, hot and buggy in that damn stinking piece of Missouri during that time of summer. And that was all it took to set off this ornery, hard-headed German...

As my two disgruntled refuge managers stomped off and left the area, it was as if a huge wet blanket had been lifted off the group of trainees. Then without missing a beat, I reassigned two more officers to perform a known dangerous car stop. One of those officers selected for that dicey detail was Bill Skar. In fact, Bill was assigned as the lead or contact officer. And in the car to be stopped were five "pretend hard cases" who eventually in the course of the exercise, were to be arrested. As Bill made his correct approach, one of the chaps in the backseat of the "suspects' car" made an assigned furtive move "spelling" danger. It was right then and there that I decided if I ever made Special Agent in Charge, Bill would be in my squad... In an instant, Bill recognized the "danger" and in one fell swoop, reached into the open driver's front door of the car and brought out all five men in his massive right arm at the same time! And in so doing, brought all of them out the driver's side! Think readers, all five men in one man's arms and all brought out the driver's side opening at the same time! Never in my life had I seen such a move... He was so big and strong, he just reached into the backseat, scooped up all three actors and brought them en masse over the front seat and out the door...! Then without missing a lick, Bill proceeded in the same movement in sweeping the two occupants in the front seat, with the bad guys from the backseat still in his arm, all out the driver's door at the same time! That was my first introduction to my gentle giant named Bill Skar.

The following year, I was promoted and assigned to the Washington Office and left the Dakotas. But I managed to keep an eye on Bill. Shortly thereafter, Bill put in for an agent's position, was selected and sent to the academy in Georgia. When he returned from the academy, I was informed that he and his wife had gone through an extremely bitter divorce. In that divorce, Bill's wife received custody of their only child and that broke Bill's heart. I heard sometime later that he took to drinking and slid a patrol vehicle under a semi while driving intoxicated. It was readily apparent that the divorce and loss of his little one had really affected Bill. Years later, I was selected as the Denver Special Agent in Charge and upon my arrival, I happily discovered that Bill was my North Platte, Nebraska, Special Agent. I also learned that after his accident with the government vehicle, he toned down his drinking. But the loss of his only child stayed with him in more ways than he cared to admit...

Throughout the following years, Bill served the Service and me very well. In fact, when we had dangerous raids and arrests to perform, Bill and I would team up. Especially if we had a very hard case or bad hombre to arrest. Instead of sending smaller officers in the front door to make the arrests and chance a fight or injury, Bill and I would perform those duties. Can you imagine two officers, both over 6′ 4″ in height and weighing in at over 600 pounds collectively, knocking on your front door? If that left anything in the bad guy's tank when it came to the "fight" department, I never saw it... Yet when we went through those doors of "destiny," I never once ever saw Bill lose his cool, become overbearing or careless. Here was an officer and real man who treated all others as he himself wanted to be treated.

To illustrate a little more about Bill the man, how about this instance. He loved little kids, especially after losing his only child in the divorce. On one particular day in South Dakota after arresting several bad guys who were commercial market hunters in the Wind Cave National Park area, it was lunch time. And believe you me, Bill and I could really put a hurt on any buffet unfortunate enough to be placed in front of the two of us after we had worked hard making arrests. I remember stopping in some South Dakota off-breed town for lunch that day, after our legal work was done.

Exiting our patrol truck, we started walking down the sidewalk towards a restaurant. Then we chanced passing an old red beat-up pickup truck. In the back of that truck were a couple of chain saws, several old chain saw bars and chains, numerous cans of oil, a gas can, a couple of axes, and three small children! Those kids probably ranged in age from four to maybe eight years old. Bill immediately stopped and walked over to the truck. There he asked the kids where their folks were. The oldest boy, a youngster about eight, said they were in having lunch in the adjacent restaurant. Bill then asked why the kids weren't with their parents. The boy responded by then saying their folks didn't have enough money to feed all of them lunch, so they were left behind! I could tell from the look on Bill's face when he heard those words, it just broke his heart! Hell, for that matter, mine, too! Then Bill just looked hard at the parents sitting at a nearby table eating lunch in the eatery through its front window. They in turn just ignored the presence of a large unknown man standing and talking to their kids in the back of the pickup...

Turning, he said, "Wait right here for a moment,

Boss." Walking across the street, Bill disappeared into a Red and White Grocery store. A short-time later, he reemerged carrying three cartons in his massive arms. As he got closer, I could see he had three cases of Snickers candy bars! Then he proceeded to give each small child a case of candy bars. Talk about excited kids, you sure had three hungry but now happy little ones in the back of that pickup that fine day so long ago. Hell, for all I know, eating all that candy probably gave those three kids a good case of the trots. But from the looks on their faces, those candy bars went a long way in filling up the cracks in their small and neglected hungry little carcasses... Then I had to laugh. Just imagine what following that pickup was like as it headed down the highway when they left town. I'll bet there was a snowstorm of empty candy wrappers flying out of that truck and into the wind and windshields of any following vehicles...

Then Bill and I went to eat lunch and nothing more was said about the episode. That was until the bill came for our meals. That was when Bill asked me if I could pick up the tab. It seemed he had spent all his money purchasing those kids three cartons of candy and now, did not have cash on hand to pay for his own meal.

Then there was the time Montana Senior Resident Agent Joel Scrafford needed another rider for his next backcountry grizzly bear horseback patrol in Yellowstone National Park. Joel called me and asked if he could request Bill as his partner for the rugged and dangerous (to me, horses are a wreck waiting to happen) trip. I advised he had better ask Bill and see if he was available. But no matter, I knew Bill would be a good one for such a dangerous detail in grizzly bear country. He already owned horses and was an accomplished rider. Plus, he was a fine pistol shot with good survival sense. So I figured he would be a natural. Well, I can hardly write these lines as I am already starting to laugh over this one... Or should I say, giving one another close in look at my gentle giant?

Joel called Bill and the two of them made arrangements for the grizzly bear backcountry detail. Then later, those two got together and met their two National Park Service officers who would accompany them on the same rugged detail. I heard no more about that detail until they returned. Now to those of you who did not know Joel, my Billings Senior Resident Agent, well, he could be a bit of a turd. You know, let one know that he was better than the next Yogi Bear in the park when it came to horseback details in front of his Park Service buddies (Joel had come to the Fish and Wildlife Service from the National Park Service). As I later discovered, the first thing Joel did was have the group ride 35 miles in Yellowstone that first day of the trip! Now to any of you horse people, that can be a bit much if one has not ridden such a distance for some time. Thirty-five miles on any horse is sure to give one a sore backside and a pair of knees who were wishing they belonged to someone else. But that was one way Joel would try to show you who was the toughest son-of-a-bitch in the valley. But Bill held his own on that grinder of a trip as I later discovered.

Joel, unable to unhorse and embarrass Bill with a killer of a ride, went to pushing Plan B even harder. During their long truck ride to Yellowstone National Park from Billings pulling their horse trailer, Joel unleashed Plan B. All the way, Joel did nothing else but tell Bill, who had never ridden in grizzly bear country before, what dangerous animals they could be. Then on the first half of that seven-day trip patrolling the backcountry for bear and elk poachers within the park's external boundaries, Joel told evil story after evil story about the very dangerous, human-eating grizzly bears. By the time the group had arrived at the Miller Creek backcountry cabin for a night's rest, Joel had pretty well sensitized Bill to the point that he was seeing hungry bears around every tree waiting to eat him at the drop of a hat. Now, I am starting to "bust a gut," so all you readers hang on to the willows for what is coming.

Having been to Miller Creek cabin in the Yellowstone outback myself on such a patrol, I could only imagine what frame of mind Bill was in. First having heard nothing but gruesome stories about the "griz" eating everyone he could catch, and then seeing that cabin for the first time. It sure as duck fat makes for excellent cooking, had to leave its first impressions... That storied backcountry cabin in the heart of grizzly bear country had bear chew marks all around the sides where the bears have tried to tear their way inside! Then on all the heavily shuttered cabin windows and the front door are numerous, very obvious, bear-proof obstacles. Each shutter and the front door had long, greatly sharpened spikes, driven from the inside out. Dozens of them to be exact! That way, if a bear tried to break into the cabin from the windows or front door, it first had to get past the vicious nest of spikes driven sharp-side out.

Knowing about Bill's gentle side, I could just imagine what was going through his mind at that point in time after seeing such a bear-proof fortress. Then while the two Park Service guys unloaded the horses and set about making dinner, Joel and Bill tended to the horses and pack animals. Currying livestock, checking horseshoes, and setting out saddle blankets to dry became the word of the day for the two men. Then into the cabin the four of them went for the warmth and the delightful backcountry supper that was to follow after a long, cold day in the saddle.

The rest of this "Bill" story I got from Joel. Now all you readers keep in mind that Joel came from the Park Service prior to joining the Fish and Wildlife Service agent core. He was also very proud of his Park Service heritage and very proud of his prowess as a horseman and back-country rider with a tough-as-nails reputation... That evening is when the wheels came off Joel's "gut full of pride" wagon. It seemed that Bill had to take a dump in the middle of that darkest of nights. A very big dump! But after five days of nonstop "a grizzly will eat everything in sight including humans" stories, damned if Bill was going to chance a long walk in the dead of night in grizzly bear country to the outhouse... So, quietly grabbing his .44 magnum pistol, Bill silently left the group of sleeping officers in the cabin. Stepping outside and remembering all the gruesome bear-eating-human stories, Bill took his big dump! And being at least 6' 9" and weighing over 300 pounds, it was a very big dump! Unfortunately, he did not go very far outside to take care of business. Or for that matter, even to the outhouse some few yards distant. As I later heard from a very embarrassed and pissed-off Joel Scrafford, Bill had gone outside and taken his very big dump on the cabin porch near the steps! Damned if any hungry grizzly bear was going to get him!

The next morning, a not-too-careful Joel had discovered the dump on the porch. Not only that, Joel had discovered the dump on the porch after finding its evidence on his custom-built cowboy boots as well! Suffice to say, my man Scrafford, "the greatest of all the American cowboys," especially when the "Parkies" saw the dump, went into orbit! He was so embarrassed over what one of his own men had done, that he was fit to be tied. In fact so embarrassed, for the next two days, he said nary a word to Bill except for chewing on his last part over the fence. Then when the riders got to the trailhead, Joel just quickly loaded up their horses and split before the "Parkies" had another good laugh over Joel's smelly cowboy boots. Then when Joel returned, he couldn't wait to call and tell his boss (me), how badly Bill had behaved in placing the dump on the cabin porch next to the steps. By then, I was laughing so hard and hysterically over what I was being told, that Joel, now pissed at me, and still not over having his badly dented pride tinted with poop, abruptly hung up on me.

Then lightning struck Bill once again. Only that time, I should have better read the tea leaves on his life to come... Months later, he spent several days training refuge officers on a refuge in Nebraska. After the training was over, the officers had a barbecue and of course some beer. About halfway through the 'good old boy' get-together, they ran out of beer. Then it was decided Bill in his government vehicle and a refuge officer would go into a nearby town and get some more beer... That was his first mistake.

Purchasing the beer, as the two men left, they noticed a hay truck full of baled hay sitting in the bar's parking lot. Well, as I came to understand it, the refuge officer bet Bill he couldn't cut the rope on the truckload of hay. Especially since the truckload of hay belonged to a county commissioner who had been a thorn in the side of the nearby refuge for years. Well, Bill being up to a challenge like going on a grizzly bear patrol, drove alongside the hay truck and cut the rope. It was then the wheels on that wagon came off as well. It seems they were seen and in their effort to get away, were identified. As it turned out, the cut rope was discovered and fixed before any harm came to the load of hay. But the damage was done.

The following Monday, the Nebraska county commissioner called my boss, the Regional

Director, in Denver to complain. Then my boss called me up to his office and in a fit of rage, ordered me to fire Bill. I told him that I would not fire anyone until I had all the facts. And then and only then would I make my decision since I was that man's immediate supervisor. That the Regional Director agreed to, and I returned to my office and called Bill. Bill told me the whole truth and from what I could discern, the rope-cutting incident had been beer and lack of common sense-driven. But no matter, it was a stupid idea on his part. Then I called the refuge manager of the refuge officer who had been with Bill and made the bet about cutting the rope on the load of hay. As I discovered through that conversation, the refuge officer who had been with Bill got one day off without pay for his part in the rope-cutting endeavor.

Then I returned to my boss's office with my verdict. I told my boss who was a bit of a coward and bully, what the circumstances were after talking to Bill. He immediately once again ordered me to fire him and I demurred. Then he immediately began yelling at me about who was the boss and that I had better obey his commands or I would be out the door as well. Like I said, my boss, Regional Director Galen Buterbaugh, was a bit of a hind end and a bully...

I once again advised my boss that Bill was

overall an excellent officer and that he had many years of excellence under his belt while working for the Service. My boss, who was still yelling at me and now up on his "high horse," ordered that Bill be fired immediately. In fact, his yelling was so loud, his secretary, Barbara Henry, got up and shut his office door. Once again, I refused. By now, Galen's face was beet red, as he continued screaming at me as he ordered me to fire my officer. Once again I told Galen I would not. But I also explained to my boss that I already knew what I was going to do with my officer for his moment of stupidity. Then still on his "high horse," Galen demanded to know what I was going to do short of firing my officer. I quietly advised my Regional Director that I was going to suspend my officer for 30 days without pay and move him from North Platte, Nebraska, to Minot, North Dakota.

Leaving Galen's office that morning, I saw the Regional Director's secretaries looking at me and giving me the high sign of support. They knew Bill was a good guy underneath all that kid he possessed in his soul. However, what I had to tell Bill was not good news but it had to be done and was certainly better than being fired. Calling Bill one more time, I could tell from the sound of his voice that he expected to be fired. When I advised him what I had decided to do in the form of punishment, he audibly breathed a huge sigh of relief.

Then Bill said, "Boss, I will never again let you down. You have saved my career and I won't forget what you have done. Especially since I have already heard through the grapevine, what you had to go through in Galen's office. You have my word on it, Chief."

Thirty days later, I moved Bill to a duty station in Minot, North Dakota. That move turned out to be fortuitous later on in his life. There he met a young lady who loved him dearly. A lady whom he eventually married. But I am getting ahead of myself...

Bill true to his word, once at the Minot duty station, tore up the bad guys. He instituted a law enforcement program second to none and I was damn glad I had given the man another chance. Especially when I found out he had discovered a new love and she was just the gal for my gentle giant.

Several years later, Bill worked his way covertly into an illegal commercial fishing operation from Tennessee, which was purchasing paddlefish eggs from the local fishermen along the Missouri River system. Working covertly, Bill soon endeared himself with the Tennessee outlaws. Soon we had a great case in the making, where paddlefish eggs were illegally moving in interstate commerce in the caviar trade. In just a short period of time, the case had expanded to such proportions that Bill needed extra help. Joel, his old friend with dump still on his cowboy boots, had by then forgiven Bill for his porch-dumping act. Together the two officers were putting together the investigation, when I got an ominous phone call from Joel. In that phone call, Joel informed me that Bill could only work until about noon and then being so exhausted, he would have to lie down under a tree and rest!

Getting a bad feeling, I drove to Minot and met with Bill. He tried to put off his physical health concerns due to the expediency of the great case he was now working. That concern was immediately put to rest with a direct order from me for him to see a doctor. That he agreed to do and did so. The news was not good. He had a rare fatal blood disease! I believe it was called "hairy cell" leukemia. Bill went quickly downhill from there. Soon he was so sick he couldn't even control his bowels. It finally got so bad he had to wear a diaper and was homebound. One can only imagine the shame going from a 6' 9", well-built giant, to just a mere slip of a man who was almost totally helpless. God, how I still hurt inside as I write these lines... On one of my last trips to Minot to see Bill, I convinced him to marry the gal he had been living with, so if nothing else, she could inherit his retirement. That he finally did, just before he died some weeks later. As I sit here writing these sad words of a dear friend lost to the ages, the tears are streaming down my face and staining the front of my shirt.

Bill was a good and unusual man, a gentle giant if you will. He had a heart as big as all of the great outdoors and a smile to match. I still miss him dearly to this day...

Sometime after his death, Bill's wife gave me his .44 magnum he had purchased for that grizzly bear patrol from what now seems "a campfire long out." His wife said Bill wanted me to have that gun to remember him by. Several years later, I gave that big magnum to my Littleton Police Officer son, Christopher. I did so since I already owned and had used such a fine firearm throughout my law enforcement career. My son gladly took the gun, understanding its history and did it proud by always shooting perfect scores with it using hard-hitting factory ammunition! Something probably only Elmer Keith, the big magnum's inventor, could have done. My police officer son, now deceased in the line of duty, was a lot like Bill. He was a large man as well with a heart as big as all the great outdoors and had a smile like Bill's to match. Today that big magnum sits in the gun safe at the house of my wonderful daughter-in-law, Lisa. She is keeping that handgun for her son Gabriel, who was born after my warrior son Chris died. I think Bill would like that, as would have Chris...

If I am not mistaken, a simple medicine with a 98% cure rate was discovered for Bill's blood disease some two weeks or so after he died. But not before that disease had taken my friend, Special Agent Bill Skar, who was just a man in his early 40's... rest in peace, my friend...

## HUNTING "THE MOST DANGEROUS GAME" ONE LAST TIME

Sitting in my office one morning in the fall of 1997, my intercom buzzed. On the other end of the intercom was my perky and wonderful little secretary, Debbie. "Scott Pearson from Sacramento is on line one," she said.

Thanking Debbie and clicking on line one, I said, "Good morning, Scottie. What can I do you in for?"

Scott was the Service's Senior Resident Agent or first-line supervisor in Sacramento. He was a big old dude like me and a damn good catch-dog officer. I had trained him earlier in his career at the national academy in waterfowl identification and during that time, we had become good friends. He was also my rookie Special Agent older son Richard's immediate supervisor.

"Terry, I need for you to come to the northern

Sacramento Valley. I recall your actions when you worked here as a game warden and later on as an agent. I also know you were deadly when it came to catching not only the run-of-the-mill outlaw duck hunters but commercial market hunters and 'duck draggers' as well. I think it is time you returned and worked with your son to show him the ropes before you retire (I was slated to retire less than one year later). Plus, being the historian that I am, just think, a Special Agent father-and-son team working together. That hasn't happened for many a year now and I think it needs to happen once again. Bear in mind once you retire, there are no more father-and-son teams in the Service. I talked this over with Dave McMullen, my Special Agent in Charge, and he thinks it is not only a good idea but a grand opportunity for you and Rich. We will pay all your expenses for such a trip and any return court costs you incur to testify on cases made, if you are interested in such a detail."

Sitting back in my office chair in surprise over the offer of a lifetime, it took a little time to ponder my response. Now don't get me wrong. I already had a full plate in running my eight-state region as the Special Agent in Charge. Truth be known, I didn't need any more of a workload to carry. But this type of an opportunity was just too good to pass up. Especially in light of the fact that I

would once again be working in my old stomping grounds one last time, hunting the lawless
and chasing those knotheads I had missed the
first time around.

"Yeah, I think I can swing it. When would you
need me to arrive?" I asked.

"Well, Chief, duck season is in full swing here
in the Sacramento Valley and now is as good a
time as any," he quickly replied.

"What does Rich have to say about having his
old man riding along as shotgun on this crazy
venture you have in mind?" I asked.

"He is all for it and could use the help. He is currently the only agent I have available to work the
valley, and the two of you could just cut a swath
long and wide as far as I am concerned. To be
frank with you, ever since you left in 1974, I think
my waterfowl outlaws have had a breather. Can
you imagine what it would be like when those
old-time outlaws from your days past see your
face grinning at them once again?" he chuckled.

"Have your secretary send some cost codes to
my Administrative Officer, Grace Englund, so
she can arrange for an airline ticket and I am your
man," I quietly replied. Two days later, I arrived
at the Sacramento International Airport and was
met by my oldest son and brand-spanking-new
Special Agent, Rich Grosz (who, currently as
these words are written, is the Resident Agent in

Charge for North and South Dakota—the same as his old man was, some 37 years earlier).

"Morning, Dad. I see you have your grubbies on and are ready to go to work. That is good. Being it is a non-shoot day here in the Sacramento Valley as is tomorrow, that gives us two days to get our feet on the ground and locate the ducks," he said with his characteristic, good-natured big grin. Then he shouldered my heavy duffel bag carrying all my gear, as I grabbed the other one full of clean clothing. "My truck and Grumman Canoe are just outside in the parking lot. We can head for the Colusa area just as soon as we get your gear stowed," he continued with his typical genuine grin. With that, I never saw that airport again until 13 long, cold, wet days later...

Arriving at his new Dodge 4x4 truck in the parking lot, I marveled at the money that my counterpart in Region 1 had. I couldn't get a Dodge truck in the region I came from. The Budget and Finance Office always told me they weren't available and I had to settle for those damn Fords. Go figure... Then I took a quick look at the Grumman Sport Boat Rich had on his boat carrier in the truck. It was an old, unsafe piece of junk!

"Neither one of us will be using that narrow-beamed, unsafe piece of junk," I sternly advised. "I used those same models in Region 3 when I was the Assistant Special Agent in Charge in 1979, and I junked all of those unsafe, no built-in flotation bastards in a heartbeat. You need to tell Scottie these are death traps and I damn near lost two officers in Illinois in these narrow-beamed crappers," I growled.

Rich just grinned saying, "Well, Old Man, you had better plan on walking a lot in that soft rice mud of the upper Sacramento Valley since now we won't be riding in a canoe when the occasion arises."

I just grinned. *He might be a 6' 7", long-legged galoot but his young spirit would be no match for his old man's deceit and guile,* I smugly thought. Boy, was I to soon get a lesson on just how old and broken down I had become compared to his youth and boundless energy. Especially when it came to hunting the most dangerous game in the northern Sacramento Valley.

Then off we went in a soft drizzling winter Sacramento Valley rain, as we headed north towards Colusa and Glenn Counties. Counties that historically, along with four other northern California counties, were haven to millions of ducks and geese during the winter migrations. An hour later, we arrived in the small Town of Williams and rented a motel room for our 13-day adventure. Then to make driving a lot easier and leaving a less noticeable "footprint" for the outlaws to fixate upon, we unloaded our useless canoe and left it in the motel swimming pool. Now, we looked just like the rest of the run-of-the-mill dirt farmers in the valley, I thought. Especially since we were driving around in an unmarked, muddy vehicle similar to all the rest of the duck hunter and farm rigs.

However, before we left Williams, we went to a store/restaurant combination named Granzella's that specialized in Italian food. Once inside, I made sure we had loaded up a mess of Italian hard salami and the like into Richard's ice chest. Then we purchased a case of Diet Pepsi, numerous pouches of Levi Garrett chewing tobacco, and we were ready to shake, rattle and roll. In fact, for the next 13 days, we only ate in a restaurant one time! The rest of the time, we lived out of that ice chest as we cranked out a continuous stream of 18-hour days. I figured if I was there to break in a rookie Special Agent on how to work the valley's duck shooters, he had better learn to run with the big dogs. And without a single complaint from the young and the old, that we both happily did. Truth be known, it was a piece of cake. Rich had been raised ever since he could ride with his old man, that a day's work was just as long as it took to get the job done, and done well.

For the next two days, we did nothing but run the valley floor in all my old known duck hunting haunts, trying to get the lay of the land, scout out the duck concentrations, and ascertain their morning and night-time flight and land-use patterns. For the most part, we left alone what few duck hunters were out and about in order to keep quiet the fact that the 'feds' were in the valley. Plus, it was a good lesson for Rich on how to really work the serious duck hunters and valley's waterfowl outlaws. First, get to know the lay of the land and the waterfowl flight and land-use patterns. Then, get to know up close and personal-like the duck shooters...

It was during that time of serious scouting that I realized the valley had changed in several major ways since I had left it some 23 years earlier. For one, there was a lot more flooding of rice paddies in the valley which really scattered out the ducks. Now it seemed almost every farmer's land was rented out as "poor boy" duck clubs. I also noticed that now instead of walking, most duck hunters used ATVs to get back and forth to their blinds or hunting areas. After two days of scouting, I also noticed that the numbers of ducks and geese had dropped off by at least 50% since my earlier days of running the valley from the late '60's to the middle '70's! They just weren't there in the numbers (millions) I was used to. I found that demise of ducks in the world of waterfowl very sad. We also discovered that, for the most part, early and late shooters of waterfowl were almost a thing of the past. Especially when I could remember catching hundreds of early and late shooters in my day every duck season. And that included many shooting up to six hours late! Especially during numerous instances when the night lights from the nearby cities were just right and reflecting off the low-hanging rain clouds, which allowed illuminated late shooting! Bottom line, the ducks just weren't there in sufficient numbers to tempt early and late shooting lawlessness. I also noticed that the length of the duck season had been shortened by about 20 days since I had left the valley. No doubt reflecting the reduced numbers of waterfowl. Lastly, hunters could only use steel shot instead of lead, like had been the norm in my earlier days. Again, a reflection and realization that the use of lead shot eventually killed the hell out of feeding waterfowl through lead poisoning when ingested as grit.

## FIRST BLOOD

It was during the last non-shoot day (normal shoot days in the Sacramento Valley had historically been Wednesdays, Saturdays, Sundays and holidays), since the beginning of our "hunt," around nine o'clock in the evening, that I was able to teach Rich an old market gunners' trick.

One that he in future days could use to turn the tables on any serious night shooters he might run across.

That Tuesday evening, just north of Williams on the west side of Interstate 5, we stopped the truck. There we attended to a call of nature in the inky black darkness of night, as a light winter rain drizzled around us. We were in that area because in times past during my earlier career, especially when the winter rains came, the waterfowl moved from the center of Colusa County to those rice fields lying west of the Interstate. When they did, the duck draggers would sneak up on feeding waterfowl at night and ground-sluice them, killing them in large numbers. That was also the pattern of the few remaining, old-time, dyed-in-the-wool market gunners.

Earlier in my career, I had discovered that waterfowl feeding in the rice fields at night did so quietly under such total darkness and winter weather conditions. In so doing, they became difficult to locate for an illegal night shoot or for those looking for a normal, next day's, early morning's duck shoot. Since our weather and duck movement was such, I figured this was a teachable moment. Then without warning, I asked Rich to locate the nearest flocks of night feeding ducks in the various adjacent harvested rice fields.

He said, "What?"

I said, "Using your finger, point out the nearest bunches of night feeding ducks." Even in the inky dark and wet of the night, I knew he was looking at his old man like I had gone daft.

"What do you mean, Old Man?" he asked with his question hanging heavy in his response.

"Simple. Point out the nearest feeding ducks in this area," I said with a knowing grin spreading across my face.

Following orders, my son listened intently as the soft rain patterned noisily off the metal of his truck. "I don't hear anything but the sound of rain drops hitting my truck and the faint sounds of cars on the Interstate," he finally responded.

"Well, that is one reason first of all that one moves away from his close-at-hand vehicle when he is interested in listening to quiet night sounds. Vehicles are inherently noisy as their engines cool down and the like. Now, let's step a few yards away from your vehicle," which we did. Realizing his absence of hearing nearby feeding waterfowl because of lack of experience, I said, "Now, cup your ears with your hands (magnifies sound) and do a 360-degree turn (I had done so earlier in the dark unobserved by Rich). When you hear a sound like flowing water, point your arm. The flowing sounds you will hear are the sounds ducks make as their bills move rapidly through the rice fields' straw, locating and eating waste grain left behind during the harvesting process."

Doing as he was told, he began his turn. Then he excitedly said, "There-there-there," as he moved in a circle from where we stood on the road, quickly pointing his arms.

Then I said, "Now, get up into the back of your truck and repeat the procedure."

That he hurriedly did and once again, he swung himself in a slow 360-degree turn with ears cupped once again. "There-there-there-there," he once again excitedly spoke out. "Dad, how come that is?" he said as he stepped out from the back of his truck.

"That, Son, is an old market hunters' trick used in the darkest of nights to locate feeding ducks. By cupping your ears, you magnify all night sounds. Getting even higher, like in the back of your truck, will increase your hearing range considerably regarding soft night sounds. Then once the night gunners echolocated on the feeding birds, they would begin their sneak, and eventually ground-sluice the feeding birds from very close ranges. Keep in mind for the future that by using your hands cupped over your ears, that acts as a magnifier of the softest of sounds. Then in the instant case when you hear that flowing water sound, you have located your ducks,"

I continued with a smile over having taught my son a clever trick. "Now," I said, "locate the biggest bunch of birds in our area by that same method." Once again my son utilized his newest trick just learned and after a careful listening period, pointed to the north and east from where we stood.

"Good," I said. "You have just pointed out where we will be come daylight tomorrow, which is a normal shoot day in the Sacramento Valley. I say that because if someone is too cowardly to night shoot those ducks, they will be on them in that spot come tomorrow morning. They will do so because many of those ducks if left undisturbed, will be there in the morning as well. And the ducks will stay there until they fill up with waste rice or get shot out. Now, what do you say we head to our motel room and get some sleep? We will have an early riser in order to get into that place just identified for tomorrow before the bad guys do. Then we will be ready for whatever the day brings."

"But before we leave, tell me what species of ducks we will be staking out come tomorrow morning?" I casually asked. Once again, I was sure I was getting a dumb look from my son over that question in the black of night. It was darker than the inside of a dead cow and now starting to rain even harder. Normally, unless one knew what he was looking for, he wouldn't be aware of anything in the way of species. I could just hear his wheels turning inside asking himself how the hell is he supposed to know what kind of ducks he had just discovered.

"How the hell am I supposed to tell you what kind of ducks I am hearing, when I can't even see them. Hell, Dad, all I can hear is the flowing water sounds you asked me to identify as feeding ducks. But I sure as hell can't see them," he replied.

"Crawl back up onto your truck and cup your ears, Son," I advised. When he did, I said, "Now, carefully listen for pintail whistles, quacking from hen mallards or the calls from wigeon. Depending on what calls you will hear, will give you a direction in which to go. If you hear a lot of wigeon, forget it. They aren't the best eating, are medium-sized and a market doesn't really exist for that species. However, if you hear mallards or pintail, that is your "mark" for the night or the next morning early. They are the larger-sized birds and considered better eating than wigeon. Those will more than likely be the ones the bad guys will either sneak at night or set up on for the next morning's shoot," I said.

Without a word, Rich cupped his ears and went silent. Moments later he said, "Dad, the birds we are going to be on come tomorrow morning are pintail. Damn, you were right. If I really concentrate, I can hear them softly calling back and forth. And what I am hearing are calls from pintail. As for the other four bunches of feeding ducks in the area around us, most of what I am hearing are calls from wigeon. What a great thing to know," he continued with excitement rising in his voice.

With that, we pulled up stakes and headed for Williams and our motel. I slept well that evening having taught my son several old market hunters' tricks on how to locate and identify feeding ducks during the darkest of nights. Tricks that had been taught to me by an old market hunter and friend whom I caught using that same trick some three nights later after he had taught me. Caught with his dark of the night kill of 183 pintail and mallards. But, that is a story for another time and place...

The next day at four in the morning found Richard and his old man safely hidden in a large grove of eucalyptus trees just south of Marengo Road. We had used a thick row of trees and some dense brush for cover in case someone else wanted to park near our area as well. Then I had Rich get out and using his hand-cupped ears, see if he could locate our flock of ducks discovered from the night before. In an instant he had located the same feeding birds some hundred or so yards south of our position. Then the two of us had a typical game warden's breakfast of hard Italian salami, crackers and Diet Pepsi to the sounds of the valley's winter rains drumming off the roof of our hidden truck.

About an hour before legal shooting time, we had company! A pickup drove into our grove of trees and quickly put out its headlights. Then the dome light went on and through our binoculars, we could see two men in camouflage clothing hurriedly exiting their vehicle.

Then it was once again as dark as the inside of a dead cow around us. It was then I got an idea...

"Rich," I said, "I have a plan. We both have portable radios and damn good binoculars. I propose just before daylight, you and your long stork-like legs take off and head in the direction our two chaps went. It will still be dark, so they won't see you coming. Remember, they probably are heading for that area where we heard that flock of feeding ducks last night. And since that area is still holding feeding ducks this morning, I would suggest you head in that general direction as well. Pick a rice check and stick to it for the cover it offers just as long as it leads towards our feeding ducks. Then when you get about fifty yards from the feeding ducks, dig in and wait so you don't spook our ducks. Just make sure you are well-hidden. Because if you aren't, new birds coming in to feed will see you and flare off. When they do, if the chaps you are stalking see that and are woods-wise, they will realize someone else is in the field. Knowing that, that may change the intent of our lads to not shoot over-limits, drag the ducks or the like. In the meantime, I will take a position from this grove of trees where I can watch the whole scene. Once I spot our two chaps sneaking the ducks, I will call you on your radio on channel one. Then I will direct you in to them so when they shoot, you can be on them before they know what hit them. Additionally, I am using a set of Steiner 7X50 binoculars with a built-in compass in the ocular. If they try hiding anything, I will get a compass bearing on that location. Then when you have them in tow, I will use that compass setting and guide you back to where they have stashed extra ducks or lead shot (use of lead shot to take waterfowl is illegal). Additionally, once they see you, if they split, I will be in a cutoff position here back at their vehicle to play spoiler in their race for freedom. What do you think?" I asked.

Hell, my plan must have sounded alright because before I could say anything else, Rich was making ready for his trip out into the rain and the still-darkened, freshly harvested rice field. Finally ready, we checked our radios to make sure they were working properly and off he slipped into the dark of the coming dawn. Grabbing the rest of my stakeout gear, I soon followed. Locating the best weedy area, I buried myself into a stand of dripping wet brush. From there, I could see the entire area from whence the sounds of feeding ducks still droned on, unawares of the soon-to-be-unfolding drama focused around them.

Just before legal shooting hours, I heard a rattle of shots in the direction of our feeding ducks. Soon I could see with a naked eye numerous strings of fleeing waterfowl heading east towards the nearest national wildlife refuge and the sanctuary it offered. Raising my binoculars, I quickly scanned the immediate area nearest from where the ducks had erupted into the air from their place of feeding in the rice field. I saw nothing for the longest time. Then I observed two heavily camouflaged individuals crawling across the field, hurriedly picking up ducks from the rice field to their front. Then into my ocular field sneaked Rich as he ran in a crouched position from rice check to rice check as he approached the area of our two shooters unseen. Then I noticed our two shooters with duck straps filled with ducks quickly leaving the field and heading back towards the direction of their still-hidden vehicle.

A few minutes later, I saw our chaps eyeballing in alarm in the direction from which Richard was approaching. It was obvious they had spotted his approach as Richard continued his sneak.

A hurried conversation took place between our two chaps and then they knelt down out of sight alongside a heavily weeded rice check. There I could partially see my chaps throwing something into the tall water grasses by the rice check where they now hid. As they did, I took a compass bearing through the built-in compass in my binoculars where their furtive actions had taken place. Finally rising, they began ambling across the harvested rice field like nothing out of the ordinary had occurred. It was then and before I could alert my son, Rich realized he had been seen. With that realization in mind, he hotfooted it straight across his rice field with his long legs towards the two shooters, who were still pretending they hadn't seen him.

Finally Richard's long ground-eating stride allowed him to intercept and confront his two shooters. Watching through my binoculars, I saw the three men talking. Then Richard began checking their shotguns for plugs (can only be capable of holding two shells in the magazine, with a third in the chamber), their ducks, and checking their shotgun shells with his magnet to make sure they were shooting only steel shot. Then I saw Richard's portable radio antenna glint in the daylight after he had moved off from the two shooters about thirty yards so they couldn't hear what was being said on his radio.

"Dad, they each have limits of ducks, no lead shot, and their shotguns are plugged. Since they didn't shoot but a few minutes early, I don't have anything on them. Did you see anything out of the ordinary that the two men did before I got there?"

"Rich, make sure you have their hunting and driver's licenses in hand. That way if they decide to run, we still can track them down. Then tell them to wait as you backtrack their earlier position. Don't say anything about receiving any instructions from me. Then head south on the rice check they were walking on until I tell you to stop. Where I tell you to stop will be in that immediate location where both men knelt down and tossed something into the weeds along the rice check. My guess what they tossed into the weeds will be lead shells," I radioed.

Rich following directions, headed back down the suspect rice check. As he did, I followed his progress until he came to the compass heading on my binoculars where I saw the two suspects kneeling once they had spotted Rich heading their way. When he hit that compass bearing, I stopped him with the radio and told him to look around the rice check in that immediate area. This he did for a few moments. Then I saw him kneel down. Soon he stood up holding a large handful of freshly killed ducks. These he held up, knowing I was watching his progress through my binoculars. That sure as hell brought a big smile to my always-handsome Robert Redford-like face. However, I was sure it curdled the milk of our two shooters, as they observed Rich finding their illegal stash.

Then Rich called me on the radio and advised he had found five extra ducks over the two chaps' limit and a handful of lead shot shells that had been tossed in the rice check grasses as well! Then he headed back to my two shooters and a short discussion soon followed. From my hidden position, I soon observed more "curdled milk looks" on the faces of our two shooters over what Rich was telling them. Then with Rich carrying all of the men's ducks, the three headed back towards where I was hidden in the brush. Not wanting to give away my hiding place in case I wanted to use it in the future, I slipped away and headed for Rich's truck. Then taking an extra set of keys that Rich had given me to use in an emergency, drove it over to where our two chaps had parked their pickup earlier in the morning. That way, they wouldn't become aware of our vehicle hiding place in case we wanted to park our truck there some future day. Once there, I waited with another big grin on my face. There is nothing like "first blood" on an operational detail. "First blood" especially in my old stomping grounds of Colusa County in the northern Sacramento Valley of California, some 23 years after I had left the area.

When my three chaps arrived, I noticed our two shooters sure had long faces. Standing on the off side of the truck, I let Rich do his thing. There he identified the problem once again, pointed out I had spotted them ditching the evidence he had subsequently discovered, and then commenced writing out the two men's citations for possession of duck over-limits and lead shot.

Me, I just stood there out of the way with a big proud grin on my face. There is never a bad feeling when one of your kids follows his "old man" into the same profession and is very successful in so doing. Especially since the sun will soon set on my time here and is just beginning to rise for that of my "seed"...

Then the unexpected happened. As I stood on the far side of Rich's truck and he wrote out the citations on the other side, another truck pulled into our grove of trees. Two lads bailed out from that truck and being dressed in duck hunting gear, just stood there wide-eyed looking on at apparently two of their friends getting cited for waterfowl violations. They said nothing nor did I, as Rich finished up with the paperwork. Then after going over the legal particulars relative to our two subjects' federal court responsibilities,

Rich's business was finished. Without saying a word, they had to be wondering how Rich had gotten onto their 'early morning little feeding duck ground-sluicing' game. Then our two lads loaded up into their truck and without another word, left the area like two whipped pups.

As they pulled away, our two recently arrived lads walked over to Rich. I knew damn well why they were doing it but I was sure Rich didn't have a clue. They were aware he was a federal since he wasn't one of their local game wardens, and wanted to get a good look at his face so they could recognize him if they ran across him in the future. I just had to grin, having been there numerous times before during my career. Then one of the lads began asking Rich all kinds of legal do's and don'ts relative to the duck hunting sport. Then he asked Rich his name.

Rich responded, "My name is Rich Grosz and I am a federal agent from Sacramento."

For a brief moment the lad asking the question just mulled over in his mind what Rich had just said and then asked, "What did you say your name was again?"

Rich repeated, "My name is Rich Grosz."

Then our questioning lad said, "Do you know Terry Grosz?"

Rich said, "Sure do. He is my dad."

Then our questioning chap broke like an egg.

"Damn, can you believe that? The son to Terry Grosz. You know, the guy who used to be our local game warden and then he turned fed. Damn, I never met that guy because I was too young to hunt. But I sure heard a lot of tales about that fellow from all my kin. They said he was as big as a bull moose, worked every day of the week, worked mostly all day and all night, and could move through these rice fields, especially if he was chasing you, like that of a locomotive. You couldn't outrun him even though he was a big man like that fellow standing over there by your truck. But I guess he could run for miles in the soft rice mud and hardly anyone ever outran him. Damn, he pinched both of our grandfathers, all our dads, most of my uncles, and every one of my cousins, some of them twice! Man, you sure have big shoes to fill and you started here today. Those two fellows you just caught have been killing ducks in this here valley like there is no tomorrow. And for years they have always bragged they would never be caught and never were. But you did it! And I can see why, if you are from that son-of-a-bitch Terry Grosz. Uh, excuse me, no disrespect meant. But he was a catching son-of-a-bitch."

"No disrespect taken," said Rich with a knowing grin, as he took a quick look at but did not identify me, to his two questioning young men.

Man, talk about a good feeling overhearing

those words. Especially after having been gone from the valley for years. Maybe all those long days and nights hadn't been for naught, I thought. Then both of those young men walked up to Rich and in a surprising move, asked to shake his hand. Then the one doing all the talking said, "I can't wait until I see my dad this afternoon and tell him I shook the hand of Terry Grosz's son. Hell, Terry caught my dad four times for taking over-limits of ducks before my dad wised up and hung up his shotguns. He has never hunted since that last time Terry pinched him in 1973. In fact my dad to this day, says Terry always treated him fairly and still considers him a friend."

Driving from our hiding spot in that grove of trees, Rich said, "Damn, Dad. That plan of yours sure worked well. If we get into another like situation, let's do it again."

"That would be alright with me," I said as I lit up one of my rancid little Italian Toscani cigars. Alright because my right knee was shot by 1997, actually bone on bone and the left one was not far behind. A direct result I suppose, from all my life's adventures logging, hunting and fishing, playing football, plane crashes, years afoot and on horseback as a game warden, and then later, many more miles rattled off as an agent. Those many adventurous miles spent on those knees were sure worth it but in the end, the devil had taken his due. I had both knees replaced several years later and still to this day, neither of them works for a damn. Oh well, sometimes one eats the bear and other times, he eats you...

But one had to admit, having drawn "first blood" on our initial day afield in the "hunt" was certainly welcome and we were just getting started. In fact, from the sounds of those two kids talking about their Princeton family's apprehension histories, it sounded like a lot of first, second and third blood had been drawn from that historic clan. Remembering back to my field days in the '60's and '70's and subsequent contacts with various members of that clan, they certainly deserved all the attention I and others like me gave them. In fact, several mean-spirited ones in that clan had been apprehended years earlier and sent to federal prison for the illegal sale of waterfowl. However, I can't take all that catchin' credit alone. I had a chap or two within that historical family group who kept me in the loop when it came to some of their little illegalities. And when they spoke, I listened. And when I listened, my catchin' went up...

## A DEAD ALEUTIAN CANADA GOOSE
## LEADS TO AN OVER-LIMIT OF PINTAIL

Trying to stay low key and keep our federal footprint light, I took to running our patrols on instinct for the next several days. After that, the word would be out that the feds were in the valley, and then Rich and I would just "flock shoot" our wayward sportsmen when it came to who was going to be checked. But for now, that meant passing up any and all sports hunting ducks and geese unless they "struck an inner chord" in my carcass. In other words, if my two guardian angels didn't focus my attention on any particular group of sportsmen, we would just look them over and then pass on by.

When I started this "hunting by instinct" operation, it frustrated the hell out of Rich. Being the young bull, he wanted to go down and check every one of them packing a shotgun or a mess of ducks. However, by operating in that fashion, it would only be a matter of a few days and the word would be out that the feds were swarming all over the valley. So because of that eventuality, I went to checking only those who "looked the part." But when we had caught 30 folks in a row over a period of a few days and issued 49 citations based on the bad things they were doing based on those instincts, Rich settled down and just let me "run." I can still hear him saying, "Dad, how do you do it? How do you know which ones to stop and check? Because every time you do, we end up catching someone."

"Rich," I said, "maybe it is just luck, instinct or warnings from my guardian angels but either way, I can't explain it. But be that as it may, we will take it. Now, let's head for the west side of Delevan National Wildlife Refuge on Four Mile Road. I used to make a lot of money on the duck hunters hunting in that area on the Newhall Farms properties and just south of the Lambertville Complex. So let's give that area a look-see and let this area cool off. Because you can bet those four lads we cited or met this morning are already spreading the "federals in the area" word far and wide on the west side of the valley.

The Lambertville Complex used to be a "nest" of shabby 30 or so duck clubhouses on the Colusa-Glenn County line. The hunters in that complex were usually a lower class of hunters who could not afford to be members of the really rich and highly successful duck clubs. Therefore, whenever many of their chaps (not all, but many) were afield, I had during my earlier career in the area found many of them willing to stray across the line of legality. Especially if we had a wintry rainy day with a little wind blowing. When that happened, the ducks and geese historically took to the air and made easier targets because of their lower flight patterns and increased airborne numbers.

Arriving at the southern entrance of Four Mile Road just off the Colusa-Maxwell highway, we turned north on the gravel. Just as we arrived south of the Gunner's Field turn in the road, I noticed a suspicious dark lump floating in a small watered ditch on the west side of the road. A lump that looked like it had just been thrown into the adjacent ditch to be rid of.

"Rich, when you get alongside that dark lump floating in the roadside ditch, let's stop," I said. "Whatever it is, it doesn't look good for the lump."

Arriving alongside the floating lump in the watered ditch, Rich stopped and got out from the truck. Walking over to the side of the ditch, he reached down and retrieved the lump. It turned out to be a freshly killed Aleutian Canada goose. A sub-species of Canada goose that was no larger than a big mallard duck and was a restricted-take species that year. During that period of time because of extremely low population numbers, that sub-species had been placed under the additional protection of the Endangered Species Act! Someone not knowing the identification markings of that bird in flight or its frequent area of use, had killed him just minutes before our arrival. In fact, the blood pouring from its wounds had not even yet begun to clot! However, there was no one to be seen. Thinking maybe the culprit was out in the extensive rice fields to our west still hunting, Rich and I decided to pull off the road into the field and have some lunch.

I figured if we did and someone close at hand showed themselves, we would have a chap to interview regarding the protected but now freshly killed, Aleutian Canada goose.

Walking around to the back of Rich's truck now that the rains had subsided, we lowered the tailgate and retrieved his large ice chest. Then we arranged on the tailgate our afternoon's fare. There were several sticks of hard Italian salami, raw sweet onion, several kinds of stinky cheeses, crackers, Diet Pepsi (is there any other kind of soft drink?), fresh jalapeños, fresh locally grown kiwi fruits, and several jars of a homemade surprise Rich had kept from me until now.

"What's in the jars?" I asked.

"Dad, Carrie (Rich's great little, and I do mean tiny, wife) made these for us. It is all kinds of fresh vegetables canned as only she can do. They are pickled and hotter than a firecracker," he proudly asserted. Now, both Rich and I liked hot things. Even some of the hottest of peppers in the world could often be found as a menu item during our tailgate "picnics." And Carrie, who is also a hot pepper aficionado, had made us some of her finest for that two-week stint, as I hunted the lawless for the very last time. Man, picture that lunch. Dark cloudy skies threatening rain, a strong wind blowing out from the northwest, cold and damp as a witch's heart, and two big galoots sitting out on a tailgate having a winter picnic eating all kinds of hot peppers. And I do mean hot peppers, if all the smoke coming out from our ears and nostrils as we ate, said anything as to the Scoville Units of "heat" we were consuming. Even a Scotch Bonnet pepper had nothing over us on that fine day so long ago! Man, I will tell you, that tiny Italian pistol Carrie that Rich married sure could cook, and her canned hot peppers were some of the best I ever ate, even to this day!

Sitting there eating our great little lunch and having such good father-and-son time together, didn't dampen any of my senses for the hunt though. While sitting there, I had noticed the north winds had picked up. That had stirred the resting ducks on the adjacent national wildlife refuge, and soon the air was filled with skeins of waterfowl looking for a place to feed in light of the oncoming winter storm. Then I noticed the freshening wind had also picked up the popping sounds of shotguns going off on the Lambertville duck hunting complex to our immediate north. As expected, the freshening wind had brought more ducks to their decoys and those gunners were making the most of it. Especially in light that their hunting area was only considered poor shooting at its best. However, with a freshening wind followed by rains, soon the ducks would be "walking" — they would be flying low and slow. And when that happened, it always seemed the Lambertville gunners would be trying to make up for lost time on all the bad hunting days they had experienced prior to the advent of stormy weather. And it was just precisely those conditions that were manifesting themselves as we finished our lunch that fine winter day. A day that was soon to get even better...

"Rich," I said, "we need to put our lunch fixin's away and head north to Lambertville. We can't do much about the person who killed this Aleutian goose, but it seems he died in this ditch to gather our attention to the fact of all the other impending gunning to our north. And if I was a betting man, I would say we could snatch a few shooters with over-limits of ducks if we made haste and our presence known. Especially in light of all those low-flying pintails I have been observing flying into that area and the resultant shooting that followed. What say we pay those chaps a visit and see if we can make any money during the rest of this day?" I said through a knowing grin, generated from my past history of success in that area.

Putting the last of our gear away just as the rains returned with a vengeance, we headed north on Four Mile Road. There, I directed Rich over to a large stand of bamboo where we could be hidden from view and yet observe part of the Lambertville complex and a mess of her now happily shooting sportsmen. Soon the two of us had locked onto an area where numerous flocks of low-flying pintail were crossing a corner of the complex as they headed for the safety of the Delevan National Wildlife Refuge. Then after another half-hour of watching, we had fastened our attention onto a large battery blind in the far corner of the complex holding at least five gunners. And it seemed from all the ducks flying into their decoy sets and the reduced numbers leaving after the shooting had stopped, that they were doing quite well. In fact, too damn well to my way of thinking.

Exiting the truck in the now-howling winds and horizontally driven rains, the two of us began our stalk towards our still-firing, like they were at Gettysburg during the Civil War in 1863, duck shooters. *Damn, they have to have killed their limits by now*, I thought. Then over the howling winds, I heard an ATV from the area of the suspect duck blind start up.

"Rich, you keep heading for the east end of that shooting area," I said. "They will have to come out from that side. And if they do and have too many ducks, you can nab them as they come your way. In the meantime, I am going to cut across this dike and try and intercept that chap on that ATV. I think he may be running some ducks and I would like to catch his ass if he is." With that and a wave of my hand, I took off at a trot in my hip boots towards where I knew from past experience my lad on the ATV would have to exit. Picking up my pace so I could get to the crossroad on the dike ahead of my oncoming ATV, I slipped and sloshed my way across the muddy trail. Coming to the crossroad, I laid down in the wet weeds so I would not be seen by the oncoming ATV driver. Soon I could hear him coming closer and closer to my hiding spot. Then careful to look in his direction through the weeds, I timed my intercept. Finally, it was time to go. Up I leaped from my muddy ditch and bank of weeds, only to slip like a buffoon and fall flat on my face in the six-inch deep gooey mud! By now, the ATV was almost on me and from the sound of his motor, he had no intention of stopping. Scrambling to my feet once again in the thick sticky mud, I arose to his front, held up my badge and yelled "Federal agent!" It was then I observed a half-filled wet gunnysack riding in his carry basket. *That has to be full of freshly killed ducks!* I thought,

True to form, my chap just lowered his head like he was in a driving snowstorm and flew down the road like he was on a mission. And he was—a flying mission to get the hell away from me! As my ATV driver roared by me, I grabbed him by the back of his collar on his camouflage duck coat and swung him off his slipping and sloshing ATV like a sack of spuds. *Whump* into the mud of the road with a loud *ooofff* went my ATV speed demon, cutting short his steel Pony Express ride. With his "glamorous exit" from his "iron horse," it stopped and sat there idling in the rain, mud, blood, and beer of the arena.

"What the damn hell are you doing?" he yelled once he got his wind and some of his "sand" back.

"Trying to avoid being run over by a mad man and do my job!" I retorted. "Federal Agent Terry Grosz," I said as I flashed my badge and credentials once again at the muddy, sputtering, and wet speed demon.

"I don't care who the hell you are!" he growled right back. "You have no right to attack me," he continued.

"Well, unless I miss my guess, I would say this gunnysack half full of freshly killed ducks says differently," I said, as I lifted the wet and half-full sack of still warm and freshly killed pintail out from the carry basket on the ATV. Opening the sack, I could see at least 20 freshly killed pintail in the sack and boy, they had to be glad to see me even though I was a little late, I thought.

"I would like to see some identification and your hunting license," I said, as a big gob of mud fell off my face and onto the muddy ground with a soft *plop*. Boy, even though I was normally one handsome son-of-a-bitch, I had to be a muddy mess that fine day. Especially after doing a full 320-pound face plant into about six inches of soft adobe mud, 10,000 years of duck crap, and tule roots. Oh well, I told myself, maybe my 'catch of the day' will think it is clever camouflage.

It was then I noticed that the rain and wind were still in full force, but all the shooting from our suspect blind had now ceased. In that, I assumed that Rich had made his presence felt and hoped his catches were as good as mine. Taking the proffered driver's and hunting license from my ATV operator, I said, "Alright, let's go." And with that, I pointed in the direction from whence he had come in such haste. After turning off his ATV, the two of us and the bag of dead ducks sloshed back towards his duck blind and his recently, just-left friends.

Arriving some fifteen minutes later, I could see that Rich had his side of the detail well in hand. He had four drowned rat-like looking chaps standing out in the elements. In front of each man laid a large pile of dead ducks. A pile whose size denoted someone had pulled the trigger on his shotgun too many times, and very successfully I might add. With that, Rich and I procured all our shooters' driver's and hunting licenses, along with their over-limits of ducks. Then with instructions for the five lads to meet us out on the main road, we began trudging back through the mud, blood and beer of the recent duck shooting arena, under loads of wet, freshly killed ducks.

"Nice camouflage, Dad," said Rich as he looked at my still fully muddy face and neck where I had taken a face plant in the soft mud earlier. "I suppose that is your idea of camouflage," he continued through a wet-faced grin as he staggered under his load of ducks. "I would say it wouldn't do you much good though, Dad. With a carcass the size of yours, you would have to use a whole lot more mud to cover it up than there is available here in this end of the valley," he continued with a grin.

"Keep it up, Rich, and I will tell your mom," I retorted. I knew he had no fear of me, but from his mom, that was different. All she had to do was go after my 6' 7", 270-pound son with a wooden cooking spoon and you would think the devil was on his tail. And to all you readers out there who are conservation officers "unfortunate" enough to have to work with my badge-toting son from North Dakota, remember that ruse. If he gets on your case, just remind him that you will tell his mom and then watch him scurry for cover under the available leaf litter.

Back at our truck, the two of us counted out 26

pintail ducks our lads had exceeded their limits by. Not bad, I thought. *We surely have paid for the gas and oil we have used this fine winter day,* I thought, as I looked at my rain-streaked muddy face in the side view mirror of the truck. Rich was right. I was one pretty damn-good-looking, muddy-faced son-of-a-bitch.

Driving down our road towards our earlier designated meeting place, we met once again our glum-faced and wet as a pack of muskrats, duck shooters. I stood outside in the rain as Rich seated each man-jack of them one at a time in his truck and wrote out their citations. I figured that way they couldn't complain to their Congressman about being treated unfairly by being made to stand out in the rain and all.

Leaving our chaps after completion of our paperwork, Rich and I headed out to the east side of Glenn County and stayed until about nine o'clock that evening listening for sounds of any night shooters. Hearing none, we finally retreated to our motel room after we had gutted our evidence ducks and had dinner once again from our ice chest. Damn, did my hot shower ever feel good that night on my tired carcass.

Oh by the way, the valuable and rare Aleutian Canada goose was not wasted or eaten. It was carefully frozen and later turned over to a taxidermist. There it was turned into a museum study

skin specimen and now resides in the Service's
Federal Law Enforcement Training Center waterfowl identification section, where it is used to
train new agents in waterfowl ID.

## MORE GUARDIAN ANGEL
## "CATCHES" ON THE EAST SIDE

Before dawn the next morning found Rich and
me on the eastern side of Colusa County near the
Butte County line. There we greeted the sunrise
amidst hundreds of skeins of flying geese and
thousands of hungry ducks. Standing up in the
back of Rich's patrol truck, the two of us surveyed
the countryside as we looked and listened. To our
south, we could hear numerous guns from waterfowl hunters having a field day on the expensive
Butte Sink Duck Clubs. But without a safe boat
of any kind, those chaps would get a free pass
during this trip. However, off to our north and
southwest, there were shotguns aplenty popping
away at the flocks of ducks and geese, who were
looking for a safe place to land and feed in the
many harvested rice fields.

To be frank, I was more interested in gathering
intelligence as to where everyone was doing it
to whom, rather than running around and just
checking everyone pulling a trigger that fine
windy and rainy morning. However, my "young
bull" Rich was eager to "spill more blood" and

was raring to go. So unable to contain his desires over just the simple gathering of intelligence in that area, off we went east down the Gridley Highway, checking out numerous hunting areas.

As I said earlier, I was just looking for those contacts once we started roaming and hunting the lawless, that "spoke" to me as to their illegality. Heading east on the Gridley Highway, we chanced passing a chap in a canoe paddling along in a small ditch on the south side of the road. I knew that ditch eventually led into the northern end of the fabulous Butte Sink private duck club hunting area. A tempting area for one in a canoe to poach, that was loaded with thousands of loafing ducks and geese.

Rich, upon seeing my chap as well, started to stop so he could check him. However, knowing our lad was still a distance from our road, I told Rich to keep driving so he wouldn't suspect us as officers. This Rich did until he could find an area in which to turn around further down the road. Then I had Rich sit for a while giving our suspect lad time to paddle up to the road and his parked car. In the meantime, my guardian angels were making noises like we needed to check this lad in the canoe.

Finally figuring our lad would be close enough to the road so he could be captured before he tossed his lead shells or extra birds, I had Rich drive back his way. Sure as God made green gumballs, there was our man just at the edge of the road in his canoe. Stopping, Rich bailed out and was heading for the ditch and canoe before he could move a muscle. Rich walked down the bank, as I watched our man for any furtive movements. Arriving, Rich finally grabbed hold of the edge of the hunter's canoe to steady it so it wouldn't roll.

"Morning, federal agents. We would like to check your ducks, hunting license, duck stamp, and shells if we may," said Rich.

From the strange look on our man's face, I knew my guardian angels had been right once again. During the field inspection of our duck hunting chap who obviously had been illegally "sniping" along the northern edge of those Butte Sink duck clubs to which he didn't belong, several things came to light. First of all, he had an unplugged shotgun, two wood ducks over the limit, and possessed pocketfuls of prohibited lead shot! Not a bad catch, I thought from up on the roadside. I also got a quick look from Rich with that "you were right once again, Dad." Seizing the illegal ducks and lead shot, Rich had our dude come up the ditch bank and into the passenger side of the truck. Once again, Rich took care of the paperwork while I continued looking and listening for other likely hunting activity as I stood by his truck.

When Rich finished with our chap, we loaded up and headed down the road once again looking for that next set of duck hunters. And the collateral inside feeling I usually felt when something was wrong. As it turned out that morning, we bypassed a dozen groups of hunters either in or exiting the duck hunting areas. Rich kept looking at me for "that" signal but none came and we did not make any moves in checking that bunch of sports.

Then as we drove by a lad sitting on the tailgate of his truck drinking coffee, I told Rich to pass him by, go down the road a piece, then turn back around. This Rich did and as we once again approached our lad on the tailgate, I advised Rich that he needed to be checked. Pulling in behind our hunter as he was rolling down his last hip boot, Rich got out. Walking up to our lad, Rich identified himself and requested to check his hunting license, duck stamp and birds. He did not ask to check his shotgun because we had not seen the lad in the field hunting. True, he had been hunting from the looks of the three freshly killed white-fronted geese lying in the back of his truck, but we did not see him afield with his shotgun in the process. Since we hadn't, we couldn't write him a citation at that point even if the shotgun had been unplugged. Hence, it was not checked. If anyone did write such a chap for an unplugged shotgun, not having seen him hunting, I always considered that a "thin pinch."

Everything checked out during that field inspection. He was not in possession of lead shot and had a legal limit of three dark geese, to wit, three white-fronted geese. Standing there by the front fender of Rich's truck and letting him handle all the contacts for the experience it provided, I noticed him quickly looking back at me. Looking at me like, "where is the violation, Dad?"

"Rich, check in his empty hip boot lying in the bed of the truck," I quietly instructed. With those words, I thought just for a fleeting second I saw a glimmer of panic flood across the eyes of our chap with the one rolled-down hip boot still on. Rich looked at me for a second as if wanting to hear that instruction once again and then did as he was told. Surprise, surprise! Out from the "empty" boot lying in the back of the truck that our lad had just taken off, tumbled two more freshly killed white-fronted geese! Two more geese meant that our lad was now two dark geese over the limit! Man, you should have seen the look in Rich's eyes of disbelief over his surprise find.

Finishing with the "over-limit" paperwork and heading back down the road from whence we had just come, Rich looked over at me with a grin. "Dad, how the hell do you do that? That is thirteen citations for nine hunters contacted in a row without a miss. How the hell do you do that?"

"I don't know, Son. I just do what my instincts tell me to do. When I do, it just seems to fall into place with a violation on the end of it. Now my instincts are telling me we need to head back to Terrill Sartain's property. When we were there earlier listening, I saw a few snow geese heading out into one of his big rice fields. I will bet by now there will be a few thousand more there with those first few. That being another large hunting club, we will have some business there before nightfall if that field is full of snow geese."

Arriving at the north end of Putnam Road, I directed Rich south. Besides, there was an old-time duck club named the White Mallard Duck Club on the western edge of the Butte Sink where I always made lots of money from those who couldn't count for sour owl droppings when it came to their limits. Especially when it came to duck limits. I figured while staking out Terrill's, I could also listen to the hunting "traffic" occurring out on the White Mallard maybe for another day's look-see.

When we arrived at the south end of Terrill's property, I saw I had been right. The air was now full of swirling lesser snow geese looking for a place to light and feed in the huge harvested rice fields on Terrill's farm. Plus, there were about 20,000 snow geese already down out in the middle of his rice fields feeding to their hearts' delight. Finding a pull-off where we could see but not be seen, we commenced cleaning our five recently seized white-fronts so they would not spoil. Then we washed off our hands with melted ice water from Rich's ice chest and lowered once again the tailgate. Sitting there, we had another picnic. Plus, anyone seeing the two of us sitting there on the tailgate having lunch all dressed up in hunting gear, would think we were just another pair of goose hunters and not the law.

It was about then we spied five lads quickly pull over on Putnam Road as they were driving south and begin eyeballing the flying and feeding snow geese. "Those five are up to no good," I said. "We need to stake out that bunch and sure as all get-out, we will be making the federal government some more money before the sun sets," I quietly said. It was then that I noticed Rich looking me over very closely with another of his big grins.

"What?" I said.

"Dad, is this another one of those feelings you are getting about those five goose hunters sitting over there watching the geese?" he asked.

"Yeah, I think so," I said slowly as my instincts began telling me those five lads were up to no good.

For the next 30 minutes or so, our lads just sat and looked at the blizzard of white snow geese amidst a constant of goose "happy calling" sounds flooding into Terrill's harvested rice fields. Then our five "lookers" slowly drove south on Putnam Road to its end and stopped once again. A short time later, their vehicle was observed heading west on a muddy farm road that led in and behind the feeding geese. Then they disappeared and all was quiet on the western front, except for the loud and incessant calling common with thousands of flocked-up snow geese and many decoying new arrivals.

As I suspected, our five chaps were pulling a sneak on the thousands of feeding snow geese. That thought was later seconded when the huge flock of snow geese went instantly quiet, followed by the roaring of five shotguns firing in unison into the blizzard of "white" feeding on the ground! Then there was a thunderous roar of thousands of frantically beating wings from 20,000 previously feeding snow geese, as they fled in panic away from the streams of deadly flying shot.

"That is our cue, Rich," I yelled, as I bailed off the tailgate where I had been setting waiting for that which I figured was sure to come. Rich, attending to a call of nature, put everything back as he scrambled for his truck. Roaring out from our hiding place, we turned south on Putnam Road until we came to its graveled end. There Rich put our truck into four-wheel drive and we also entered the same muddy farm road that headed back into where the snow geese had been recently feeding. By now, the air was full of calling snow geese as many were typically resettling further back into Terrill's rice fields. *Typical damn dumb snow geese*, I thought. They don't have the sense God gave a dying goose when it came to coming in out from a rain of deadly shot.

Arriving at what had been the backside of the rice field previously holding thousands of feeding snow geese, Rich and I saw our five chaps out in front of us running to ground numerous crippled snow geese and wringing their necks. Pulling up to where they had parked their SUV, I saw Rich was taking the wrong side of the road. In an instant, we went to our axles in soft, gooey mud! However, that new concern was secondary to our initial venture of catching the bad guys, so out from the now deeply stuck truck we flew.

Within minutes, we had rounded up our feeding snow goose ground-sluicers. Then as luck would have it, they didn't have any over-limits! The white goose daily bag limit in those days was six birds per shooter per day. Our ground-sluicing rubes were two geese shy of full limits. No matter how hard Rich and I looked along the weedy rice checks, we could not muster up another dead or dying snow goose. Plain and simple, our rubes had shot into the geese from too great a distance. In so doing, they had killed very few considering what they could have killed had they been closer. However, shooting into thousands of feeding geese at such a long distance meant many more were now carrying shot in their bodies as they winged off! Some probably hurt badly enough that they would not see another sunrise. Anyway, back to our five shooters, no over-limits there.

*Damn,* I thought, *my string of catching a long line of violators may have come to an end, all based on my gut instincts.* Then we got to checking hunting licenses, duck stamps, and for plugs in their shotguns. We did better there. Two of the lads did not have federal duck stamps and three of the chaps were using unplugged shotguns. My string was safe as Rich wrote up five more citations based on a gut instinct and a lucky hunch. We now had issued 18 citations while only checking 14 waterfowl hunters. I was on a roll!

Since our five shooters had taken 28 snow geese by illegal means (no federal duck stamps and unplugged shotguns), all their geese were seized and held as evidence. It could have been worse. I remember working that same field 23 years earlier with Special Agent Jerry Smith in which five shooters using unplugged shotguns under the same circumstances, had killed 98 geese or 68 over the limit! And untold cripples had been able to escape only to die in the Butte Sink or in the mouths of foxes or coyotes, probably that evening.

As our five disgruntled snow goose shooters left, we faced another dilemma. Rich had stuck the truck to the frame. "Dad, you are better at this four-wheel drive thing than I. Why don't you try to get us unstuck instead of me?"

Looking the situation over, I said, "Rich, get into the back of your truck. Then when I start trying to get us unstuck, you jump up and down in the back. Your weight every time you come down will help the tires get a better bite when it comes to traction. Let's see how that works. If it doesn't, it is shovel time." Getting into the truck, I put it into four-wheel drive, low range for the power I would need. Then I yelled to Rich to start jumping up and down in the back of his truck with his 270 pounds of weight. At first nothing moved. Then as our wheels spun furiously and Rich jumped up and down, we began to move. Giving the truck more power, we soon were inching along. Then all of a sudden our wheels caught and out from the sticky deep mud we came. "Nothing to it," I said as I exited Rich's truck. However, a quick "thank you" was fired off to the "Great Duck Hunter in the sky" for getting us out so we could continue hunting those doing bad things to His critters.

Later that evening, Rich and I staked out about 25,000 ducks feeding in a Butte County harvested rice field on the lookout for duck draggers (night shooters who would sneak up on the feeding ducks unseen in the dark and shoot into the masses on the ground, perhaps killing hundreds). The night proved uneventful, as the two of us cleaned our 28 seized snow geese so they wouldn't spoil. Then Rich and I had a pleasant supper from our ice chest as well as a hot time in the old town that night as we had some more of Carrie's hot canned cauliflower and broccoli.

We left the feeding ducks sometime around eleven that evening, not having any night shooting takers and headed back to our motel in Williams. The next morning around four o'clock, we headed for the headquarters at the Sacramento National Wildlife Refuge. There we dumped off all our evidence ducks and geese into Rich's evidence freezers on-site. Then we headed into Glenn County to greet the sunrise and see what kind of trouble we could stir up.

## RICH GETS A SURPRISE AND SO DOES HIS DAD

Leaving the refuge, Rich and I headed due north on old highway 99, until we came to what is known as "Riz Siding." There we turned due east on state highway 60, until we came to what

I called the "S turn." The "S turn" being nothing more than an "S" turn in an otherwise straight road. Then I directed Rich to head south on Road SS in Glenn County. Driving south for about two miles, we came upon a duck club on the west side of the road. It being a non-shoot day in the valley, the club was deserted. Rich and I got out and looked around the duck club and in so doing, discovered a huge pile of duck wings and guts on the west side of the duck club in a ditch. It was plain this was where the members cleaned out their daily kills and from the looks of it, they were doing very well. In fact, a little too well to my way of thinking. Leaving the clubhouse area, we proceeded further south to the club's shooting area proper. There we discovered a flooded harvested rice field, numerous rice check duck blinds, and several stand-alone, brushed-up blinds out in the middle of the field. Plain and simple, this club's main shooting grounds were in a 'backwater' area probably not frequently checked. And with all the evidence back at the clubhouse pertaining to cleaned ducks, I lodged this area in my mind as one needing further attention come the next shoot day.

Then Rich and I headed further east, hit Highway 45 at Cordura and headed south for another area I wanted to scout out. Continuing south for several miles, I soon directed Rich to turn off on a farm road heading due west. Driving towards the 2047 Canal and the east side of Delevan National Wildlife Refuge, we quickly arrived at another duck club I wanted to scout out. It had been many years since I had visited the place and figured I needed a refresher on its layout. As remembered, it had a nice clubhouse, dog kennels and numerous ATVs sitting behind the main residence. Then to avoid being caught by a young farmer, a fellow from the Beauchamp clan, who was allegedly a bit of a heavy trigger-fingered duck shooter in his younger days, we took a quick turn about the hunting area proper. Refreshing my cobwebbed mind on the area, we then beat a hasty retreat out of the area before being discovered by the caretaker. Then for the rest of that non-shoot day and Tuesday following, Rich and I worked parts of Glenn and Butte County scouting out other future work areas of interest. As we did, we also made note of any die-hard duck hunters who were out hunting every day they got the chance.

Come Wednesday, a valley shoot day, found Rich and me positioned on the "S" turn on Road 60 in Glenn County. While we sat in the truck prior to daybreak eating our typical breakfast, we became aware of the heavy shotgunning to our south on the Road SS duck club—the very club we had scouted out earlier on the previous non-shoot day. As usual, our hunters had come into the clubhouse the night before. That way they would be ready for the Wednesday morning's shoot. And with that now happening, we quickly finished breakfast. I had arranged for us to be at that site on Road 60 during that morning to allow for the arrival of any latecomers to the duck club. If so, they would have caught Rich and me parked along their main road and the "cat would have been out of the bag." Or even worse, they would have caught us sneaking in on their blinds. So, we had made sure everyone was in on the club and shooting before we made our final move. Finishing breakfast, we drove down Road SS for a mile or so until we came to a locked cable running across the road that had not been there during the previous non-shoot day.

Looking over at Rich, I said, "I would take that locked cable across the road as an indicator those folks do not want us or anyone like us, on their property nosing around today. What do you think?"

Looking over at me in the gloomy light in the truck, Rich said, "I would not bet against you, Dad, on that one. Are you getting that feeling that we might make some money on this duck club today?"

"Why do you think we are here? That is why I wanted to scout the area to see the amount of hunter and duck use on the club. Besides with this stormy weather and the wind, today bodes well for a good shoot day just about anywhere we went. And in my earlier days, the real club out-laws many times would come in on a Wednesday to shoot their over-limits of ducks because there were less folks around to witness their illegal acts," I said.

"How do you want to work this, Dad?" Rich asked.

"Well, Son, there is an old heavily weeded ditch running south from the clubhouse that leads directly to the north end of their shooting area. From there we can set up and see the whole shooting area and not be seen. I would suggest we walk to the clubhouse, then work our way down that ditch out of sight. Once we get to its end, we set up out of sight and just watch the ac-tion. That way we don't have to expose ourselves if some of the members get cold or wet and de-cide to come in early. If we lie down in the ditch, they will walk right by us and never know we are there. Then at that time, we can decide what we want to do. One way to do it is to work out into their shooting area after we have observed them breaking the law and put the grab on them. By doing it that way, we would be on them before they could toss or hide their over-limits. Or we can waylay them from the end of this ditch as they walk by or wait until all of them are back at the clubhouse. If we wait until they are all back at the clubhouse, we can corral the whole bunch at the cleaning site. And that would be easy to do, because we would have each member's exact kill figures, based on what we saw on our earlier stakeout in the ditch. No matter which way we decide to go, we ought to be able to gather in the whole lot before they even know we are in country. What is your thinking on the issue?" I asked, wanting Rich to learn and think out the "capture" scenarios himself in the process as well.

"I like the ditch idea and waiting until they have played out their hand. That way we will know the full extent of their illegality," he quietly said.

"Well, let's get with it. We are just burning daylight sitting here on our last parts over the fence," I said with a grin normally found on a hunter's face.

Backing up our truck and parking it near some recently placed idle farm equipment hoping it would just look like the local farmer's rig, we grabbed our gear, dressed for the weather, and headed out. Twenty minutes later found us at the end of the deep and weedy ditch. We were wet as a couple of muskrats from the incessant rains and sweating heavily from the difficult walk in rain gear, but at long last we were in position.

Then setting up our gear, we soon discovered we had ten hunters out on the hunt club, all banging away at the ducks. Ducks which because of the windy and rainy weather, were all riled up and flying low and every which way. And with the worsening weather, the day's shoot looked to get even better with all the low flying and heavy numbers of waterfowl.

For my readers' benefit, ducks and geese have a heck of a time when flying in the rain. First of all, they have to fly with their eyes open. Imagine what it would feel like zipping along at forty miles an hour with pellets of rain hitting your sensitive bare eyeballs. It plain and simple has to hurt. So our ducks and geese were flying low and slow during such inclement weather periods trying to avoid the eye pain. And in so doing, became better targets. Keep in mind in the winter, ducks and geese have to eat high-energy fuels like grains and seeds in order to maintain their health and keep warm. The gunners knew that, hence their presence out in the rice fields on such days during inclement weather patterns. Good game wardens know that as well and react appropriately if they are worth their salt.

Then for the next four hours, the two of us laid in that damn wet ditch as Mother Nature poured cold rain drops down the back of our pants and down our necks. Our shooters to our front were a mess. They were some of the worst shooters I had ever seen. They had the opportunity all morning long to kill a million low-flying ducks due to the howling wind and driving rains. As it finally ended up, nine shooters out of the ten finally managed to take over-limits. Small over-limits due to their poor shooting abilities but over-limits just the same. The most any one shooter had was three ducks over. As for all the rest of the sports, they had from one to two over the limit on ducks. Since the geese had not over-flown our shooters that day, none were taken. Then our lads began straggling in one or two at a time, having had enough of Mother Nature's soaking or because they ran out of shells. That made it difficult to bag the whole bunch in the field so Rich and I decided we would let all the shooters gather at the clubhouse. Then we would spring our trap and have all of them in the bag at the same time. Plus by so doing, we would not have to give away our hiding place in case we wanted to use that ditch once again in the near future on these same chaps.

When the last of our shooters had disappeared into their clubhouse to dry out, Rich and I arose from our hiding place and walked down the road towards the clubhouse. Our plan worked perfectly. Since all our lads had taken a damn good soaking while out in the field, they were all in the clubhouse changing their clothing. But before so doing, they had hung all their ducks in the club's hanging facility to cool out with nary a tag on them as is required by federal law. Tags meaning those ducks left in temporary storage had to be tagged with their names, date birds were taken, species of birds killed, and the signature of the hunter taking such birds. Walking by their clubhouse undetected, I headed to retrieve our truck and the citations. Rich on the other hand, peeled off and headed for the duck-hanging facility so he could control the now-intact bunches of evidence birds. There he camped out also undetected in the perfect ambush position after looking over all the untagged birds. Retrieving our truck, I drove back to the locked cable running across the road. Once there, I took out the huge ring of keys I had amassed from working seven years earlier as a warden and Special Agent stationed in the valley and soon had the lock open. Driving down the road, I parked the truck in the middle of the road alongside the clubhouse in front of God and everybody. Then getting out, I waited on the far side of the truck to see how Rich handled the situation himself with ten soon to be cranky duck shooters, once they learned they were to be cited for their morning's duck hunting transgressions. He was soon to get a real surprise in pissed-off human behavior.

With the arrival of the truck, several members bailed out from the clubhouse to see who had come through their locked gate. In so doing, they ran smack dab into Rich who was sitting on an old 55-gallon barrel by their hanging facility. Upon spying Rich, the word boiled immediately out, and soon ten lads roared forth from their clubhouse like a mess of mad bald-faced hornets. First, mad because I had breached their locked gate and was now sitting on their private road. (Remember, we were dressed like normal duck hunters and our patrol truck was unmarked. In fact, Service agents since 1973 have been without uniforms and have used unmarked patrol vehicles.) Second, they now realized that Rich, a total stranger, was on their property and by their hanging facility with all the evidence of their over-limits! Third, they all immediately developed a huge case of the vapors, when Rich announced that he was a federal agent and was there to check their ducks, hunting licenses, duck stamps, shot shells for lead, and shotguns for plugs! This time even though they were back at the clubhouse, we checked for plugged shotguns. We did so because we had observed all of them earlier in the rice fields shooting at ducks. So checking their plugs in their still-wet shotguns at that time, even though they weren't hunting, was appropriate.

Nine of the ten men went through the roof, claiming Rich was trespassing and had conducted an illegal search contrary to their Constitutional rights! Not so. Service agents have the right of trespass under the Supreme Court's "Open Fields Doctrine" to inspect hunters and their bags. The tenth lad, an older gentleman (the one without an over-limit and properly tagged ducks hanging in the picking facility), oddly enough, just quietly stood by and watched the hell-raising unfolding. Fourth, everyone realizing Rich had them with obvious over-limits, did their "first stage of a violator" infantile bit of screaming and hollering. That was followed by their "second stage of a violator," namely profuse denials of any illegality even in the face of the obvious over-limits hanging in their cooling facility. Then that mood change was once again followed by the "third stage of a violator," with much moaning and grinding of teeth as Rich stubbornly held his ground in the face of such surprising behavior. Even though the odds were nine to one, he held his ground pretty well and I was proud of him. And since he was handling the situation well, I didn't bring "my" reinforcements.

Note I didn't say he didn't lose any tail feathers in the "hoorah." He did and a few hind end feathers as well. But he professionally prevailed doing what he was supposed to do, surprising behavior on the part of the duck shooters or not. Remember, I was still quietly standing by the off side of his truck out in the road just sweetly minding my own business. I just figured if Rich wanted to be a real agent, he had to "eat" some "horse pucky," as well as learn to ride the horse. It was all part of growing into the profession. And if one wanted to succeed and be among the best on a "vision quest," he had better learn to be an "eager eater."

Then Rich got down to the paperwork at hand. And I do mean a mound of paperwork. There were nine citations issued for over-limits, six for possession of lead shot, two for unplugged shotguns, one for non-possession of a federal duck stamp while taking migratory waterfowl, and a citation issued to the duck club proper for having untagged birds in temporary storage. Surprisingly, the tenth man still quietly standing, remained square with the law. And as Rich and I were soon to discover, he was also the owner of the duck club. And that oddly enough didn't end there. He was still holding one more surprise for both Rich and his old man.

Finished with all the paperwork, Rich advised the still-sullen group that since all of them had taken over-limits of ducks, all the birds would be seized and held as evidence. That was when the wheels came off the wagon! Damn, you talk about a genuine hell-raising and hand-wringing. Between Rich and the nine violators, it now became a cat fight with one feline and nine mad as hell dogs. However, Rich was one BIG 270-pound cat!

When the fur had softly drifted off into the wind after that cat fight ended, Rich started gathering up all the ducks as the sullen crowd stormed back into their clubhouse to lick their wounds and see if "Jack Daniels" had any words of wisdom. That was when the quiet tenth man approached Rich.

"Say, Officer, do you have a minute?" he asked.

"Sure do," said Rich as he turned and faced the man.

"What did you say your name was?" he asked.

"My name is Rich Grosz and I am a federal agent based in Sacramento," he replied as he shook the man's now-extended hand.

"Damn, did you say your name was Rich Grosz?" he asked in surprise.

"Sure did," said Rich.

"You any relation to Terry Grosz? The one who used to be a California Department of Fish and Game Warden?" he continued.

"Sure am. He is my dad," said Rich.

"Well, I'll be jiggered," said the man. "I used to be a captain in the Fish and Game some years back. A marine captain on a patrol boat off the coast of California. I heard of your dad when he worked for us. He had a reputation for being a catching son-of-a-bitch and was a legend in the Sacramento and San Joaquin Valleys, as well as up on the north coast. I always wanted to meet that guy. What is he doing now?" asked the man.

"He is the Special Agent in Charge for the Rocky Mountain Region for the Fish and Wildlife Service stationed in Denver," said Rich who by now was getting a sly grin on his face and a mischievous twinkle in his eyes.

"Like I said, I always wanted to meet him. He was hell on wheels on the north coast when it came to catching those taking too many abalone. And what really amazed people, especially those he was hunting, according to all the tall tales I heard, was his ability to not be seen until it was too late and he was upon them. Especially so, because he was such a big guy and yet it seemed he could hide under a leaf," he continued talking excitedly. "As I said earlier, I really wanted to meet that guy and shake his hand for being such a fine hunter of humans. He sure put the fear of God in many of those folks breaking our fish and game laws."

"Well, that is no problem if you want to meet him. He is standing right over there by my truck," said Rich with a mischievous twinkle in his eyes and a grin to match.

"You have got to be kidding! Hell, I thought that guy was just a senior officer in the Service, letting you get your wings and experience working angry but guilty as hell duck hunters," he exclaimed. Then with that, he quickly walked from the duck club over to where I was standing. Extending his hand and shaking mine like it was an old-fashioned pump handle on a water pump, he said, "Damn, I am glad to meet you. I am Captain Swan, marine patrol, and I used to work for the Fish and Game Department on a 110-foot patrol boat named the Albacore," he excitedly said as he continued vigorously pumping my hand.

"Glad to meet you, Captain Swan," I said somewhat embarrassed over all the outright fussing from the man.

"Terry, we in the duck club are having a crab cioppino dinner this evening. I brought up a mess of good old Dungeness crabs and they are cooking in the pot with all the rest of the tomato stuff for our annual dinner this evening as we speak. You and your son have to come and be my guests," he excitedly continued.

I just chuckled and said, "I don't think so, even though I love that dish. After all the hell-raising your lads did over being pinched, I just don't think sitting down and having dinner with them would be too wise. If we did, I would think most of them would choke on a crab claw."

"To hell with them! I own this duck club and if they don't like it, they can go to hell," he exclaimed strongly, "you and your son will be my guests."

"No, I think Rich and I will pass. We thank you for your generosity but we have other plans relative to a stakeout this evening (I lied)," I said.

"Well, suit yourself, but I am one hell of a cook and you two would sure be missing out on a great feed," he said with genuine disappointment rising in his voice.

I was disappointed as well. Crab cioppino is a wonderful dish with all the associated fish, oysters, shrimp and the like in its spicy tomato and onion broth. That and some good garlic San Francisco sourdough French bread, a crisp green salad, along with several glasses of "Dago" red wine, would sure go a lot better than the "ice chest" dinner we would be having that evening in some cold and damp rice field, I thought.

Later that evening Rich said, "Dad, I have been figuring. We have only been out and about for several days now. And as near as I can figure it based on your gut hunches, we have issued 37 citations and have only checked 24 hunters. That is unreal. Will I ever develop any such instincts like that down the line?"

"Son, most good officers from every profession will develop those gut hunches as you call them.

A good officer learns to 'listen' to them from within and then religiously follow them. They will not only many times point you in the right direction but like in my case, save your life many times over as well throughout your professional life. I to this day, don't know how or why they develop but I thank God every day I put on the badge and gun for those instinct-based 'helping hands.' For I know without them, I would have been dead a long time ago."

Sitting there in the dark and damp on another nameless rice field, listening for any night shooting activity, Rich finally said, "Damn, Dad, this repast sure isn't a hot bowl of crab cioppino soaked in a gob of good sourdough French bread, is it?"

## CAPTURING A CALIFORNIA
## STATE PARKS "HIGH CHIEF"

I slept little that evening. The north wind howled all night long and the rain kept pelting our motel window, noisily keeping me awake. Finally, by three in the morning, I had enough of Mother Nature's hell-raising and woke up my son and fellow Special Agent, Richard.

"Rich," I said, "let's roll out, get some grub in our bellies. Then we can see what awaits us in the rice fields in the way of feeding ducks and maybe some dark of the night drag shooters."

Without a word, Rich unfolded his 6' 7" inch frame and headed for the bathroom. Me, I peeked out the motel window. Out in the courtyard of the motel under the parking lot lights, I could see the rain coming down in wind-driven sheets. Just right, I thought. On a morning like this, one would more than likely not find any game wardens out and about. That is just what I would want any market gunner or local "duck dragger" to think. *Rich and yours truly weren't "just" any standard bomber turned game wardens,* I thought with a sly grin. I had been here before and had learned that the most hardened duck draggers or market gunners would brave such weather, because they knew most sane men would still be in bed during such weather conditions.

Then I began hustling into some warm, layered clothing and my chest highs. I figured instead of wearing my regular hip boots that ended at my thighs, I would wear my chest highs.

That way instead of eventually getting wet legs and a hind end, I would stay drier with chest highs and enjoy the warmer protection they offered. One couldn't run as fast in chest highs but at least we would remain dryer and warmer, I thought. Besides, I would just have to find a way to head the bad guys off at the pass instead of chasing them out across the sloppy rice ground in bulky chest high waders if they did something really stupid requiring a chase. Especially in light of my ruined knees, a chase would not be what I favored, if I had my druthers. One must remember that the world of wildlife has two faces. One in which you can plainly see, and the other which you must look for. If I got into a running situation, I thought, I would find that second face, somehow.

Emerging from the bathroom, Rich took a glance at my "uniform of the day" and dressed accordingly. Then the two of us sat on our beds and quietly ate our standard fare of dried, high in fat Italian salami, crackers and Diet Pepsi for breakfast.

"Where we going this morning?" asked Rich.

"I kinda figured we would head out on the east side of Colusa County and see if we can slip in on the backside of the Boggs Ranch. Garvin Boggs and some of his clan have been known in past years to hunger for a duck dinner or two now and then. That in mind, I thought we would slip by his farmhouse and head out into the rice fields behind his residence. Then if any of his clan arose to the occasion and came out to shoot the ducks, guess what they would reap in the offing? Plus that area is pretty remote. Since the Butte Sink and its great reservoir of resting ducks and geese are so close at hand, they will probably bleed off into that area to feed come daylight. That being the case, I figured that would make that a good place to start our day. Plus when I was a young officer stationed in Colusa, the duck draggers used to pound the hell out of the night-feeding ducks in that area. So we can start our day there and then drift northward into that area just north of the Colusa-Gridley highway in those newly flooded rice fields we spotted the other day. Ducks, teal and pintail in particular, just love a freshly flooded rice or weed field. And so does a certain strain of "heavy on the trigger finger" duck shooter familiar with freshly flooded rice fields and the penchant for ducks to swarm into such eateries. So those newly flooded areas are a must for our patrol this fine wet and windy day as well," I continued.

Rich nodded as he helped himself to another chunk of salami as did I. Then a fresh gust of wind sprayed our motel shaking the structure as a wave of raindrops followed noisily against our windows. "Dad, are you sure someone is nuts enough to go out into this kind of weather in the dark of morning to shoot some ducks?"

"To be frank with you, Rich, this kind of weather only beckons to the really dyed-in-the-wool duck killer. First of all, the feeding ducks are harder to find on a night or morning like this. They typically quickly and quietly feed up, then head right back to the safety of their loafing areas. Secondly, it is a lot harder to get around in this kind of weather and is hell-on-wheels trying to keep your vehicle unstuck. So most shooters will have someone drop them off in the vicinity of the intended shooting area. Then the shooters will sneak and shoot the ducks as they feed. Following a historical pattern, they will either stack them in a ditch or alongside a rice check out of sight or grab up what they can. Then they will meet their drop-off at a different location to avoid being caught. That way they are harder to catch. Then the really greedy ones will come back in the morning during daylight hours and glean the fields they just night-shot, for any dead or crippled ducks they missed after their first shoot. So if a "tule creeper" (game warden) is sharp and misses the shooters during the first go around, he might just get lucky and snare them if they come back for seconds and clean up their area just shot. Lastly, hunting ducks in this kind of weather is a bitch. If you aren't wet or cold, you are both. So other than the dyed-in-the-wool killer, most folks stay home during this kind of weather and that includes game wardens. Especially those who would rather be warm and dry instead of wet, cold and dedicated. Your chances of catching such a bad weather night shooter are slim and none. But I have done it and you might as well as get used to all the types of action prevalent in a heavy duck and hunter use area. Especially one like here in the Sacramento Valley, that has a rich and bloody history of killing ducks at night during any kind of weather."

Rich just grinned and said, "If my old man can do it, then so can I. No one ever said either of us were sane," he grinned through a mouthful of crackers, cheese and salami.

Around three-fifty in the morning, Rich slipped his truck quietly by the darkened Garvin Boggs ranch house running without lights and in four-wheel drive. "Try to keep your wheels in the puddles on hard or sandy roads in order for the water to clean out your tires' cleats. By so doing, that will give you fresh traction when you hit the mud. But when you are in the mud, if there is a grassy area in the middle of the road or along its sides, keep your tires there. By so doing, you once again give your vehicle the best traction," I cautioned, as Rich slid and sloshed his way along into the harvested rice fields' roads on the back-side of the Boggs Ranch. As one could plainly see, Rich's training by his old man included that of operating correctly a four-wheel drive as well as catching those a mite heavy on their trigger fingers.

Finally locating a likely looking parking spot in a small grove of trees where we would be out of sight, we slid to a stop in the mud and shut off the engine. Getting out, I cupped my ears during a lull in the wind and rain and carefully listened. As for finding any ducks, it was pretty quiet. Then climbing up into the back of Rich's truck, I listened once again through cupped hands.

Off to our northeast, I faintly heard a mess of feeding widgeon giving out an occasional call. Getting back into the truck, I said, "All I hear are a small bunch of widgeon feeding off to our northeast. That really isn't the kind of duck a night shooter will go after unless he is uniniti-ated. They are smaller and not as good eating because of their high intake of grasses in their diets, as compared to the larger seed-eating pin-tail and mallards. But they will do for starters because they will attract other hungry ducks as the morning goes along."

As the renewed rains and winds discovered us once again, we partially rolled down our win-dows and sat there quietly listening for any signs of night shooting activity. I had done so many a time in the past when I was a younger officer, only then usually by myself. That morning it was different. I was now sitting there in my official capacity with another officer, my son. A lot of good feelings came from that outing to my way of thinking. Especially when it came to "passing the torch" from my old and tired hands to his strong as steel and younger ones...

Like a lot of other patrols, that one produced nary a chap stepping across the line of reason. However, it sure gave Rich a different perspective when daylight came. Unbeknownst to both of us, ducks of all makes and models had silently flooded into that area all around us throughout the morning's dark. Come daylight, there were about 25,000 quietly feeding ducks all around us! I could see the look of amazement on Rich's face with that revelation. Then another "deeper" look soon prevailed. One that indicated he had just realized in the Sacramento Valley during duck season, he needed to be out and about even during some of the worst weather Mother Nature could throw at him. That was followed by another even richer look on his face. He had just imagined the look on a night shooter's face after he pulled the trigger into such a mass of ducks and then discovered a couple of previously staked-out badge-toting galoots hot on his trail. Yes, I thought, the die had been cast on this lad for all time during his vision quest in the world of wildlife...

After realizing no one was going to "blow up" our feeding ducks, Rich and I pulled up stakes as the last of our previously feeding ducks headed back to the sanctuary of the Butte Sink in the daylight. Picking our way north on farm roads towards the Colusa-Gridley Highway, we finally turned off the mud and onto the blacktop. All around us there were duck and goose hunters doing what they did best when given the opportunity. Combined with those crazy enough to be out and about in such weather, were hundreds of skeins of hungry ducks and geese trading to and fro in the rain-darkened skies as well. With those two combinations, there was shooting and killing galore. However, none "reached out and touched my soul," so we continued heading east towards the small Sacramento Valley Town of Gridley.

Gridley is a town richly steeped in the illegal market gunning history of ducks around the turn of the century. And to be quite frank, for years thereafter in the teeth of new laws passed in 1918 forbidding such unmitigated killing and sale as well. Case in point. I personally knew one market gunner who killed and shipped by train over 100,000 ducks and geese to the markets in the San Francisco Bay Area! He was still alive in 2007 when I saw him last. Even then, he still had that glint in his eyes when he talked about killing the duck in the "golden" old days. He died shortly thereafter from prostate cancer. Wherever he is now, I hope there is a pintail or two for him to shoot on a windy and rainy day...

With the stout winds and heavy sheets of rain, the overall duck and goose hunting in the area that morning was outstanding. Rich and I saw lots of birds decoying into various decoy sets and heard the shooting. But once again, nothing piqued my interest. Then slowly moving east on the Gridley Highway, I observed three lads leaving a rice field and heading towards a small parking lot at the edge of a larger rice field.

"Hold it, Rich," I said. Rich quickly stopped in the middle of the highway. Then I quickly told him to get off the main road so we wouldn't appear so suspicious. He was then directed into the nearest parking lot at the edge of the rice field. "We are less obvious that way, Son," I said as I continued watching my three chaps with more than a "fatherly" interest. They were about three hundred yards out in a flooded rice field with another flooded rice field right next to where we were parked. For the longest time I watched my three chaps as did Rich. I could clearly see through my wonderfully clear Steiner binoculars that each man carried numerous ducks in his duck strap. Then as they continued walking down a muddy dike road alongside the far side of our rice field, I noticed the lad on the outside edge of our hunters looking our way more than intently. In so doing, I could see him in my binoculars saying something to his two companions. Then all three men as they continued walking, began eyeballing our truck in their parking lot along with the other vehicles. And in so doing, were doing so with more than a normal pedestrian interest.

"Rich," I said, "that outside guy in that pack of three duck hunters is rotten. I can clearly see through my binoculars that he is very concerned over our recent arrival and presence in this parking lot. I am getting that feeling that he needs your attention."

With that Rich said, "How do you want to work this one, Dad? They are so darn far out in that other field that they can ditch anything they have that is illegal, when it is obvious we are moving in their direction. And in so doing, we will never see them ditch whatever they are carrying that is illegal because of the distance separating us."

"I need you to grab your portable radio and head out to those men. Keep your radio on frequency one. I will sit here in the truck where they can't see me and continue watching them with my binoculars. They are so far away that even if they can make me out in the truck, they can't see what I am doing. Besides with your long legs, you can cover the ground between us and them much faster than I can with this bad knee. So take off and head for them. I will keep in contact with you like we did that first morning when we started this operation. Remember when we caught those two shooters with over-limits and lead shot? If any one of these lads we are now watching decides to break the law, I will see them and just report their actions to you." However, as I spoke, I kept my binoculars on the one who looked to me like he needed a little attention.

Grabbing his gear without a word being said, Rich was out the door and on the hunt. When he exited the vehicle and took off down the farm road that led over to our three chaps, I saw the men in the field on the other side of our rice field slow their walking to a stop. For a second they looked at my long-legged son hotfooting it down the farm road that would eventually lead to our three suspects. Then there was an obvious hurried conference taking place on the far side of our flooded rice field as Rich continued on his way. Finally the lad I figured that needed a little of our attention started walking out from the field with his two companions like before. However, I noticed that he now was walking on the outside of the farm road on which the three were traveling near a cattail-filled ditch. It was then I saw my targeted chap just ever so slightly move his right hand which was on the far side from me and down by his side. When he did, I marked his hand movement with the corresponding degree setting on the compass inside my binoculars. Keeping that compass bearing in the back of my mind, I continued closely watching my "sleight of hand" chap until about five minutes later when Rich walked into view in my binocular field of vision.

I could see Rich identify himself to our three chaps with his badge and then begin checking their ducks, shotguns for plugs, licenses, shot shells for lead, and their duck stamps. Then Rich walked down the farm road from whence our three had just trod and once out of their earshot, called me on his radio.

"Dad," he said, "they are 'slick as hounds' teeth when it came to any violations. Their licenses and shotguns checked out and are OK. I can't find any lead shot on any of them. Do you have anything for me?" he asked.

"Walk back on their farm road from whence they just came. Stop when I tell you to because that is where I saw the one chap dressed all in camouflage gear hesitate, slightly move his off hand at his side like he dropped something, then moved on," I said. That information in hand, Rich began walking back on the farm road as I watched my degree settings on the built-in compass in my binoculars reversing backward. Taking a reverse compass sighting, I stopped Rich when he was at the exact bearing where the man in full camouflage clothing had hesitated earlier. "Stop right there, Rich," I instructed. "Whatever that guy did, he did right there alongside that cattail ditch," I once again instructed.

Rich began looking down and all around where he now stood, and soon he went out of sight into the densely lined cattail ditch. After a few moments, he re-emerged holding up for me to see a handful of ducks and something else in his other big paw hand that I could not make out.

Trotting back to his three chaps, there was a short conversation and then Rich began marching them to where we were parked. Back at our vehicle, I got out as Rich and the three men arrived. Then Rich walked over to our truck and tossed five freshly killed, green-winged teal ducks up onto the hood. Then he laid out a handful of lead shotgun shells he had discovered in the ditch full of cattails where the teal ducks had been discovered. Taking the chap I had suspected of being in violation, Rich brought him over to the hood of the truck. Then he got out his citation book and began writing our suspect chap for possession of an over-limit of ducks and lead shot. It seems when Rich had walked back to my directed compass setting, he could see where my chap had walked over in the mud and had thrown the extra ducks and shot shells into the ditch. From then on, it was all downhill for our violator.

Then as Rich commenced writing him up, he complained that his dog had brought those ducks to him, saying someone else had taken them. Hearing those words and wanting to forego any such defense in court down the line, I took out my thermometer. Then I checked each duck from his duck strap and the ones he had tossed into the ditch. All the ducks were within five degrees of each other in body temperature. That fact was pointed out to our chap "now in tow." He had nothing to say countering that temperature revelation. Then another quick look in their immediate hunting area with my binoculars revealed not another soul duck hunting. That observation was also brought forth to the light of day. Once again, our chap in tow had no counter to that fact, namely, the lack of anyone else close enough to do such a thing as he now claimed. So much for the theory of another chap killing the ducks that he had just "found."

When Rich asked for the man's occupation, he waffled for a few moments in consternation. Then he professed to being a very high chief in the California State Parks Division! Then he asked if Rich could cut him some slack and give him some professional consideration; i.e., not issue the citation because of the man's exalted, badge-carrying position with the state.

Rich looked over at the chap now squalling about being caught red-handed saying, "You of all people should know better about breaking the law. Especially since you are also a badge carrier in the state park system. I will give you the same break you would give me if I violated any of your state park conservation laws."

With those stone cold words, our man quit his squallin' and took his citation in sullen silence. Finished with the paperwork, Rich advised the man of his responsibilities in the federal court system and then handed him an evidence receipt for the seized shot shells and the ten ducks removed from his possession.

To my readers, as stated, ten ducks were seized. Those he claimed he lawfully killed as well as those he "found." All birds were seized because in any court of law, you do not have an over-limit in your possession as charged, if you give back a legal limit to the shooter. You only have what is left. Then if someone wants to make an issue in court that you do not have an over-limit as charged, well...

After our three lads left, Rich looked over at me and said, "How the hell do you do that, Dad? We have driven by several bunches of hunters today and you have not said a thing about checking any of them. Then we see this bunch and you tell me to pull over because one of those three chaps needs checking. Dad, they were over 300 yards away in a rice field. What the hell did you see that made you want to check those three chaps?"

"I don't know, Son. They just looked too good to pass up, plus something inside me told me they needed checking. So we did and bingo."

"Well, no matter how you did it, that is citation number 38 I have written from the 25 folks you wanted checked based on your 'gut feelings,'" he said slowly as he gave me that look of disbelief once again.

Looking back at my incredulous-looking son, I said, "Rich, it is almost lunch time. What say we celebrate our latest catch with some of Carrie's wonderful canned, hot vegetable specials?"

I never had that same feeling, that someone needed checking, no matter how many groups of duck hunters we passed the rest of that day...

## RICH BLOWS AN ASSIGNMENT AND LEARNS A LESSON

After a wonderful lunch from our ice chest once again, we headed easterly. Passing the Butte County line, I had Rich pull over. I had observed a small bunch of pintail flying over a flooded rice field in a grove of trees to our south and heard shooting. Not one duck from that little bunch that had just flown into that area had escaped. Figuring we might have some further business, Rich and I began watching our duck hunting chaps located at the far edge of trees in a harvested rice field. The more we looked, the more we observed that there was only one way into those shooters without being discovered by those chaps. That is unless we wanted to swim some damn deep irrigation ditches in the dead of winter. About that time, four duck hunters from two different blinds finally began walking out from the hunting area in question. Once up on a muddy farm road leading towards what appeared to be a duck club, our four hunters moved right along in the pelting rain. Not seeing anything out of the ordinary or having that feeling that those guys needed checking, I advised Rich we would give them a pass. However, they were located in a good area if the numerous flights of ducks passing over their previous hunting area said anything. That plus each of our four men appeared to be carrying limits of ducks on his duck strap as they walked back towards their clubhouse. If nothing else, those full duck straps meant our chaps had a "honey hole" and probably deserved a closer "look-see" by a couple of now highly interested lads like Rich and yours truly.

Then I got an idea. "Rich," I said, "we need to check those lads in the field tomorrow. We need to get up early, walk in on their shooting area in the morning and get hidden. Then we can "count drops" (count the number of shots fired, the number and species of ducks killed) on our lads and see if they are behaving. When finished with that chore, we can follow them back to their clubhouse and see if they are in possession of over-limits. Especially since they have a goodly number of ducks on each of their duck straps today. And if they shoot like-limits of ducks tomorrow, they will be over their possession limits. What do you think?" I asked.

"Sounds good to me, Dad," grunted Rich as he continued looking over our four chaps' hunting area with his binoculars for evidence of any other duck blinds in the vicinity our shooters might use.

Little did Rich know, I had an important learning task for him to perform on the morrow. One in which he would learn a great deal about stalking humans and doing it right. Moving on, Rich and I spent a great day together in which I was able to teach him many other things about the duck hunting public, hunting the lawless, and how waterfowl behaved during the winter season in the northern Sacramento Valley. All of those lessons were spiced with the history of the area and her peoples. Peoples I had learned about as a young man when stationed in that same area as a game warden, and later as a federal agent. It turned out to be a great day with father and son out and about in the world they both dearly loved. Loved for the look of nature in its element, the ability to learn new things, and the thrill of hunting your fellow man.

Returning to our motel later that evening, after

many long hours afield ending with a stakeout of feeding ducks at night, we called it a day. In fact, since I had arrived in the Sacramento Valley that winter, we had not worked one day less than 18 hours. But who was counting?

Eating a quiet dinner out of the ice chest, I said, "Rich. Tomorrow, I want you to set up our venture onto that duck club we spotted earlier in the day in which we saw those four chaps with duck straps full of ducks. I want you to set up our detail from start to finish. In so doing, I will not offer any words of change or encouragement as to what you set up, unless you have us heading for a wreck. Then we will see how you do and take it from there. Any questions?"

Rich just looked over at me with a smile and said, "Sounds OK to me, Old Man. Just you remember, if we score big from my detail, you owe me a hot dinner at any restaurant of my choosing."

"Sounds OK to me, but you remember, if you screw up, you owe me dinner at any restaurant of my choosing, as well," I said with a grin. Now for you folks who know me through reading my many earlier books, you know I am built like a brick outhouse and of the same size. To get that big, one doesn't lose many bets involving some free chow. After all, I didn't get as big as I am by being last in the line...

The next morning Rich got us up at five in the morning. He had chosen that time because legal shooting hours were not until around seven in the morning. And since it was still raining out, that meant a cloudy beginning to our day. In such situations, because of the darkness the low-hanging rain clouds created, shooting time would be even later than legally allowed. So to Rich's way of thinking, that meant we could sleep in a little later since we didn't have to be afield until later. Finishing our out-of-the-ice chest breakfast, we loaded up our gear and set out for the long drive to Butte County. Finding a hiding place for the truck, we finally departed for the suspect duck club from the previous day's discovery. Rich, worrying about the truck being seen, had hidden it a long way from the club in question. Then we had a long, wet walk to the duck club. Surprise, surprise! By the time we arrived at the duck club which we had to walk by in order to get into their shooting area, their lights were on and the members were up! Additionally, the duck club members had released all their hunting dogs into their outside pens and there was no way we could now walk by the duck club without the dogs raising hell and alerting the club members to the presence of strangers. I never said a word, just waited for my learning rookie son to give me some directions after our "we are here too late" discoveries.

"Dad," said Rich, "I don't think we can sneak by the duck club without the dogs creating a fuss and giving us away. Now what?"

"Well, I agree. If we run by the clubhouse, the dogs will explode into barking and let the members know that something is not right. And this is not the first rodeo these guys have been on with the 'tule creepers,' if I don't miss my guess. If we run by and the dogs explode, they will know someone is on their duck club. If that happens, they will be little angels all day today would be my guess," I said with a grin, having been there before.

"What the hell we going to do?" asked Rich.

"Well, since this one has gone in the tank and not wanting to waste our remaining time together, I say we move on and learn from our mistakes," I quietly advised.

"Damn," said Rich, "I figured these guys for over-limits today with the weather and all."

"Son, we are burning daylight. There is no way to work these guys from the field side because of all the deep ditches we would somehow have to cross in order to catch them red-handed. And they would see us too far in advance if we tried to cut them off as they headed to the clubhouse. When that happened, that would more than likely give them time to quietly and without our notice, dispose of any irregularities. So to avoid any pissing contests in this unique situation, let's save this one for another time. I say that because the best way to work this club is catching them right in the field red-handed."

With that, we pulled up stakes, tucked our tails between our legs, and headed back to our truck. On the long, wet walk back to our truck in the wind and rain, I said, "How about a fatherly and fellow officer critique?"

"Sounds OK with me, Dad. Especially since I now know I screwed up. We have been working such long days, I figured it would do both of us some good to get a little extra sleep. So in defense of my actions, that and the foul weather is why we started so late. But I guess that will be covered in my lesson to come. Right?" he said, with a big-hearted grin.

"One thing I learned a long time ago was to set a time in which to leave for a detail. Then, add an extra two hours onto those departure and arrival times. That way if someone in the group breaks down or the members turn their dogs out into the front yard, you are ahead of the enforcement curve. On this particular detail, I would have been on-site by three in the morning even though legal shooting hours didn't begin for another four hours after that hour. That way, you have time to fix any mistakes that may arise. Plus, at three in the morning, I have found that duck club members are sleeping their hardest during those hours, and you can usually get away with anything because of that. Then it will take you time to look over the area you want to observe, give you a chance to look around for any bait or lead shell caches, and find a place where you can hide, and see and not be seen. That also allows one to get into a hidden position whereby you can also avoid their shot patterns as they shoot at any ducks and still see what is going on. Lastly, everything is different-looking in the darkness. I have discovered over the years, that in such darkness, it always takes you a little longer to get your bearings and get set up for the detail at hand. So, getting there more than early is a boon. The other thing I noticed was that you hid your truck a thousand miles away from the work-site to avoid detection. Son, you will find out that if the bad guys aren't looking for you, they won't see you. Hell, park your vehicle close at hand to the work-site. Then write out a note that you broke down and have gone for a wrecker. Slip your note under the wiper blade on your vehicle and that makes for a short, unnoticed walk into your work area. That way if anyone is suspicious of your vehicle, they will see the note under your wiper blade. If they are really suspicious, they will stop and read the note. Not suspecting anything because of what you wrote, they will replace the note and you are in like "Flynn." That is what I would do. Especially since you are driving an unknown and unmarked vehicle. Now, what say we write this one off to a good lesson in learning and go looking for some knotheads who do not know any better and need some of our attention," I said with a grin.

Of note to my readers, Rich still works for the Service as an agent in North and South Dakota. He is one damn fine officer having learned from his earlier mistakes like all of us had to do. However, as I have discovered over these many years, Rich has not forgotten that lesson learned so long ago in the rice fields of the Sacramento Valley. Today when he sets his sights on someone's miserable carcass, he is spot-on, time-wise and otherwise. In short, the only way to get him off your trail if you are the target of his intentions, is to shoot him. However, that I wouldn't recommend. He is a crack shot with a handgun, rifle and shotgun.

For the rest of that day, Rich and I scouted out places in Butte County that we could examine more closely in the days that followed when looking for those stepping outside the laws of the land. And we discovered one really good looking potential site that needed a closer look. It had only one blind on a long dike in the middle of two large flooded rice fields. All around that blind were about 1,500 decoys! When we finally set up to watch its operation, we had a break in the weather. Out came the blue sky and the wind dropped to zero. The kind of a day that wasn't really good for duck hunting and the shooting of over-limits. However, every lone duck that came by as we watched that area, ended up dropping right into the trap and was killed. I could only imagine what a killing field that set-up would be on a crappy day weather-wise. Sadly, my time in California came to an end before we had a chance to really work that area in bad weather. That is one thing about my profession. Sometimes you eat the bear and sometimes he eats you. One can't catch them all... Damn!

Driving back towards Colusa County en route to Glenn County for some night work, we found ourselves approaching Butte Creek. But not before stopping at a roadside stand and purchasing about a ton of fresh kiwi fruits that were locally grown in the area. Suffice to say, during the rest of our stay together, we ate a lot of those devils. Damn, one can't beat California for its fresh fruits and vegetables. Especially when one lives clear out in the middle of the Rocky Mountain country like I did.

Continuing working our way from Butte County, I said, "Rich, what say we stop at an old public fishing spot I remember alongside Butte Creek and have some lunch?"

"Sounds good to me, Dad," said Rich as he swung his truck off the Gridley Highway and onto a much-used parking spot alongside Butte Creek. Parking alongside an old Ford pickup, we dug out our trusty ice chest and laid out our fare for lunch. We had no more than done so when I heard an outboard motor coming up the creek towards our picnic site. Soon a ten-foot homemade wooden boat hove into view. In the boat were two chaps obviously dressed for duck hunting. They had two shotguns in the boat along with several fishing poles. As they pulled onto the bank near our parking spot alongside Butte Creek, they waved. Waving back with a mouth full of crackers, kiwi and salami, my eyes as a matter of course, swept the length of the old boat. It contained several old Winchester model 12 shotguns, three fishing poles, several tackle boxes, a five-gallon live bait bucket with a lid, and several bedraggled and wet dead mallards lying in the bottom of the boat. Having seen such a get-up before, I surmised we had a couple of fishermen who had taken along their shotguns just in case a duck came flying by so they could fish or duck hunt.

Looking over at Rich, I said, "Those chaps need to be checked. They have a couple of ducks in the bottom of the boat and more than likely were sniping off a private duck club somewhere along the lower portions of Butte Creek. Plus, something is wrong according to my two guardian angels," I said. "They seem to be having a bad case of the 'flutters.'"

Rich just nodded as he gave me "that" look, finished his mouthful of food and then hopped off the tailgate and walked over to our two fishermen. There he grabbed the bow of their boat and steadied it so the lad in the bow could exit the unsteady craft safely. Then Rich pulled their boat further up onto the stream bank.

When both men had disembarked, Rich identified himself and requested their hunting licenses, duck stamps, shot shells, ducks, and shotguns. Me, I just sat there on my lazy duff on the truck's tailgate and watched. In so doing, I noticed that both men appeared to be a bit nervous. And sure as shooting as Rich checked the two men, he began discovering a lot of things wrong. The men were using lead shot, had unplugged shotguns, and neither possessed a federal duck stamp! Man, talk about "flock shooting" citation-wise a couple of very unlucky errant gentlemen out for a good day's adventures.

As Rich explained to both men about the laws they had violated, I still had the feeling that something was not right in the henhouse. However, I let Rich continue. Walking back to his truck, Rich retrieved his paperwork and set about squaring away the sails of our two unfortunates, who just happened to interfere with the lunch of a couple of having one hell of a damn good time federal officers.

Sitting there watching Rich write out the citations on our two chaps, I spotted something amiss... There on one of the chap's elbows was a single crane body feather! With that observation, I sat bolt upright. We always had a zillion cranes loafing and feeding on the Butte County side of Butte Creek in the many vast milo fields. During that era, the taking of cranes was prohibited in California! In short, they were flatly totally protected by state and federal laws! Yet, here one of our chaps had what appeared to be a crane feather on the sleeve of his hunting jacket. Without moving, I took a closer look at the back of his hunting jacket which contained a game bag, or pouch for holding game. The back of his jacket was as flat as a flounder and it was obvious there was nothing in it. But damn it, there was the issue of that single crane feather stuck to the elbow of his hunting jacket! And after working the bird identification arena for many years, I sure as hell recognized crane feathers whenever I saw them...

"You boys have anything else in your boat in the way of game birds?" I asked as I walked over to the citation-writing action alongside the truck.

With those words, Rich looked up. He knew when I asked questions in that tone of voice, that I was on to something. Both men turned and answered together the only birds they had were the two mallards now on the hood of Rich's truck. Looking closely at the men's faces for a long moment, I could tell by the way they were looking away from my stare that they were lying. And the imaginary movement inside my miserable carcass, be it either instinct or my two guardian angels slopping around, indicated something wasn't right. Leaving the men, I walked back to the men's boat and took another gander. Sure as God made Democrats one gene shy in the common sense or having a strong military department, there was another crane feather lying in the muddy water in the bottom of the boat! Leaning over and picking it up, I took a long and hard look at it. It was a crane body feather alright. Turning, I saw my two men were looking at me like a couple of African meerkats looking at a large fat grub. Well, I was large and fat but certainly not grub-like. More like a young Robert Redford... Looking again at the insides of the boat more closely, I saw nothing else out of the ordinary. Then I took a look at the lidded, five-gallon live bait bucket. There was a bloody fingerprint on the side of the bucket and on the top of the lid! Then I looked back at my two chaps once again. They both now looked like they had just seen a ghost, realizing I was now looking closely at their bait bucket! Rich on the other hand, just quietly looked on with a knowing grin on his face. We had worked together enough over his many years as my son when I was on patrols and recently as officers, that he recognized the signs when I was on the "hunt."

Walking over to the bait bucket, I lifted it out from the boat and sat it on the bank. Then lifting up the lid, I was greeted by several dozen fat head minnows swimming around looking up at me. Oh, by the way, there were two sets of breasted out chunks of illegal crane meat in the water of the bucket as well!

"Rich," I said. "You need to add the illegal possession of two sandhill cranes to their mess of violations as well." Then reaching into the bait bucket, I lifted out a chunk of crane breast meat to show him my discovery. To his credit, he never showed any emotion. He just went back to his paperwork and amended the citations to read, "illegal possession of two sandhill cranes" as well...

After our bout with the Butte Creek sandhill crane killers, we headed across the Sacramento River en route to Glenn County for some more night work staking out and protecting fields of feeding waterfowl. Lighting up a couple of cigars along with a pinch of "chew" between our cheeks and gums, Rich finally spoke.

"Well, it happened once again. How you bagged those two with the crane meat has to be a first," he continued looking over at me. "No matter. We have now written 46 federal citations and have checked 27 sports based on what you are calling 'instincts.' Come on, Dad, that 'gut feeling' stuff has to be horse pucky. How are you doing this?" he asked with a disbelieving grin.

"It has to be all that Diet Pepsi, kiwis and Italian salami we are eating," I said with a grin. "That aside, you now owe me a hot dinner at Granzella's in Williams," I said with a triumphant grin. "Since you blew our duck club operation this morning, you have a bet to square with your old man. And if you don't, I will tell your mom," I said over the in-house family joke.

"Yeah, I know. You don't have to rub that small mistake on my part in any deeper," he said with a grin.

Then our radio came to life. On the other end of the transmission was Williams State Fish and Game Warden Charlie Jensen. He had replaced Warden Chuck Monroe in Williams about the time I had resigned my state commission as a state game warden and had accepted one with the U.S. Fish and Wildlife Service as an agent. Charlie was a good guy whom I playfully called "the hippie puke" because of his modern-day liberal attitudes and long hair. Well, what else does one call a friend who has a real honest-to-goodness barber chair sitting in his living room as a bona fide piece of furniture? I had broken Charlie in as a game warden trying to show him the ropes when he entered the Williams Fish and Game Warden duty station. In so doing we had become good friends and now that he knew I was back in the valley after all those years, wanted to hook up. That we agreed to and decided we would meet at Granzella's for dinner that evening. As it turned out, that was a major mistake. True, it was the very first time Rich and I had eaten at a real live restaurant since my arrival. Having skipped the restaurant scene the whole time while living out of an ice chest, so we could maximize our limited time together in the field chasing duck hunters. Just something a couple of "crazies" would do when having a grand time.

Eating bland food for days on end, three meals a day and later a hot, rich meal of all-you-could-eat of creamy clam chowder at Granzella's, turned out to be the perfect storm! Especially after having been introduced to fresh kiwis just a day or so before and having now consumed about a ton of them in the offing. My quicker readers should see this one coming, being familiar with my writing and eating style, as it leads into this next winter waterfowl adventure...

## LEAVING A "PAPER TRAIL"

Exiting our motel room at three in the morning, we headed east to the small valley Town of Colusa. Once in Colusa, we headed north on the Princeton Highway until we came to a small muddy farm road I was looking for that headed due west. "Turn left here," I instructed Rich, which he did. Then heading west towards the backside of where Gunners Field is located, we finally arrived at the 2047 Canal. "Get up on the 2047 Canal dike and follow it south," I instructed. Following that muddy dike road south, we finally arrived on the backside of the Delevan National Wildlife Refuge, located on the opposite side of the canal. Stepping out into a drizzling rain, we could hear the tens of thousands of ducks and geese across the 2047 Canal on the refuge "buzzing" as they awoke to another day. Since the weather for that day was forecast for just rainy drizzle, we had chosen to wear our hip boots for the ease in the long walk that lay before us. Hiding our patrol truck in a large thicket of bamboo alongside the canal, we began a long walk south on its muddy dike. After a walk of about half of a mile, we walked down off the 2047 Canal levee and onto a farm road heading due east. South of that road lay the external boundaries of a large duck club that Rich and I had scouted out earlier on our trip as a future work candidate to check out. Walking about an eighth of a mile due east, we came across another small farm road heading south into the center of the duck club in question. Turning south, Rich and I walked about another quarter of a mile into the very center of the suspect duck club. Here we decided to set up our stakeout so we could move either way towards where the sounds of the guns would lead us once the day's duck shooting started. Scouting out that hiding area as hundreds of ducks fled from our intrusion, we finally decided to lie at the edge of a small flooded rice field. We had picked that area because, as we soon discovered, we were right in the middle of the area where the duck club members parked their ATVs. Then from there, they would begin their travels out into their assigned duck blinds. Plus, it was a constriction point for any hunters returning from their morning's hunt. In short, if a busy shoot day was the word of the day, the hunters would be within easy reach no matter which way we walked. And if they had gotten carried away with the numbers of ducks killed, they would be in our clutches in a heartbeat.

Standing there in the dark and a lightly falling rain, I said to Rich, "See why we are getting into this club so early? Notice all the racket the ducks made as we flushed them off their watered roosting areas when we came onto the club. Also notice the lights are still out in the clubhouse even at this time of the morning. By arriving when we did, the men are still sleeping and did not hear the disturbance we created with all the ducks being spooked off their roost sites. Additionally, by getting here so early, that will allow the ducks to slowly return. Then when our duck club members come out to hunt, they in turn will also scare off the ducks once again. In so doing, if the club members are 'coyote' to game wardens, they will think because there were so many undisturbed ducks on the club, that the wardens are nowhere nearby. Like having the sun to our backs, Son. It will give us a psychological edge when it comes to surprising these chaps," I said with a grin of anticipation.

Exploring the flooded rice field and its dikes where we were to hide, we laid out exactly where we could burrow into the weeds to avoid discovery when the anticipated members arrived later that morning. We also made sure we were far enough away from the ATV parking area that if the members came out in daylight, we would be unseen as well.

Standing there in the morning's dark and drizzling rains, we finally saw the lights come on in the clubhouse some several hundred yards distant. Then we could hear their kennel of dogs waking up and begin barking in excited anticipation of the morning's hunt to come. Even in the darkness, I could see the two of us grinning over having pulled the wool over our potential targets' eyes...

Then the crap hit the fan and the wheels came off my wagon! The night before as mentioned earlier, Rich and I had met Charlie Jensen at Granzella's restaurant in Williams for dinner. That was our first hot meal in days and we both looked forward to our upcoming treat. After visiting for a bit among the three of us, we ordered our dinners. Rich and I chose the all-you-can-eat soup and salad bar. We both figured eating something hot for a change would be wonderful. And with a garbage can full of salad under our belts, that ought to move all that salami, cheese and crackers we had been eating almost non-stop since my arrival. However, in our calculations, we had forgotten the fresh kiwi fruit factor... Then as it turned out, the soup of the day at Granzella's was rich, creamy, buttery, slick as cow slobbers, thick and gooey clam chowder. Needless to say, one of my absolute favorites! When Rich and I had finished "bore pigging" that soup bar, there wasn't a single clam left on the entire west coast of California! If I remember correctly, I had 12 bowls of that rich and creamy chowder and Rich had something like 15! And they weren't any wimpy-sized bowls of chowder either! Barely able to wiggle, Rich and I bid Charlie goodbye because of our next morning's planned detail and left for our motel. Well, the rest is history...

About six in the morning as I stood knee-deep in water alongside a rice check, my guts began grumbling and growling. When I say growling, not like some mangy fox's den full of little ones, but more like the lion cage at feeding time at the San Francisco Zoo! Soon there was no denying it, near arrival of shooting time or not, I had to "see a man about a very large horse," if you get my drift. Realizing there wasn't a whole hell of a lot of high ground in a flooded rice field to take care of business, I chose the water. Moving carefully out into the knee-deep water of the rice field, I opened up my rain coat, put my side arm in a large side pocket so it wouldn't be lost when I bent over, and slowly lowered my hip waders and britches. It wasn't a moment too soon as a deluge of the previous evening's creamy, rich, thick and gooey clam chowder exited my carcass at a very uncontrolled and high rate of speed like the back blast on a jet engine...

Just imagine that picture out in the darkened rice field—a rhino-sized chap, balancing himself like a stork in the deep mud and water of a flooded rice field. All the while trying to keep the water out of my partially rolled-down hip boots and out of my pants. And yet staying bent over enough so one wouldn't fill his hip boots and pants with "once-used clam chowder." Then to top it all off, it now began raining heavily, and that cold water was going places it shouldn't have! Finally, after getting rid of all that clam chowder, salad, kiwis, my liver and gizzard in a rush, I was finished. Then once again carefully balancing my rather massive bulk, I attended to the "paperwork." I did all that all the while to the incessant taunting of my son. Rich was a few feet away down the rice check laughing at me and letting me know how foolish I looked in the pre-dawn. His comparison was that I looked like a huge chicken laying a rather large egg from an ugly perched up-in-the-air position. Then if that wasn't bad enough, he had the audacity to inform me that I was smelling up the neighborhood. Pulling up all my now-wet clothing from the heavy rainstorm, I replaced my pistol in the holster and returned to my hiding spot along the muddy rice check.

Then the gods smiled on my miserable carcass. I hadn't been lying there more than five minutes when Rich said, "Dad, you have any more toilet paper? I think I have to get rid of all my clam chowder as well," he moaned.

By now, guess who was laughing? Not wanting to miss an opportunity of that nature, I informed Rich that I had just a little toilet paper left and I was saving that for myself.

"Come on, Dad. Do you have any toilet paper left? I really need to go and I mean right now," he groaned out.

Guess who was now in the driver's seat? Yep, you guessed it. I gave Rich just enough toilet paper to wipe a small bug's hind end! So much for taunting the old man... But even a drowning man will catch a spear thrown at him if he had a choice. That was Richard's case as he headed for my previously used spot in the flooded rice field. And from the sounds of it, not a moment too soon!

Gross as it is, imagine this. Two huge piles of once-used clam chowder covered with toilet paper dumped out and floating in the water of a rice field. But the action didn't end there. Oh, no! It was just beginning as this plot thickens...

Rich returned to his hiding place on the rice check not a moment too soon. Down from the clubhouse raced two ATVs coming in our direction. Both Rich and I burrowed into our rice check's mud and weeds, hoping not to be seen in the pre-dawn light by our oncoming duck hunters.

Soon they arrived and parked their vehicles not twenty feet from us! Then the bad news arrived. They had a dog! *Damn*, I thought. *I hope he doesn't wind us and start barking*. If he did, he more than likely would give away our positions and ruin our chances at catching any bad actors.

Then the plot thickened further. As our two duck hunters made ready for the hunt, they let their dog run around and take care of business. Unfortunately, he began running down the rice check Rich and I had chosen to hide alongside. There we were. Lying half onto the rice check to stay as dry as possible and the leg portions of our bodies wearing hip boots, lying out into the water of the flooded rice field. Twice that dumb dog ran over the both of us as we lay on the rice check trying not to be "dogged."

Then the dog winded something "wonderful in a dog's world." With that, he ran right over Rich's long legs and then mine as he headed out into the water of the rice field. *Splash-splash-splash*, went our dog on a mission. When he did, he ran right to my wonderful pile of liquid essence. Then all I could hear over the steadily falling rain was *slurp-slurp-slurp...* That damn dog was making a rich breakfast from my dinner the night before! Then I could hear him rolling in whatever essence remained uneaten! Then off that goofy dog went to Rich's pile of essence, and once again the loud slurping "feast" began in earnest!

Holy cow! I was laughing so hard, I almost broke a gut. I later found out that Rich was doing the same. Then that damn dog rolled in the remainder of Rich's floating breakfast with great vigor as well. Finally, here he came once again right over my legs as he headed back to his master who was now calling him! But this time when he came by, this black Labrador of profound size, inbreed intelligence, and even more profound smells, was covered with strips of toilet paper and brown chunky matter where he had rolled in our leavings! As this dog roared by my hiding place, I tried with my feet to kick off the toilet paper so his master wouldn't be on to us lying just a few feet away. But my efforts were to no avail as the dog thundered by. Then he ran over Rich. Here again as I later discovered, Rich had tried to do the same in kicking off the toilet paper as the dog thundered by him. Once again, all to no avail.

It was then I heard the dog's master call his dog one more time. Then I saw the man kneel down as if to welcome his dog with a big hug. In so doing, I heard the dog's owner saying, "Here he comes. Good dog, Charlie. Good dog, good dog—Auggh! What the hell did you roll in? Get away from me! Auggh!"

By now, I was a hopeless, laughing mess. Hell, the man could have had a 100 ducks over the limit and I was in such a hapless state, I could never have caught him I was laughing so hard.

Then I heard the other duck hunter with the dog's owner bellow! "Auggh! Get away from me, you son-of-a-bitch. What a stinking son-of-a-bitch this dog of yours is, Jim. You have got to quit feeding him that brand of dog food," I heard him say as they walked out of sight to their duck blind...

By now, Rich and I were both complete messes from laughing so hard. Boy, I had to think that dog sure took care of the paperwork as he sped by me. Even with the two of us trying to kick off the toilet paper so we wouldn't be detected, the dog had foiled our best attempts. And in the process, the two duck hunters didn't see anything wrong either with a toilet paper-covered dog as they headed out to their blind. Rich and I later discovered that those two duck hunters had quickly placed their dog outside their duck blind because he was still stinking so badly. Hell, I didn't think my clam chowder had tasted or smelled so badly going down that first time. And what a poor mournful dog that turned out to be. Run out from the dry duck blind because he smelled badly. Then having to sit outside all day in the rain... That poor dog sure looked down in the mouth when we saw him later on in the day.

Oh, by the way, Rich and I caught those same two duck hunters without a federal duck stamp in their possession. Here you had a couple of wealthy duck hunters who could afford to belong to an expensive duck club and couldn't afford a federal duck stamp... I guess it takes all kinds.

However, the later cost of their citations would have purchased four federal duck stamps, each truth be known.

Walking back to our truck that fine day now that we weren't carrying that load of clam chowder, Rich got real quiet and then finally said, "You did it again. To my way of calculating, we now have written 48 citations having only checked 29 duck hunters all on a 'hunch.' I really hope someday, Dad, I can do that as well."

"I wouldn't worry about it, Rich. If you have the gift, you will soon develop that trait yourself. If you do, follow it religiously. It will serve you well. Now I think we need to pay the White Mallard Duck Club a visit. For some reason, I have a feeling that we need to hustle over there and see if we can catch some of their duck hunters coming in with something they shouldn't have in their possession."

## IS THERE A DOCTOR IN THE DUCK CLUB???

Leaving our area, we once again headed north on the Princeton Highway. Crossing the ferry over the Sacramento River, we turned and headed south for Putnam Road. Once on Putnam Road, we headed south directly to the lower portion of the White Mallard Duck Club.

The White Mallard Duck Club had been established as a shooting preserve in the Butte Sink since before the turn of the century. Like other Butte Sink duck clubs, they catered to the rich and powerful. And ever since my days in the Sacramento Valley as both a state game warden and later as a federal officer, they had given me fits. Fits in the forms of over-limits of ducks and geese, unplugged shotguns, untagged birds temporarily stored, and "floating duck stamps."

For you "unwashed," a "floating duck stamp" was a loose federal duck stamp kept handy at the front door of an outlaw duck club. Many times on a stand in an empty cigar box by the front door, as a reminder to those sports not possessing one. Then anyone who did not have a required federal duck stamp would just grab one as he left to hunt. When that sport returned, if not checked by a game warden or federal agent who made the holder then sign in ink across the face validating it, the duck stamp was returned to the box for someone else to use at a later date. In my seven years of service in the valley during duck season, I never visited that club in my official capacity in which I did not write someone a citation, many times more than one! They just never seemed to get the conservation message. A classic example of where money and power generated a form of arrogance that I was never able to correct or control. Even when the fines ran into the hundreds of dollars, the next trip out on that club produced more of the same violations discovered just the week before. I guess after all, it was only money to their way of thinking...

It was that form of arrogance that I hoped to tap into that fine day, as Rich and I headed their way. Parking our vehicle at the end of Putnam Road like any other duck hunter hunting on Terrill Farms, we hotfooted it over to the White Mallard's lower boat landing. My plan was to wait until any of their hunters returned from the flooded portion of their duck club. When they did and tied up at their boat dock, we would spring into action and see if we could catch any unfortunate with too many ducks, no federal duck stamp or in possession of lead shot.

Slinking in behind their duck club, the two of us just waited at the corner of the building where we could see when anyone returned to the boat dock after their morning's shoot. Then our plan was to let Rich make the contact and I would remain off to one side, letting him handle the action. Then when finished, I would critique his actions if any was needed. Pretty simple plan, right? Well, we had no more than gotten into position, when we could hear a motorboat coming our way.

Waiting in position where we could see and not be seen, we observed a duck boat coming down the canal towards the duck club. Then it slowed and finally landed at the boat dock.

It was then that I noticed the old gentleman riding in the bow of the boat was an old 'friend' from previous contacts. He was a wealthy doctor from the San Francisco Bay Area and as crusty as any old barnacle you would ever want to scrape off the hull of a ship. Additionally, I had cited him numerous times over the years I had served as a badge carrier in the valley. In short, he would do it his way and damn the torpedoes! As a result of this hell-bent-for-leather arrogance, I had torpedoed his "barnacle-clad" ship many a time. In fact, one time for possessing 13 white-fronted geese over the limit and another for 33 mallards over the daily bag limit.

Waiting until the men had tied up their boat and could not easily escape back onto their deepwater duck club, Rich finally made his appearance. Walking up to the two surprised men, Rich identified himself and requested that he check their hunting licenses, duck stamps, shotgun shells (for lead), and their ducks. Without a word, both men dug into their hunting gear somewhat sullenly for the required licenses and duck stamps. When they were produced, Rich examined them and both men were legal as the law required in that department. Then Rich asked to check their shotgun shells for any lead shot. The first man produced his shells and Rich checked them with a magnet. All proved to be steel shot as required by law. Then when it came time to check the "good, older than glacial dirt, barnacled doctor," Rich discovered he had his pockets filled with lead shells by the score! It was apparent from the number of shells the good doctor was toting, that he did not believe in using the new steel shot over his trusty, tried and true lead shot when it came to killing ducks.

Rich then advised the good doctor that he was in violation and would be receiving a citation for hunting waterfowl while having lead shot in his possession. Rich also advised the good doctor that since he had taken his ducks with lead shot, they would be confiscated as well since they had all been taken illegally. That set the good doctor to grumbling in fine fashion.

Now for my readers' sake, at that time in his life, the good doctor had to be in his late-eighties! And remember, I had pinched him every time I had contacted him during my seven years in the valley. Pinched for 33 ducks over the limit once, numerous times for no federal duck stamp, four times for using an unplugged shotgun, and numerous times for lesser over-limits of ducks and geese. Oh, and twice for killing totally protected swans because he said when they fed on the duck club, they had dug deep holes while eating roots and tubers. Deep holes which several members of the club had fallen into when they had walked into the duck ponds to retrieved their downed birds. Suffice to say, the good doctor came to know me very well...

As the good doctor rummaged around in his wallet for his driver's license as Rich had requested, all of a sudden he stopped dead still. Then looking up intently at Rich he said, "What the hell did you say your name was?"

Rich replied, "My name is Rich Grosz and I am a federal agent from Sacramento."

"You any relation to that son-of-a-bitch Terry Grosz?" asked the good doctor.

"Yes, Sir," Rich replied. "I am his son."

For the longest moment, the good doctor just looked long and hard at Rich as he processed what Rich had just said. Then he spoke saying, "God damn it! Now I suppose I will have to put up with another one of you 'Grosz' son-of-a-bitches for the next twenty years of my life!"

Standing there at the side of the duck club unobserved, I just had to laugh. That crusty old barnacle son-of-a-bitch was at least 88 or 89 at that time! And now he was concerned that he had another twenty years of his life to put up with us "Grosz son-of-a-bitches." Well, that "Grosz son-of-a-bitch the younger" wrote out that good doctor a citation for having lead shot in his possession. And in so doing, opened up a new chapter in the good doctor's life of lawlessness. He now had been written up for waterfowl violations by a father-and-son team. To my way of thinking, that made the whole trip to California in 1997 well worth it. I now had "closed the ring" to my way of thinking. From father to son, the younger Grosz was now picking up the torch from his old man's tired hands and carrying on with the tradition. Who says there isn't an all-knowing and all-seeing God? I have discovered over all my years of wildlife law enforcement in the world of wildlife, that there are always two faces looking back at you. One you can see quite readily and the other you must look for. I guess we found the second one that fine day so long ago in the good doctor. I wonder if he is still illegally shooting ducks?

By the way, that was 49 citations for 30 hunters checked, based on instincts by an "old man and wildlife warrior" about to step out the door. A warrior who was on his last "hunting" trip, hunting the lawless...

For the remainder of our days as a father-and-son team in the Sacramento Valley hunting the lawless, Rich and I at last just had fun and checked everyone we came across. After a 49 citations-issued to 30 sports-checked run based on instincts, I gave my guardian angels the rest of the "time" off. "They" had earned it. "Time," because I was sure by then, they were a tad feather-thin and worn on their wings!

That was the last time a father-and son-team worked together in the Service at the time these words were written. There were glorious times within my career and vision quest. And if I could do the same all over again, I would do so in a heartbeat. However, hunting the lawless or the "most dangerous game" is a young man's adventure. I am a broken-down old man now and waiting "for my next assignment." I can only hope it will be as wonderful as was my first...

# CHAPTER SEVEN

## "BECAUSE OF THE GRACE OF GOD GO I"...

I WAS BLESSED during my 32 years of state and federal wildlife law enforcement service, and my whole life for that matter. Right off the bat, I was "given" two guardian angels in which to watch over me from "The Old Boy" upstairs. "Given," because He created all the critters in this great land of ours and I was blessed with the opportunity to watch over them for Him... In short, when the critters needed help, I was there for them. And when the bullets and fists were flying in the world of wildlife, He was there for me. Plus, since I was bigger than the average "Yogi Bear" and prone to professionally discover the "bad and the ugly" on a frequent basis, I was given two guardian angels to watch over me. It was a good thing that He provided those two, because as you will soon see, they were needed by me more than "often" and as a result, ended up feather-thin after their many years of excellent service...

June 22, 1941, was the date and year of my birth. By no means was it an ordinary birth. My mother was in hard labor with me for 48 hours before my arrival. During that time, the doctors had administered two quarts of ether to help her with the pain of childbirth. As a result of those chemicals, she lost all her hair in the process. She regained her hair afterwards but what a wild ride she had in the birthing process of yours truly.

Finally, in the birthing process as she got weaker, the doctors decided the time had come to do something or they would lose the mother and child. They had waited too long to do a C-section. It was then the doctors had advised my biological dad that they could save the mother or the child. Which did he want to save? My dad advised that they needed to save the mother. Whereupon in order to save her life, they put clamps on my head and just jerked me out! I ended up being injured at birth (left shoulder) and was black and blue from head to toe. In fact, when my mother first saw me, she told the nurses to take me back. Take me back because I wasn't her baby because I looked so awful. On the day of my birth, my mother weighed only 105 pounds, was jaundiced, and I weighed in at over a whopping nine pounds! But I made it... Because of the grace of God go I...

I was born in the United States of America, the finest Nation and land in the world. I have truly enjoyed "her" blessings and many freedoms throughout my life, as do all of my family members. My heartfelt thanks to all our military folks and their families for their sacrifices for the freedoms my family and I enjoy. Because of the grace of God go I...

At age six, my best friend was a boy named David Hagger. He contracted polio in the days when no one knew what caused such a disease and there was no known medical remedy. Soon he was bedridden and abandoned by his friends for fear if they visited him, they would contract the much-feared and little-understood polio as well. However, he was my best friend. I visited him daily in the face of all the warnings from those around me because of my love for my friend. He died shortly thereafter and is buried in the Quincy Cemetery under some stately oaks. I never contracted polio even though I visited him daily. Because of the grace of God go I...

I was raised for many years by a devoted, divorced mother with my sister during extremely hard times (read my book, Genesis of a Duck Cop). Mom worked in a local box factory and I babysat my little sister when my mother was working on Saturdays. (I was six and my sister was several years younger at the time!) For years we had very little and were hungry most of the time. That forced me out at the age of seven to do any kind of job I could for the money it provided. Mom took half of the money I earned to help out with family expenses. She placed the other half in the bank for my college.

During those hard times, I learned to be independent, hardworking, honest, and driven with an excellent work ethic. Those traits served me well in later years, aiding me in paying my way through six-and-a-half years of college, to working successfully in a fine profession in the world of wildlife law enforcement. Because of the grace of God go I...

I was raised by the finest mother in the world, who instilled in me many strengths including that of survival. Her love for her children was boundless. She later remarried a man named Otis Barnes. No finer stepfather could one ever have. Between those two parents, I had the best guidance in life, ever. Because of the grace of God go I...

In junior high school, I met the love of my life. Years later she became my wife and at the time these words are written, still is the love of my life. Donna is the mother of my three children Richard, Christopher and Kimberlee. She is still my best friend and has always been at my side throughout our varied and interesting lives. She was truly a heaven-sent blessing. Because of the grace of God go I...

At age 15, because of my height and size, I lied about my age and went to work in a local box factory to earn money for my college. At age 16, I once again lied about my age (one had to be 18 to work in dangerous professions) and went to work in the woods of Plumas County as a logger. One of the jobs I did as a logger was to help in fighting the many forest fires that occurred in those days. Loggers did so because in those days, the U.S. Forest Service did not have large fire-fighting contingents of firefighters like they do today. So they hired loggers to fight forest fires because we were woods-wise and physically able to do the strenuous and dangerous work of running chainsaws and operating bulldozers.

On one of those fires, the Mosquito Creek fire in the rugged Sierra Nevada Mountains, I got my first helicopter ride. Five of us loggers running chain saws, felling timber back into the fire, got trapped on a rocky ridge by a crown fire (one that races through the treetops at a high rate of speed). Trapped by the blazing fire at the edge of a rocky ridge, a passing Forest Service helicopter spotted our plight. Swinging back, the small observation helicopter dropped down and took two men off the ridge with promises to return for the others. When it returned, he took two more men off the ridge, leaving me, the largest man, behind for last. When it returned for the third time, there was danger.

As the chopper approached my getting-hotter and more dangerous by the minute position, the pilot yelled, "Hurry and get on board. I am almost out of fuel!" As he drifted by without touching the ground, I jumped for the passenger seat on the two-man helicopter. Hanging half in the chopper and the rest of me standing in the rescue basket, over the edge of the ridge we went.

Then we began dropping the thousand feet or more to the Feather River Highway far below! It was at that moment that we ran out of gas! "Hang on!" yelled the chopper pilot as we began our thousand-foot free fall! Man, that was not what I had anticipated my first helicopter ride would be like... Finally, the last few feet before we crashed, the chopper pilot auto rotated the chopper to the ground. However, he misjudged the distance and we hit hard enough to smash the landing skids, allowing the chopper to tip over and smash off its still-whirling blades! Man, when the smoke and dust had cleared, I was one happy lad that my first helicopter ride was over without any injury, other than a very sore back, to my carcass. Standing there on the highway,

I heard the far-off explosions of our left-behind chain saws and their gas tanks blowing up on the high ridge top. Because of the grace of God go I...

In 1957, I lied about my age and went to work for the Meadow Valley Lumber Company in Plumas County as a logger. My first job was on the brush crew where we cut logging leftovers and placed the limbs onto exposed soils to slow erosion. After working in such a position for a month, Cal Cole, the woods boss, had me walk down to the landing. There he assigned me to another job as a "powder monkey" (one who works with explosives) because my predecessor had just blown off his hand! In that work, I assisted the head powder monkey blowing stumps and building road so the logging show could proceed through the timber-harvesting process.

It was during one of those days that I came in close contact with my mortality. A careless delivery man had stacked ten cases of 20% dynamite at the edge of our new road near a tall dirt bank. On top of the dynamite the thoughtless delivery agent had stacked ten boxes of electric blasting caps!

Walking back to our truck for more dynamite and blasting caps, I met my woods boss Cal Cole. As we walked back to the powder cache, we observed Clyde Barnes, an expert bulldozer operator, dragging a skid of logs towards us. He was on a side hill coming in our direction and heading right at the unseen ten boxes of dynamite and electric blasting caps! Both Cal Cole and I ran towards Clyde trying to warn him to stop before he hurtled over the high bank with his dozer and the skid of logs right onto the explosives! However, the loud noise the bulldozer was making precluded Clyde from hearing our warnings and he just waved back. Seeing he was going to land his cat onto the unseen explosives, both Cal and I dove behind a large pine stump for whatever protection it would offer when they blew up!

Clyde sailed off the bank and landed his D-9 bulldozer right onto the pile of explosives then turned his cat and headed down the road dragging his skid of logs and mangled explosives behind him! Nothing happened! The explosives did not go off, and when we finally got Clyde to stop and he saw what he had done, he had to sit down on a nearby stump... Especially in light of the fact that the very sensitive electric blasting caps had not gone off and set off the ten boxes of dynamite!

After that, we brought up a fire truck and hosed down all the smashed sticks of dynamite and then carefully unhooked the skid of logs and drove the cat off the smashed explosives. Even though Cal and I had laid down behind the stump as Clyde came our way, we were so close to the cache of explosives we would have been blown to bits had they gone off! Because of the grace of God go I...

Then there was the Tahoe-Truckee forest fire. Arriving with the rest of the logging crew, we were directed into a canyon in front of the still-advancing fire. The experienced logging crew supervisor dissented over being placed below and in a steep canyon in front of the advancing forest fire. However, we were once again directed by the Forest Service Fire Boss to head for the canyon and begin cutting a trail down to the mineral soil. That we began doing. Then sparks from the oncoming fire drifted across the canyon, setting the other side on fire as well. Soon our entire logging crew was trapped in the canyon between two raging forest fires! Running down the side of our canyon, the logging crew raced for the safety of a small creek at the bottom. However, the fire now raged down both sides of the canyon and we were in danger of all being burned alive!

Then we heard the sound of an approaching airplane. It was an old Stearman biplane carrying borate fire retardant. The brave pilot raced into the lower end of the canyon and dropped his load of borate on one canyon side creating an escape trail. All of us loggers raced for our lives along that borate trail to safety. However, the brave pilot, now trapped in the canyon's swirling winds, crashed and burned to death before we could reach him!

Our woods boss later located the fire boss and beat the hell out of him for exposing his fifty-man logging crew to being burned alive in the bottom of a canyon. Later, all our woods crew donated $100 each from our pay, and the company matched our donations, which were delivered to the brave pilot's widow. Because of the grace of God go I...

I was raised in the high Sierras in the small Town of East Quincy, California. Ran Slaten was a boyhood friend and a neighbor. He, like me, was crazy as a loon and soon got into flying while still in high school. Soon his family owned a J-3 Cub and a Cessna 172 aircraft. It was just a matter of time before Ran and I spent many an hour flying all over the Sierras. Then the devil took over, as we two crazy-as-loons kids began dropping things from the airplanes while in flight to see the explosions the falling objects caused upon impact when hitting the ground... One day we were flying over Snake Lake, a high mountain lake in Plumas County. As usual, we had loaded the plane full of large rocks to drop. Spying a beaver lodge out in the lake, Ran made a pass and I began dropping rocks out the window to see if I could hit the beaver lodge. Flying over the far lip of the high mountain lake just over the trees, we hit a down draft! The bottom of the plane smashed into the high treetops of the ponderosa pines at the far end of the lake! In so doing, it pitched the nose of the plane downward... Fortunately for us, there was a steep drop-off at that end of that lake and we dropped over the cliff and gained airspeed once again. When we landed at the Quincy airport a bit later, we had to clean pine branches out from our landing gear! Because of the grace of God go I...

In 1961, my friend Ran Slaten flew down to Santa Barbara, where I had just completed my first two years of college. I met Ran at the airport. There we loaded my college gear plus sacks of oranges, lemons, grapefruit, avocados and potted flowering plants for my mother. Suffice to say, the Cessna 172 was grossly overloaded! In fact so overloaded, we had to have the airport manager hold the tail up so we could start and taxi the airplane at the Santa Barbara Airport.

Once airborne, the airplane struggled to get enough air to fly over the Coast Range as we headed for Bakersfield. Once over the Coast Range and en route to Bakersfield, the gas gauge read, "no safe takeoff!" Too late to turn around, we looked for Plan B. Looking for any safe spot to land as we continued along, we finally reached the Bakersfield Airport.

Taxiing over to the gas pumps, we filled up the aircraft. It was then we realized we had landed with only three gallons of gas in our tanks... However, that didn't deter us two goofy aviators! Now very grossly overloaded, we managed to take off in 90-degree heat! As it turned out, we used up the entire 5,000-foot length of the commercial runway before we got up enough airspeed to lift off. Then when we did, we couldn't gain any altitude. Then the engine began overheating! Realizing we were soon to crash unless we lightened up the airplane, I began dumping things out the window. Sacks of oranges, lemons, grapefruit, avocados, potted plants, schoolbooks. and the like were dumped just as fast as I could. We finally lightened up the plane enough to gain altitude, where the cooler air prevailed. That in turn cooled down our engine from its "red line." Then like a couple of dummies, we flew back to Quincy in northern California using highway maps to chart our way... Because of the grace of God go I...

In 1962, I went to Alaska as a seasonal employee for the Bureau of Commercial Fisheries, U.S. Fish and Wildlife Service, to gain work experience in my major field of study at college. There I was assigned to a sockeye salmon tagging crew on the Naknek River. After tagging over 34,000 salmon, our tagging crew was sent into the upper reaches of the river system to recover our tags on the spawning grounds.

Throughout our travels, we used rugged and heavy 18' Cordova skiffs, oak-planked boats, powered by twin 40-horse outboard motors. We used 1,000 feet of nets evenly divided between two boats, which was carried in the bows for our salmon tag recapture efforts.

Our technique was to locate a large school of resting salmon in the pools of the rivers or lakes. Then we would motor below them as one boat slipped up along one side of the salmon and the other boat slipped up alongside the opposite side of the fish. As we did this, we played out our connected 1,000 feet of net. Then once the salmon were surrounded by the net, we motored to shore, hauled in the net and recovered our tagged fish located in the overall school of fish. Then we would take blood and scale samples from our tagged fish and release all the fish to continue their migration.

On one occasion we were located where two huge rivers met. Where the two rivers joined were about 1,500 resting fish in a large pool. In order to trap those fish, one boat would have to motor up one river in reverse playing out its net and the other boat come up the other river in reverse doing the same thing. The plan was to trap the fish in the net, motor the boats to the shore and recover our tagged fish. However, that is where the wheels came off the wagon!

One of the swift flowing rivers was the Savanoski. It was a glacier river and one could not see the bottom because of its milky mixture (typical of all glacier-fed waters). The other quieter-flowing river was the Grosvenor where one could see the bottom. However, directly below where the two rivers joined was a raging whirlpool about 35 yards across full of dead trees and one dead moose! The object was to net the salmon and sideslip above the dangerous whirlpool. Being we were "knowing everything there is to know" bull-headed college kids, none of us were wearing life preservers that day! Plus, the rivers were running about 36 icy degrees temperature-wise!

Starting off like we had done a hundred times before in other river systems, we began our run. However, the swift-flowing glacier river was really carrying the mail and the man operating that boat had both 40-horse engines going full throttle in reverse! The other boat backing up the Grosvenor which was less swift, was only running one motor in reverse. As the two boats backed up the rivers, the lads in the bows were playing out the 1,000-foot net. However, we had not anticipated the extra drag the fast-flowing glacier river was putting on that chap's net. Then in a flash, that boat was ripped out into the swiftly flowing river! Then that boat operator hit a rocky bar and busted both shear pins on his roaring motors, rendering that boat powerless. Now we had one boat wildly flying down the glacier river out of control and the other boat gamely trying to hold itself in place against the pull of the net from the other boat. In the process, the two lads in the bows of the boats tried gathering in the fast-spilling-out nets before their drag rendered both boats totally out of control...

Zingo, out into the lower reaches of the glacier river both boats were quickly dragged and hurled into the maw of the raging whirlpool! Now we had two boats madly whirling around in the whirlpool, rapidly filling with icy water! In the meantime, both bow men tried gamely to retrieve their portions of the 1,000 feet of rapidly playing out net. In so doing, the boats smashed together, knocking one bow man senseless and back down into the bottom of his boat. Then with no one to control his net, it began spilling out once again. I was in the bow of the other boat. As such, I tried bringing in the rest of my net and now my knocked-out partner's portion of the net as well. Then our nets whirled back under my boat, fouling the propeller of my still-functioning outboard motor. That killed the one remaining running motor! Then a log sailed into our net, whirled around and dragged both boats out of the whirlpool. Then down the raging glacier river both of our boats went!

Finally smashing into a low overhanging riverbank, I grabbed an exposed root from a tree and holding on with all my strength, my partner began gathering in the rest of the two nets. Finally we recovered the nets and I managed to untangle the net from our propellers. Then starting up the outboards in my boat and holding both boats together, we motored back into the Grosvenor River and finally banked our boats on a sandbar. Exhausted, all four of us just laid down on the sand and thanked our lucky stars for not drowning or freezing to death in either of those two rivers. Which sure as all get-out would have occurred had we gone overboard with our hip boots on and wearing no life preservers... Because of the grace of God go I...

Later that same summer in Alaska, three of us were transporting our belongings from the Coville Lake smolt camp to the Grosvenor Lake camp. An easy trip of just 20 miles across the lake. However, before we arrived at the Grosvenor camp, we were trapped on the lake by a violent storm. It was a freak Arctic storm with 80-mile-an-hour winds that caught us out in the open water.

We knew it was coming as it was at that moment hitting our main camp on the west coast in King Salmon, Alaska. The radio operator at that camp advised since we were 100 miles inland, we could beat the storm running across Grosvenor Lake if we left at that very moment. Since our boat was loaded with all our gear, extra gasoline, all our belongings plus a month's food supplies, off we went for our normal one-hour trip.

Larry, our most experienced boat operator, ran the twin 40-horse motors as Doug and I sat on the middle bench seat enjoying the scenery. We hadn't been gone for more than 20 minutes when we noticed that the entire daytime sky had now turned ashen and it looked like we were in a dense fog. As it turned out, the fog was volcanic ash! Mt. Trident in Katmai National Monument (now a national park) had chosen that moment to erupt and now we were in the middle of a dense ash fall. Then behind us to our west, we could hear a roar. Up until that point, Grosvenor Lake had been as smooth as a sheet of glass. Then the roaring gale-force winds caught us and as it turned out, it was the tremendous winds from the Arctic storm off the Bering Sea now hitting us. In an instant, our glassy lake turned to twenty-foot rollers with an 80-knot tail wind turning us every which way but loose!

Soon our heavy 18' oak-planked Cordova skiff was smashing into twenty-foot rollers and sending walls of water over the bow! Doug and I began bailing out our overloaded boat like crazy, as Larry fought to keep us heading into the waves and not getting blown sideways. Because to be blown sideways meant we would roll over and sink! Once again, we weren't wearing life preservers over our heavy rain gear and hip boots! Hell, it wouldn't have mattered anyway. If we had gone overboard, the water was 34.6 degrees! And if we had survived that swim to shore, we now had a twenty-mile trek along the shoreline inhabited by hundreds of hungry brown bears...

For the next two hours, the three of us fought to keep from swamping. Normally the trip only took one hour. Now we were into our second hour and our gas tanks were running dry since our engines were running full bore to maintain headway. That forced me to pour fuel from our spare tanks into our boat's gas tanks. In so doing because of the howling winds and smashing waves over the bow, I managed to spill some fuel into the bottom of our boat. Soon we were slipping in gasoline and oil (our outboards were two-cycle, not four-cycle).

Two-and-a-half-hours later, we finally arrived safely at our Grosvenor River camp totally exhausted, soaked to the bone like a bunch of wet muskrats, and "prayed out."

There we soon discovered that all our dry food

on board had been ruined by the spilled gasoline and waves coming over the bow. We tried to radio our base camp at King Salmon requesting an airplane with new food but were refused. With all the volcanic ash in the air, it was not safe for aircraft operations in the area. So for the next three days, we ate nothing but fish taken from the nearby river and lagoons.

Three days later, the ash "fog" lifted and we were finally resupplied with food from the air. But a huge brown bear found our recently arrived food cache, broke into our cook shack and ate everything. Then habituated to the food source, it kept trying to break into our little plywood shack for more food while we were inside with only one rifle and three bullets! But, I will save that bear adventure for another time. I will tell you this, it got so dangerous that the giant brown bear had to be put down. Then he was so big we couldn't move him from in front of the door to our cook shack. Within one day, all the other bears in the area (and there was a pile because the salmon were running) came and feasted on our dead bear as we sat in our shack some six feet distant with only one bullet left for our rifle. But that story can be read in my book titled, Genesis of a Duck Cop. Suffice to say, our boat trip was something out of the ordinary. Because of the grace of God go I...

Upon graduation from college with my Bachelors and Masters of Science degrees, I successfully competed with 1,400 applicants for one of 25 California State Fish and Game Warden positions and was selected. That action was fortuitous, because my young Bride Donna was now pregnant with our first child and I needed a job. Because of the grace of God go I...

On my first investigation as an undercover officer for the Fish and Game Department in 1966, I was confronted by eleven construction workers who were snagging and selling white sturgeon on the Klamath River in northern California. When it came time for me to make the arrest by myself, since I had no backup, they tried to jump and kill me. I brought in all eleven culprits and successfully prosecuted the same. I did so with only one set of handcuffs (read the first chapter in my book, Wildlife Wars)! Because of the grace of God go I...

One night while on stakeout by myself in Trinity County near the Hoopa Indian Reservation, I ran into a carload of night-hunting deer poachers. Having to make a head-on car stop (the most dangerous kind), I ordered everyone out the driver's side of the car. In the dark and bedlam, I did not see one Native American drop to the road and slide out of sight under his car! Finally getting the remaining four Indians out of the car and standing them in the middle of the deserted logging road, I noticed something wrong. All my prisoners were quiet as a bunch of mice pissin' on a ball of cotton... That was unusual behavior for Indians I had discovered early on as a game warden. They would usually be getting all over me for being a white man and stopping them from their God-given hunting rights and the like. Not realizing what was the problem, I slipped between the two cars to affect my arrest for their illegal spotlighting activities. When I did, the Indian hidden under their car came out but was immediately attacked by my eight-month female Labrador who had been sleeping in the back seat of my patrol car! Yeah, you got that right... Attacked by a gentle, eight-month old female gentle as all get-out Labrador pup! By the time I got the dog and Indian untangled, she had eaten most of his face off. As he laid in a pool of blood, I noticed a ten-inch Bowie knife lying by his hand. He had intended to stab me in the back when I confronted my group of four silent Indians. No wonder they were so quiet... They were all booked just the same in Willow Creek, bad face from the dog or not. Because of the grace of God go I...

In 1966 while working along the Trinity River as a game warden, I noticed a large group of men doing something along a quiet pool below me.

702 | TERRY GROSZ

The salmon run was in full swing and suspecting some kind of illegal fishing activity by the men, I quietly parked my patrol car on a side road. Finding a trail leading down to the river, down I went just as it was getting dark. The men had set up their camp on a small sandbar, had a roaring fire going and were preparing their evening meal. Looking over the group of men from my hiding place in the river's dense brush, I observed they were drinking whiskey and having a good time. A closer look with my binoculars revealed all my men were of Indian extraction. Then I noticed why they were there. Out into a quiet pool of the river, I observed a string of floats from a 50-foot illegal gill net!

Watching all of my lads, I soon came to the conclusion that only five of the men were actively tending or fishing the illegal gill net. The remainder of the men were just drinking and having a good time. I was sure all were going to share in the illegal spoils, but for now, only five were involved to my way of thinking. Since only five of the men were in violation of running an illegal net, I decided to join their little party before more men got involved in illegal activities, which would complicate my arrests and before any salmon had been unnecessarily taken.

Realizing I would probably have a fight on my hands if I tried to end their fun and games,

I slipped on my leaded sap gloves just in case. Emerging from my cover, I approached my group of 13 Indians. Once they realized who I was based on my uniform, the group got instantly quiet. Walking into the group, I advised the men as to who I was and further advised that five of their number would be cited for running an illegal net. As if on command, all the men rose from their sitting positions and charged me. Realizing my degree of trouble, I took a defensive stance and faced off on my charging men. It was then that I realized they had screwed up. Although they were small in stature (they were Yurok Indians) but large in number, they were coming at me one at a time.

When the first Indian arrived within striking distance, I hit him in the face with the force of an angry mule's kick. One can do that when he weighs 320 and is using sap gloves weighing a pound each! Down went my first arrival, which was soon followed with another within striking distance. One hard "whang" on his pumpkin head put him out like a light just like the first man. But I had no time to gloat for the third man was now upon me. He followed the first two sailing into the sand with a knot on his head and lights out as well. Then came the fourth and fifth in line. They, too, suffered the same fate as had the three previous Indians arriving within striking distance.

With five punches thrown and five Indians knocked out, the remaining bunch of Indians slewed to a stop and wanted none of what their buddies had received. Man, was I ever happy they chose that route. Because I knew if any of them ever got me down, that would be my last day on this earth...

Backing off the remainder of the group and having them sit down, I cuffed my five with the still-ringing heads. And I had learned my lesson. When working the salmon rivers in Indian Territory, bring extra handcuffs just in case.

I then had two of the non-impacted Indians retrieve the illegal net and wrap it up. When my five chaps with the knots on their heads were able to get back onto their feet, I had them carry the gill net up to my patrol car. Then loading up my charges, we headed to Willow Creek and the nearest jail. There they were booked for fishing with an illegal net.

It was only on the way home that I realized just how close I had come to meeting my Maker. Especially if I had tried to arrest all 13 of my river chaps... Thank God they had come at me one at a time where I had time to deal with each one individually. That and a damn good pair of hard-hitting sap gloves will work miracles every time. Because of the grace of God go I....

One summer in 1966 while working the illegal Indian gill net fishery on the Klamath River with my friend and fellow game warden Herb Christie, I almost got killed. That evening while running the river in a rowboat checking for illegal gill nets without using any lights, we came across one strung across a quiet pool. Staking out the net until the next morning, we finally apprehended the illegal fishermen. It was an old Indian and his young grandson. They were issued citations for running the illegal gill net and the net was seized. Since it was now daylight (gill nets are only effective at night when the salmon can't see the mesh of the net), we went back upstream to our camp.

Shortly afterwards we noticed a boat coming upstream with a lone operator. It turned out to be the young Indian we had just seized the gill net from. He rammed his boat onto shore next to where I was getting coffee water from the river. Then he ran out from his boat with a cocked and loaded 30-30 rifle and pushed it into my stomach. Then in a fit of rage, he said, "I want my net back!" By then Herb had drawn down on the Indian from a few feet away and seeing that, I was able to talk the young man down and after a time, take away his loaded rifle. There was a live shell in the chamber and we were miles away from any medical help had he chosen to pull the trigger! My being calm with a rifle shoved in my belly and Herb holding firm on the man's head with his revolver saved my life (read that story in my book, *A Sword for Mother Nature*). Because of the grace of God go I...

In the winter of 1967 while working alone in Colusa County chasing commercial market hunters night-shooting ducks, I was ambushed by a "drop-off" (read the story, "Taking a Load of 4's" in my book, *Wildlife Wars*). When I arose to chase three night shooters who had just killed hundreds of night-feeding ducks that I was staked out on, I was shot in the back by their "drop-off." A market gunner "drop-off" who was there to provide cover for the shooters who had just shot into several thousand night-feeding ducks in front of me. That "drop-off" shot me three times with a shotgun from about 35 yards distant using number four shot.

Over 200 pellets were later removed from my carcass by Sheriff Alva Leverett. He did the removal in the quiet of his home without the services of a local doctor, so other commercial market hunters were not aware I was stove up and not 100% effective. Then I returned the next day to the ongoing wildlife wars in the rice fields of Colusa County. Sure, I wasn't moving too fast, but I was alive and managed to keep the shooting episode away from my wife so she wouldn't worry. I did so by working all night and sleeping during the day while she was away working. That way she did not see all the leaking bandages. She did not find out about that episode until about five years later... I was never to catch that shooter and prosecute him but I made sure he knew I knew who had pulled the trigger. I also made sure he realized if I ever got the chance, his turn in the barrel was next... Because of the grace of God go I...

One evening while working night fishermen on Butte Creek, I chanced upon two men fishing from a boat in violation of the fish and game laws. Being in a canoe and in the darkest of nights, I silently paddled towards my subjects. They did not see me coming and I could see little in the darkness as well. As I quietly approached their boat, one of my two guardian angels advised that I should "duck!" Not seeing anything to worry about, I kept paddling silently towards my unaware subjects. Then my guardian angel told me again to "duck" and the warning that time was almost physical! So, I ducked. As I did, an unseen frog gig sticking out from my subjects' boat grazed the side of my head, taking my hat off in the process! If I hadn't ducked, I would have lost my eyes to the sharp prongs on the frog gig. Because of the grace of God go I...

In the spring of 1970, I received a call from Agent in Charge Jack Downs of the U.S. Fish and

Wildlife Service. In that call, he offered me a job as a United States Game Management Agent. That had been the job and profession I had always wanted. Jumping at the chance, I later resigned my commission with the California Department of Fish and Game for one with the Fish and Wildlife Service as a U.S. Game Management Agent. As it turned out, one of only 178 such positions in the Nation! With that, I went on to a magnificent career with the Fish and Wildlife Service. One that took me across the Nation and into some foreign lands. Because of the grace of God go I...

Early on in my federal career as a U.S. Game Management Agent, I was assigned to backfill for Agent Ben Crabb who was on vacation. His duty station was San Francisco and the rest of the Bay Area. A large part of his duties involved checking all imports of wildlife from foreign lands arriving by passenger, boat or airplane cargo. On one of those inspections at U.S. Customs, I seized a sackful of illegally imported poisonous black and green mambas, along with some other bad character biters. While transporting the snakes to the San Francisco Zoo, the bag came undone because of a near car wreck and they escaped in the front seat of my patrol truck. I noticed the open bag when I came to a stop at the near-wreck site and bailed out from the truck and shut the door. A snake handler from the zoo eventually arrived and managed to capture all the snakes except one... No matter how hard we looked, one's absence remained a mystery until I lifted up the hood somewhat later and discovered the last snake. He had escaped through the dash and rested on the still-warm engine. Then when starting up the car, he was struck in the head by the moving fan blade... Because of the grace of God go I...

One fall in 1970, I headed into Humboldt and Trinity Counties in northern California to work band-tailed pigeon shooters who were allegedly taking over-limits of that migratory game bird. Stopping in one mountainous area and after citing a lawbreaker, I heard more shooting further below me in the forested bottoms. Driving in that direction and then hiking towards the sounds of shooting in the dense, second-growth coniferous forest, I stumbled onto an unusual area. It appeared to be a small campsite near a stream so far back in the bushes, that the chickens had square faces in that neck of the woods. Small plastic water hoses led from the dammed-up stream to a garden plot hidden under a dense growth of Douglas fir trees. Walking down a heavily used trail and wondering why the hell someone would be camping and gardening back there, I got the surprise of my life. Lying in the middle of the well-used trail on which I was walking, lay a shiny silver dollar. Starting to walk down the trail to recover the dollar, my guardian angels inside my miserable carcass went wild. Pausing and looking around for the danger they were foretelling, I saw nothing out of the ordinary. Then striding down the trail for the silver dollar once again, my angels blew up! Stopping once again and realizing from past "angel experiences" there was danger at hand, I looked around once again. It was then I observed a heavy wire leading from a sunken "T-Post" in the dirt along the trail just shy of that shiny and alluring silver dollar almost within reach. But once again, I held back out of deference to my time-worn instincts. Backing up and grabbing a loose fir limb, I began poking the middle of the trail as I continued walking. All of a sudden, snap went a huge New House #5 bear trap, which had been buried in the middle of the trail attached to the suspect wire! Had I not listened to my guardian angels and continued walking down that trail like a dingbat, my leg would be in the trap instead of that stout fir limb. As I later came to realize, that was my first discovery of a hidden and booby-trapped marijuana grow-site. And the silver dollar was nothing more than a lure for the unsuspecting to walk down the trail and right into the trap! Had that happened, my skeleton would still be there in that trap to this day... Because of the grace of God go I...

One fall duck season while working in the eastern half of Colusa County, I ran across a strange character in his duck blind. As he killed his ducks, he would pick, clean, and cook them in the blind on a small stove. Never before had I seen such an operation. Walking over to check the fellow, I ran into one of his bodyguards and that damn near ended up in a shooting before I checked my strange fellow in his duck blind. However, everything checked out and I left the strange chap and his bodyguard for another day. Later that summer, I once again ran across this strange little man while banding dove on a ranch in the western Colusa County foothills. That time we met in the kitchen of the old ranch house in which I was staying and got to trading cooking secrets since both of us were cooks. However, he still had his now-two bodyguards and once again I found that to be strange.

That was until the feds from another agency met with me one morning at my home and advised that my new friend was one Art Columbo, allegedly head of Murder Incorporated of the Mafia! They wanted me to watch him and since I was also a federal officer, I advised I would. I ran across Art deer and duck hunting numerous times after that and soon, regardless of his background, we became fast friends. He was a lonely man and much older than I. I would say our relationship was more like father and son regardless of his alleged past and present, especially when it came to our mutual cooking activities.

Later that next spring while checking "froggers" in the early morning hours, I caught an Italian fellow with 198 frogs over the limit! As I was soon to discover, my "frogger" was a Made Mafia man from south San Francisco. And pissed over me citing him and releasing all his frogs back into the marsh, put a contract out on me!

Meanwhile my friend Art got wind of the Mafia contract on me and quashed that deal. But not before having his bodyguards beat the hell out of the man who originally put the contract out on my head... In fact, the fellow taking out the contract on me, later floated up in San Pablo Bay near the mothballed Navy fleet. My friend Art and one of his bodyguards were later assassinated in 1978. But not before he had saved me from a deadly fate because of the previously issued Mafia contract. Because of the grace of God go I...

In December of 1970, I had occasion to work the San Joaquin Delta area on some of the ultra-rich duck clubs. One morning early in a dense fog while navigating several islands trying to get to the one on which I wanted to work, I got the surprise of my life. Steaming right along in the dense fog trying to hurry to my work destination before my outlaws got into place, I careened off the side of an oceangoing ship in the ship channel! Skidding off the side of the vessel before I could stop in the dense fog, I was thrown clear of my patrol boat. However, not before sailing over the windshield and tearing my chest high waders open in the process (as usual, I was not using any flotation gear again!). Now without any flotation of any kind and a ripped set of chest high waders filling with water and dragging me down, I immediately sank to the bottom of the 30-foot deep channel. In winter's icy cold water and on the darkened bottom, I struggled to cut myself free, right myself, and swim back to the surface. However, the large oceangoing vessel chose that very moment to glide over me, and the wash from its huge propellers bounced me around on the bottom once again. Finally cutting off the last of my chest highs, I once again struggled to swim to the surface. In so doing, I was now losing my ability to swim due to the lack of oxygen, lactic acid buildup in my musculature, and the effects of the icy cold water. Luckily, I finally broke the surface and managed to swim back to my patrol boat now stuck in the cattails a few yards away. Then once back again on board and after almost freezing to death as I motored back to the marina for some warm and dry clothing, I finally arrived. Taking an hour to warm up from stage one hypothermia in the front seat of my patrol truck after a fresh change of clothing, I finally returned to my boat. Then in broad daylight and lifting fog, I went back after my suspected illegal goose shooters. Those I later caught with 56 white-fronted geese over the limit! You can read this story in my book, *For Love of Wildness*. Once again during my adventures, I escaped the avenging angel. But just by a little bit on that occasion... Because of the grace of God go I...

In 1974, I received a phone call from my boss in Sacramento, Agent in Charge Jack Downs. He informed me that the Service was reorganizing and soon there were to be a slew of new first-line supervisors or Senior Resident Agents. He also informed me that he was nominating me for such a position. Hell, I only had four years on with the Service and knew I did not stand any chance of being promoted so early in my career, so I shrugged it off. Two months later, I was advised that I had been selected as the new Senior Resident Agent for North and South Dakota! I was now one of 45 new Service supervisors. Because of the grace of God go I...

While stationed in Bismarck, North Dakota, as the new Senior Resident Agent, my wife called. We had stopped having children after the birth of my second son because it had been so hard on my Bride and dangerous. Now she wanted to adopt a child, namely a daughter, since we only had two sons. She had a lead with an adoption agency so I told her to go ahead and see if we could be considered for adoption. Weeks later we went through the home study program to see if we would be suitable parents and passed. About then Vietnam was collapsing and the opportunity arose for adopting a Vietnamese War Baby. So, we threw our hat in the ring. Several weeks later, we were advised we had a bouncing baby girl coming in on the next plane to Bismarck! It all happened so fast our heads were swimming and we had not yet planned for her arrival. After a stay in the hospital recovering from all her ailments, Donna and I received one Kimberlee Dawn Grosz from Vietnam and began the lengthy process of adoption. She is now part of our family, has three children of her own, and is living in Arizona. Because of the grace of God go I...

While stationed in North Dakota, I was fortunate in running across and befriending two of the finest Assistant United States Attorneys in the Nation. One was David Peterson in Bismarck and the other was Lynn Crooks in Fargo. Together the three of us instituted numerous new programs in

North and South Dakota that are preserving our wildlife heritage even to this day. Programs protecting our wetlands from drainage by the dirt farmers, cattle trespass programs that stopped illegal grazing on Service lands, and new border enforcement programs with our Canadian counterparts that made it hell on wheels for smugglers of wildlife. All that and much more became just part of our new wildlife law enforcement initiatives. Those programs were soon followed with covert programs that slowed the illegal trade in eagle parts and products, resulted in seizures of numerous aircraft violating the Airborne Hunting Act, and initiation of interstate big game inspections now led the pack. In short between those two attorneys and my agents, we made life hell for those destroying the natural resources of those two states. Because of the grace of God go I...

During my short tenure in the Dakotas, protecting the very rare whooping crane was a number one priority in those days for Service officers. And since I had those rare cranes passing through the Dakotas during their spring and fall migrations, it became a priority for me and my officers as well. One day in the early fall as I worked in my Bismarck office, I received a call from one of my refuge counterparts in western North Dakota. During that call he advised we had nine whooping cranes out in the middle of a huge wheat field near a small, northwestern North Dakota town. He requested assistance in protecting the migrating birds while they fed and I immediately complied. I called my Minot agent and had him head towards the cranes' location immediately. Immediately, because duck season was in full swing in the area as was sandhill crane season, a look-alike bird! By that time my angels were really raising hell, as I headed out the door of my office. Why they were raising hell was a mystery to me but I kept their flutterings in mind as I quickly headed north for the rare cranes' location.

Forty minutes later, I arrived at the wheat field in question and quickly had my Minot officer take the east side of the field as I took the west. Then I had my two refuge officers on hand take the north and south side of the field in question as well. Now we had all the bases covered and if anyone tried to hunt in that area, we could caution them and have them leave the area. That way the exhausted, migrating whooping cranes could rest and refuel for the rest of the trip that lay ahead of them on their way to southern Texas.

However, my guardian angels were still fluttering around inside me like there was no tomorrow. What the hell is going on? I thought. Then way out in the wheat field by an old abandoned farmhouse, I saw a woman standing by the side of the building by herself. What the dickens? I thought, as I put away my binoculars, jumped into my patrol car and headed her way. Arriving, I pulled up alongside a woman who was watching the wheat field holding the nine rare whooping cranes. Stepping out from the car, I said, "Federal officer, may I help you, Ma'am?"

She said, "No. I am just watching my son sneaking up on those cranes to see if he gets any."

"What!" I said in alarm.

"My son. He is out there with his .22 trying to sneak up on those cranes to see if he can kill one for our dinner," she said in surprise over my explosive reaction.

"Ma'am, those are the totally protected whooping cranes. There are less than 200 of those birds left in the world!" I said in rising alarm. "Where is your kid?" I yelled.

Now concerned because I was so concerned, she said, "He is over there in that draw leading up to the cranes."

With that, off I flew in my patrol car across the harvested wheat field like a striped-assed ape! Soon I overtook the surprised lad and hustled him into my patrol car. Then we hurriedly left the field so as not to further disturb the cranes. On the way back, you can bet my lad's ears got blistered over the fact he was hunting migratory game birds illegally with a .22! And even more so over the fact he had been stalking one of the rarest birds in all of North America! Then when I got back to the boy's mother, I filled her ears with the word of the day if you get my drift. But in a fully professional way I might add... And now thankfully, my angels had settled down.

After all, they had feathers just like the cranes and were somehow kindred spirits. Thank God none of those birds had been killed. If any had, I would have had to dig a deep hole and throw them into it. Then jump right in behind them myself... As it turned out, the birds fed the rest of that day and into the next morning, then safely left. Because of the grace of God go I...

During the winter of 1975, I was heading towards Fargo to pick up a handful of arrest warrants. I was in a BIG hurry because a blizzard was predicted for the late afternoon and I sure as hell didn't want any part of that. Especially being out on the highway when it arrived. In order to get to Fargo and back before the arrival of the storm, I was speeding just a bit. Well, I would call going over 80 miles per hour more than just a little bit. However, to avoid the local law, I had taken a highway that was mostly abandoned that time of the year, so I moved right along. By then, winter had been a fixture for several months and snow was everywhere. However, the highway was dry but I did have the usual North Dakota winds blowing from the north. And when they did, they would blow loose snow up and over the raised roads and down the other side. However, when that happened, as the snow blew across the highway and cars drove over it, it soon became packed ice.

Steaming right along as I said earlier on the deserted highway in the now-blowing snow, I came to a series of rolling hills. It was at that point I noticed that my two angels were starting their "takeoffs" like they didn't want to be there. Coming to a rather high hill, I chanced a look at my engine gauges and speedometer. I remember to this day I was going 84 miles per hour when my eyes left the gauges and I looked back up onto the highway. As I did so, I topped the rise on the hill and there before me was a patch of ice about 40 yards long, in the middle of the road! And right in the middle of that long patch of ice was a sharp, 90-degree turn! I was on the ice before I could do anything and I knew the last thing I needed to do was to hit the brakes. Letting up off the gas, I said a quick prayer and hung on knowing when I hit that patch of glaze ice in the turn, I was going to fly off the road and roll about ten times before I stopped because of my excess speed.

Holding my breath, I hit the patch of ice at the top of the knoll and began my turn...

I came through that turn like it had been nothing but dry road! Stopping down the road, I said a quick prayer of thanks and then backed up. There was no way I should have traversed that patch of icy road without flying out across the windswept North Dakota prairie countryside in a series of flaming rolls. Stopping short of the patch of ice, I got out and walked back to where it ended. It sure looked like glare ice, I thought as I took a step out on it. When I did, I went flat on my hind end in a flash! It was plain and simple glare ice. There was no way I should have traversed that patch of ice without flying off that portion of the road and hurtling across the countryside in a flaming wreck. But somehow, I had... Because of the grace of God go I...

In 1976, I was promoted and fast-tracked into the Washington Office as the new Endangered Species Desk Officer. Additional duties included being the new Foreign Liaison Officer as well. Duties I would soon come to realize, that would drag me out of the fifteenth-century way of thinking and introducing me to how the Service operated. My two-plus years in the Washington Office really allowed me to mature as a wildlife officer, as well as qualifying me for later senior promotions. Several other promotions in the field soon followed. One of which allowed me to become the youngest Special Agent in Charge at age 39, in the Fish and Wildlife Service at the time. Because of the grace of God go I...

In 1978, as the Washington Office Foreign Liaison Officer, I was sent to eight countries in Asia to assist those governments in understanding U.S. import/export laws and for me to learn their laws as well. However, on a side trip in Kwai, Thailand, to check a crocodile farm, my contingent of Thai wildlife officer got ambushed by Burmese insurgents! Before the shooting had stopped and the ambushers driven off, my driver was killed, and five other officers were either seriously wounded or killed (read that story in the "Asia, 1978" chapter of my book, *Defending Our Wildlife Heritage*). Once again, I had managed to escape the Grim Reaper through some fast thinking and accurate shooting on my part. Shooting I might add, for which I had no authority to do. But I would be damned if I was going to just sit there amidst flying bullets and be picked off like a big old fat and happy toad in a lily pond... Because of the grace of God go I...

While stationed in Washington, D.C., in 1978, I initiated the idea that the Service needed its own national wildlife forensics laboratory. Our laws were species-specific and the only way to determine the accurate species on parts and products of illegal goods being smuggled into the United States, as was required by our federal laws, was through good forensics. It took me ten years of battling my own Division of Law Enforcement and the Service to convince folks that not only was our lab needed but legally necessary. Ten years after I started my battle for the lab, the National Fish and Wildlife Forensics Laboratory now resides in Ashland, Oregon. The only one of its kind in the world created specifically for wildlife forensics, as these words are written! Because of the grace of God go I...

In 1979, I was fast-tracked to Minneapolis, Minnesota, as the new Assistant Special Agent in Charge under Special Agent in Charge, Bob Hodgins. As it turned out and lucky for me, Bob was one of the finest human beings, officer, and administrator I met in my entire career with the Service. Never had I seen such a forward-thinking officer, whose officers and natural resources of the land came first every time. He was absolutely uncanny as a teacher and I learned more from him on how to manage officers, resources, and politics than will ever be known. He in part, was responsible for making me the officer and administrator I became, as I moved up through the ranks of the Service. Once on-line, I hardly ever had a day off because of Bob's forward-thinking policies and our hard-working officers. I found myself in a two-year whirlwind of learning never before thought possible. I learned so much at times, I had to just sit back, relax, and try to absorb what was coming at me in such volumes. As I was later to learn, there was a reason for me being there. It was called the Special Agent in Charge position—one of only eight such positions in the Service at the time...Because of the grace of God go I...

In the spring of 1981, the position I had dreamed of was coming open. Soon the position of Special Agent in Charge in Denver, Colorado, was to be advertised! With Bob's help, my training and previously held positions as qualifying experience, I submitted my application. Submitted my application for one of only eight such regional positions soon to be left in the Service. In June of 1981, I was selected as the new Special Agent in Charge in Denver. In fact, at that time, I was the youngest Special Agent in Charge in the Nation! I was now as high as I wanted to go and at the peak of my career... Because of the grace of God go I...

In 1984, I was awarded the Conservationist of the Year Award by the Colorado Wildlife Federation for work performed protecting Colorado's big game herds from commercial market hunters. Only one of two such awards presented that year. Because of the grace of God go I...

One morning before daylight in 1986, I was

hurrying to work. Leaving my home in Evergreen in the dark of the morning (I was always at work by six in the morning), I thundered down from what are called the foothills heading for my office in Lakewood. Stopping at the only stoplight in town at that time, I checked the gauges on my truck's engine. Then the light turned green and I began going through my four-speed set of gears. Hitting fourth gear at 55 miles per hour, I sat back with my window open and welcomed in the pre-dawn morning Colorado cool mountain air.

Then all of a sudden from over the bank, a huge cow elk ran right in front of my speeding truck no more than twenty yards distant! At 55 miles per hour and at that range, a collision was in store! Whipping my steering wheel to the left violently and hoping I didn't roll my truck in the process to avoid the elk, she did the worst thing she could do. She ran right in front of my still-speeding but veering truck. Wham, slammed my truck violently into the elk! I hit her so hard and fearing she would be vaulted through the front window of the truck, I ducked. However, at the last moment before impact, I let up on the brakes as I had learned over the years in like situations and the hood of the truck rose slightly. In so doing, it hit the running elk square on instead of being nose down. As a result of that action, the cow did not roll over the hood and come through the front window of the truck.

However, the impact slammed the elk's head through the front windshield of my truck with such force, it blew open my glove box and tore the passenger door clear off the truck. Then she slid the full length of my truck smashing in the entire side, killing her instantly! Glancing off the elk, my truck careened across two lanes of oncoming traffic. Fortunately, at that time of the morning, there was no other traffic and I soon had my truck or what was left of it, under control. Since the engine was fine, I backed up to the dead elk lying in the highway. Dragging her off the road so no one else would hit her, I continued on to work like nothing had happened. However, my windshield was gone, the hood was smashed in, the passenger door had been blown off with the elk's impact, and the entire right side of my truck was smashed in.

It was an old truck but because of budget problems, I could not replace it. However, a hood, door, and windshield from another junker was soon added and I had a truck as good as new. What really made it funny was that my truck was now of many colors. And it made one hell of a good undercover truck when later working elk, deer and sage grouse hunters. Since no one believed such a multi-colored truck could be the law, they did as they wanted and more often than not, I was able to catch them with my truck of many colors. But truth be known, that elk should have come through my windshield at that speed and killed me! But, she didn't... Because of the grace of God go I...

In 1989, I received the second Guy Bradley Award to be awarded. A national award named in honor of Guy Bradley, a conservation officer killed in the early part of the century by poachers as he tried to protect migratory birds in Florida from plume hunters. That award honored me for my lifetime commitment to those in the world of wildlife who have little or no voice. Because of the grace of God go I...

In 1995, during an annual law enforcement physical, the doctor discovered that two of my arteries were totally blocked. Double bypass surgery followed. Since I had an Interstate roadblock upcoming in Nebraska several weeks later, I made sure I was up and running shortly after surgery. However, in the recovery room, my three roommates all recovering from bypass surgery as well, did not make it... Suffice to say, I was damn glad to get out of that recovery room! I was not at my best on that roadblock two weeks later because I was still healing, but I made it... Because of the grace of God go I...

In 1995, I was awarded the National Wildlife Federation's Conservation Achievement Award for career excellence in protecting those in the world of wildlife who have little or no voice. At the time of this writing, I am the only conservation officer, county, state, tribal or federal in the United States, to be awarded such an award. Because of the grace of God go I...

During the 1996 elk season in North Park, I caught two chaps breaking the big game laws. Sitting out in the middle of a logged-off clearing, I wrote up my two offenders. After they took their citations and left, I was sitting in the front seat of my patrol truck of many colors. There I was in the process of writing up the events of the apprehension, in case I had to testify later in court. I had my driver's side door open as I sat there writing in my diary of the previous events, when I heard a barrage of shooting by at least two heavy rifles some distance away. Figuring elk shooters, I shifted my body on the seat and then heard a *zzzipp,* followed by a loud *whunk*...

Startled, I got out from my truck figuring a bullet from the last barrage of shooting had struck my truck. Looking all around, I discovered no bullet holes. Getting back into my truck, I finished writing in my diary. Then closing my door, I looked over at the inside of my closed passenger side door. There was a bullet hole in the side panel! It was then I realized that when that barrage of shots had been fired, a stray bullet had come my way. When it did, from the angle and height of the hole in the passenger door, the bullet had entered my open door, zipped just over the top of my legs as I sat there writing notes, and hammered into my passenger door! If my legs had been just two inches higher in the seat, that rifle bullet would have struck me, causing considerable bodily damage! Because of the grace of God go I...

In 1996, I was awarded the U.S. Fish and Wildlife Service's Meritorious Service Award. That award is the Service's second highest award. In the present instance, it was given for career excellence in the world of wildlife law enforcement. Because of the grace of God go I...

In 1998, I retired from wildlife law enforcement after 32 years of state and federal service at the mandatory age of 57. My years of service in the world of wildlife law enforcement were nothing short of a vision quest. During those years of service to the American people, I was shot at six times and only hit once. I also managed to make it through many boat and horse wrecks, fights, chasing outlaws over cliffs at night, attacked by mad animals, numerous physical altercations, and several plane near-crashes. And, I was able to make a difference... Because of the grace of God go I...

In 1998, I began my second career in the world of writing. Since that date, I have had eight wildlife law enforcement books published based on my 32 years of experience, with four more awaiting publishing in 2012 and 2013. My first book, Wildlife Wars, won a National Outdoor Book Award. My second book, For Love of Wildness, also won a National Outdoor Book award as First Honorable Mention. Other true-life adventure books based on my career as a wildlife law enforcement officer are Defending Our Wildlife Heritage, A Sword for Mother Nature, No Safe Refuge, The Thin Green Line, Genesis of a Duck Cop, and Slaughter in the Sacramento Valley. Because of the grace of God go I...

The books, Wildlife Wars, For Love of Wildness and Defending Our Wildlife Heritage, were also placed under contract by the Discovery Channel. In 2003, a two-hour movie was produced for TV's Animal Planet based on some of the short stories from the above three books. That movie, titled, Wildlife Wars, is still being played today on specially selected TV channels. The monies derived from that film were donated to St. Jude Children's Research Hospital. Because of the grace of God go I...

In 2009, I began my third career, this time in writing historical novels. Turning to historical novels dealing with the fur trade, there are five books currently under contract for publishing. The first book, Crossed Arrows, was published in 2009. The second book, Curse of the Spanish Gold, was published in 2010. The third book in the series, The Saga of Harlan Waugh, Mountain Man was published in 2011. The remaining two novels, The Brothers Dent, Mountain Men and The Adventures of Hatchet Jack, Mountain Man, are to be published in 2012. Three more novels are on the drawing boards... Because of the grace of God go I...

As I said earlier in this chapter, I have been blessed. My entire life has been chock-full of many more blessings than I can ever remember or account for here. But the blessings I have never forgotten are those that came from my faithful pair of guardian angels, my loving Bride of 50 years, my family and dear friends.

As I see it, we are all travelers in this great world of ours. From our genesis to the exodus of our souls, we are travelers in time between the two great eternities. And during those travels, we are all looking for the grace and Almighty hand of God... Because of the grace of God go I...

# WILDLIFE'S QUIET WAR

IN *WILDLIFE'S QUIET WAR*, Terry Grosz continues to chronicle his remarkable career defending America's voiceless wild creatures. Since his first days as a California game warden in 1966 to the end of his career as a special agent for the U.S. Fish and Wildlife Service in 1998, Terry Grosz has been fighting to put those in the business of extinction out of business. In this, his tenth memoir, he recounts his adventures as a United States Game Management Agent, where he matched wits against cold-hearted killers: commercial market hunters, run of the mill poachers, sportsmen run amuck, outlaw Native Americans, politicians and law enforcement officers straying over the line of legality, wealthy land owners, and just plain dedicated killers of wildlife. Some set out to poach wildlife; others are recreational shooters who become possessed by their primitive blood lust. The effect is always the same, however-the destruction of a natural resource that should belong to those Americans who are yet to come. Bad weather is thrown into the mix, and a ton of side-slapping funnies that can only can happen to those who offend Mother Nature. Terry Grosz made every effort to stop these heartless criminals, and Wildlife's Quiet War book has all the excitement, drama and sometimes lively humor of his adventures.

*Available Now from Wolfpack*
*Publishing on Amazon.com*

# ABOUT THE AUTHOR

TERRY GROSZ EARNED his bachelor's degree in 1964 and his master's in wildlife management in 1966 from Humboldt State College in California. He was a California State Fish and Game Warden, based first in Eureka and then Colusa, from 1966 to 1970. He then joined the U.S. Fish & Wildlife Service, and served in California as a U.S. Game Management Agent and Special Agent until 1974. After that, he was promoted to Senior Resident Agent and placed in charge of North and South Dakota for two years, followed by three years as Senior Special Agent in Washington, D.C., with the Endangered Species Program, Division of Law Enforcement. While in Washington, he also served as Foreign Liaison Officer.

In 1979, he became the Assistant Special Agent in Charge in Minneapolis, Minnesota. Two years later in 1981, he was promoted to Special Agent in Charge and transferred to Denver, Colorado, where he remained until his retirement in 1998.

He has earned many awards and honors during his career, including, from the U.S. Fish & Wildlife Service, the Meritorious Service Award in 1996, and Top Ten Award in 1987 as one of the top ten employees (in an agency of some 9,000). The Fish & Wildlife Foundation presented him with the Guy Bradley Award in 1989, and in 1993 he received the Conservation Achievement Award for Law Enforcement from the National Wildlife Federation.

Unity College in Maine awarded Grosz an honorary doctorate in environmental stewardship in 2001. His first book, *Wildlife Wars*, was published in 1999 and won the National Outdoor Book Award for Nature and Environment. He has had ten memoirs published since then — *For Love of Wildness, Defending Our Wildlife Heritage, A Sword for Mother Nature, No Safe Refuge, The Thin Green Line, Genesis of a Duck Cop, Slaughter in the Sacramento Valley, Wildlife on the Edge, Wildlife's Quiet War, and Wildlife Dies Without Making a Sound* (in two volumes) — and his Mountain Men Novels — *Crossed Arrows, Curse of the Spanish*

*Gold, The Saga of Harlan Waugh, The Adventures of the Brothers Dent,* and *The Adventures of Hatchet Jack.*

Several of Grosz's stories were broadcast as a docudrama on the Animal Planet network in 2003.

Terry Grosz lives in Colorado.